Wildlife production systems
Economic utilisation of wild ungulates

Cambridge Studies in Applied Ecology and Resource Management

The rationale underlying much recent ecological research has been the necessity to understand the dynamics of species and ecosystems in order to predict and minimise the possible consequences of human activities. As the social and economic pressures for development rise, such studies become increasingly relevant, and ecological considerations have come to play a more important role in the management of natural resources. The objective of this series is to demonstrate how ecological research should be applied in the formation of rational management programmes for natural resources, particularly where social, economic or conservation issues are involved. The subject matter will range from single species where conservation or commercial considerations are important to whole ecosystems where massive perturbations like hydro-electric schemes or changes in land-use are proposed. The prime criterion for inclusion will be the relevance of the ecological research to elucidate specific, clearly defined management problems, particularly where development programmes generate problems of incompatibility between conservation and commercial interests.

WILDLIFE PRODUCTION SYSTEMS

Economic utilisation of wild ungulates

Edited by
Robert J. Hudson,
University of Alberta, Edmonton, Canada
K. R. Drew,
Ministry of Agriculture & Fisheries, Mosgiel, New Zealand
L. M. Baskin,
Academy of Sciences, Moscow, USSR

Section subeditors
David H. M. Cumming,
Department of National Parks & Wild Life Management, Harare, Zimbabwe
Neil Fairall,
Cape Provincial Administration, Stellenbosch, South Africa
David R. Klein,
University of Alaska, Fairbanks, Alaska, USA
Richard A. Luxmoore,
IUCN Wildlife Trade Monitoring Unit, Cambridge, UK
J. Brad Stelfox,
University of Alberta, Edmonton, Canada

The right of the
University of Cambridge
to print and sell
all manner of books
was granted by
Henry VIII in 1534.
The University has printed
and published continuously
since 1584.

CAMBRIDGE UNIVERSITY PRESS

Cambridge

New York   Port Chester   Melbourne   Sydney

CAMBRIDGE UNIVERSITY PRESS
Cambridge, New York, Melbourne, Madrid, Cape Town, Singapore,
São Paulo, Delhi, Dubai, Tokyo

Cambridge University Press
The Edinburgh Building, Cambridge CB2 8RU, UK

Published in the United States of America by Cambridge University Press, New York

www.cambridge.org
Information on this title: www.cambridge.org/9780521349147

First published 1989
This digitally printed version 2009

A catalogue record for this publication is available from the British Library

Library of Congress Cataloguing in Publication data
Wildlife production systems: economic utilization of wild ungulates
edited by R. J. Hudson, K. R. Drew & L. M. Baskin: section subeditors,
D. Cumming . . . [et. al.].
p.   cm.
Includes index
ISBN 0 521 34099 3
1. Big game ranching. 2. Big game hunting. 3. Ungulata—Economic
aspects. I. Hudson, R. J.   II. Drew, K. R.   III. Baskin, L. M.
SF400.5.W55 1989
333.95′9—dc 1988–20418 CIP

ISBN 978-0-521-34099-1 Hardback
ISBN 978-0-521-34914-7 Paperback

Dedication

Jack R. Luick

(photograph by Don Borchert, Institute of Arctic Biology, Fairbanks, Alaska)

This book is dedicated to the memory of Dr Jack R. Luick, who served with distinction as Professor of Nutrition at the University of Alaska (Fairbanks) until his death in 1983. Through many years of innovative research on large herbivores, numerous scholarly publications, and unfailing efforts to promote the international exchange of information and ideas, Jack contributed substantially to many of the scientific insights presented in these chapters. We are hopeful that this volume reflects adequately the advances in our biological knowledge – a sluggish, often disorderly process which Jack constantly endeavoured to expedite. Although he did not live to serve as convenor and subsequently editor, this is his creation in that he laid the groundwork for the International Theriological Congress symposium from which it grew.

CONTENTS

CONTRIBUTORS

Bai, Q., Department of Animal Science, Jilin Agricultural University, Jilin, People's Republic of China

Baskin, L. M., Institute of Evolutionary Animal Morphology & Ecology, USSR Academy of Sciences, 33 Leninski Prospect, Moscow 117071, USSR

Blyth, Charles B., Chief Warden, Nahanni National Park, Bag 300, Fort Simpson, Northwest Territories, Canada X0E 0N0

Bubeník, A. B., 10 Stornway Crescent, Thornhill, Ontario, Canada L3T 3X7

Cumming, David H. M., Department of National Parks & Wild Life Management, Box 8365 Causeway, Harare, Zimbabwe

Dezhkin, V. V., Central Scientific Laboratory, Glavohota RSFSR, Losinoostrovskaya lesnaya dacha, kvartal 18, Moscow 129347, USSR

Drew, K. R., MAFTECH, Invermay Agricultural Centre, Ministry of Agriculture & Fisheries, Private Bag, Mosgiel, New Zealand

Fadeev, E. V., Department of Vertebrate Zoology, Moscow State University, Leninskie Gori., Moscow, USSR

Fairall, Neil, Department of Nature & Environmental Conservation, Cape Provincial Administration, Private Bag 5014, Stellenbosch 7600, South Africa

Fennessy, P. F., MAFTECH, Invermay Agricultural Centre, Ministry of Agriculture & Fisheries, Private Bag, Mosgiel, New Zealand

Fletcher, T. John, Reedie Hill Deer Farm, Auchtermuchty, Cupar, Fife, UK KY14 7HS

Gates, Cormack C., Department of Renewable Resources, Box 1196 Fort Smith, Northwest Territories, Canada X0E 0P0

Green, Michael J. B., IUCN Wildlife Trade Monitoring Unit, 219c Huntingdon Road, Cambridge, UK CB3 0DL

Groves, Pam, Institute of Arctic Biology, University of Alaska, Fairbanks, Alaska 99775-0180, USA

Hawley, Alex W. L., Alberta Environment Centre, Bag 4000, Vegreville, Alberta, Canada T0B 4L0

Hudson, Robert J., Department of Animal Science, AgFor 310, University of Alberta, Edmonton, Alberta, Canada T6G 2P5

Klein, David R., Alaska Cooperative Wildlife Research Unit, University of Alaska, Fairbanks, Alaska 99701, USA

Lebedeva, N. L., Institute of Evolutionary Animal Morphology & Ecology, USSR Academy of Sciences, 33 Leninski Prospect, Moscow 117071, USSR

Lewis, Henry T., Department of Anthropology, University of Alberta, Edmonton, Alberta, Canada T6G 2P5

Luxmoore, Richard A., IUCN Wildlife Trade Monitoring Unit, 219c Huntingdon Road, Cambridge, UK CB3 0DL

Marks, Stuart A., Department of Sociology & Anthropology, North Carolina State University, Box 8107, Raleigh, North Carolina 27695-3180, USA

Mossman, Archie S., Department of Wildlife, Humboldt State University, Arcata, California 95521, USA

Payne, C. Harvey, Manitoba Wildlife Branch, Box 24, 1495 St James Street, Winnipeg, Canada R3H 0W9

Renecker, Lyle A., Department of Animal Science, University of Alberta, Edmonton, Canada T6G 2P5

Rogacheva, E. V., Institute of Evolutionary Animal Morphology & Ecology, USSR Academy of Sciences, 33 Leninski Prospect, Moscow 117071, USSR

Scotter, George W., Chief, Wildlife Conservation, Canadian Wildlife Service, Twin Atria 2, 4999-98 Ave, Edmonton, Alberta, Canada T6B 2X3; and Adjunct Professor, Faculty of Agriculture & Forestry, University of Alberta, Edmonton, Alberta, Canada T6G 2P5

Skinner, John D., Mammal Research Institute, University of Pretoria, Pretoria 0002, South Africa

Skjenneberg, Sven, Nordic Council of Reindeer Research, Box 378, N-9401 Harstad, Norway

Sokolov, V. E., Institute of Evolutionary Animal Morphology & Ecology, USSR Academy of Sciences, 33 Leninski Prospect, Moscow 117071, USSR

Stelfox, J. Brad, Department of Zoology, University of Alberta, Edmonton, Alberta, Canada T6G 2P5

Syroechkovsky, E. E., Institute of Evolutionary Animal Morphology & Ecology, USSR Academy of Sciences, 33 Leninski Prospect, Moscow 117071, USSR

Taylor, P. G., Severn Hills, Stanthorpe, Queensland 4380, Australia

Tiplady, B. Ann, Institute of Arctic Biology, University of Alaska, Fairbanks, Alaska 99775-0180, USA

White, Robert G., Institute of Arctic Biology, University of Alaska, Fairbanks, Alaska 99775-0180, USA

Yorks, Terence P., Department of Range Management, University Station Box 3354, University of Wyoming, Laramie, Wyoming 82071, USA

PREFACE

The World Conservation Strategy calls for a more pragmatic approach to conservation arguing that conservation must be part of development rather than at odds with it. To some, this seems like a desperate compromise, but to others it represents a welcomed maturation of the conservation movement. One of the most controversial aspects is that dealing with sustainable utilisation of ecosystems, particularly the priority requirement which invites us to:

> Utilize indigenous wild herbivores, alone or in combination with livestock, where the use of domestic stock alone will degrade the land ... The potential of wild herbivores for subsistence and commercial use should be given priority attention. Two main actions are needed:
>
> assessment of social and economic potential of game ranching, looking at commercial utilization, subsistence utilization, and domestication options, as well as the market potential for products.
>
> assessment of current and potential ecological impacts of trypanosomiasis control in Africa, including consideration of new developments in control techniques. (*World Conservation Strategy Sec. 7.12*)

This book was written in response to that challenge. Through international comparisons, we attempted to chronicle the changing role of wildlife, define the main production strategies, establish their ecological and economic basis, and reflect on implications of these trends.

We adopted the required international scope in its geographical, cultural, and political diversity, but narrowed treatment taxonomically. Because of their potential on international markets and pressing conservation needs, we emphasised wild ungulates. Marine mammals, fur-bearers, rodents, birds, fish, and invertebrates were excluded except where necessary to provide perspective. We felt justified in this prejudice because broader coverage already is available in excellent recent reviews by I. L. Mason (1984, *Evolution of Domesticated Animals*, London: Longman), and S. K. Eltringham (1984, *Wildlife Resources and Economic Development*, Chichester: Wiley). We also have limited evaluation to the more controversial aspects of ungulate management; namely, consumptive utilisation to the exclusion of game viewing in parks and zoological gardens.

The evaluation was conducted by professionals engaged in research and development relevant to wildlife production systems in representative parts of the globe. Most of the authors presented preliminary invited papers at the Fourth International Theriological Congress (ITC) held in Edmonton, Canada, in August 1985. But gaps were subsequently filled by additional invited contributors.

Numerous people contributed to this project. William Fuller, as chairman of the ITC organising committee, laid the groundwork for the symposium, and Robin Pellew of the IUCN Conservation Monitoring Unit suggested its publication. Alan Crowden and Karin Fancett of Cambridge University Press guided it through to print. Brad Stelfox assisted with copy-editing. Yasmin Hudson provided moral support and cheerfully lightened the load of indexing. The following served as external referees: J. Anderson (Mafeking, South Africa), Wendy Arundale (Institute of Arctic Biology, University of Alaska), J. du P. Bothma (Eugene Marais Chair of Wildlife Management, University of Pretoria), R. Brown (Caesar Kleberg Foundation, Texas A & M University), R. J. Bunnage (Alberta Agriculture), C. Challies (New Zealand Forest Research Institute), B. des Clers (IUCN Ethnozoology Commission, Paris), M. Freeman (Department of Anthropology, University of Alberta), S. C. Gerlack (University of Alaska Museum), J. Hanks (WWF-International, Switzerland), T. Haynes (Division of Subsistence, Alaska Fish & Game), M. Hoefs (Yukon Renewable Resources), R. Hofmann (Department of Veterinary Anatomy, University of Giessen), P. A. Jewell (Research Group in Mammalian Ecology & Reproduction, University of Cambridge), J. F. Jooste (Nature and Environmental Conservation, Cape Provincial Administration, South Africa), R. N. B. Kay (Rowett Research Institute, Scotland), E. L. Kozicky (Caesar Kleberg Wildlife Research Institute, Texas A & I University), J. A. Lalouette (Union Ducray, Mauritius), H. Meissner (Department of Animal Science, University of Pretoria), M. Meldgaard (Zoological Museum, University of Copenhagen), M. Mentis (Department of

Botany, University of Witwatersrand), M. Pattison (New Zealand Game Industry Board), H. Reynolds (Canadian Wildlife Service), D. Rowe-Rowe (Natal Parks Board), J. M. Suttie (Invermay Agricultural Research Centre, New Zealand), J. G. Stelfox (Canadian Wildlife Service), R. Taylor (Department of National Parks & Wild Life Management, Zimbabwe), E. S. Telfer (Canadian Wildlife Service), T. Tennessen (University of Nova Scotia, Canada), C. Tisdell (Department of Agricultural Economics, University of Newcastle, Australia), N. Vietmeyer (National Research Council, Washington), and M. Wilhelmsen (Royal Agricultural College of Sweden).

Robert J. Hudson
March, 1988

APPROXIMATE CURRENCY CONVERSIONS

Country	Currency	Notation	Equivalent US$
Australia	dollar	A$	0.71
Britain	pound	£	1.65
Canada	dollar	Can$	0.75
Finland	markka	FIM	0.23
Nepal	rupee	Rs	0.05
New Zealand	dollar	NZ$	0.57
Norway	krone	NOK	0.15
South Africa	rand	R	0.49
Soviet Union	rouble	Rb	1.57
Sweden	krona	SEK	0.16
Tanzania	shilling	TSh	0.02
United States	dollar	US$	1.00
Zimbabwe	dollar	Z$	0.61

Conversions at mid 1987.

SECTION A

Introduction

ROBERT J. HUDSON

For much of human history, animals have served our material needs for food, bone, sinew, and hides; a dependence which has played an important role in our cultural and perhaps even biological evolution (Harding & Teleki, 1981). Some rural communities still supplement their diets with bushmeat although estimates of this nutritional dependence are difficult to obtain (Prescott-Allen & Prescott-Allen, 1982; Martin, 1983). However, the largest contribution to game meat supplies, at least from large mammals, apparently comes from sport hunting (Prescott-Allen & Prescott-Allen, 1986). With escalating demands for alternate uses of land and urbanisation of attitudes towards wildlife, such opportunities are disappearing. But recent trends suggest that wild ungulates may regain some of their importance in an expanded economic role.

Commercial production of wildlife has earned respectability as an agricultural strategy (Fennessy & Drew, 1985; Kyle, 1987; White, 1987). Net international trade in game meat now exceeds 35,000 tonnes (Chapter 2) representing perhaps 7% of world production (Krostitz, 1979). Supplies are projected to rise sharply; the largest exporter of farm venison, New Zealand, expects to increase output to about 20,000 tonnes by 1995. Compared with exports of domestic meats (over 13 million tonnes), these volumes are modest, but not insignificant in economic terms since the value of game products is high with the legal world trade approaching US$ 150 million (Chapter 2).

In general, these developments have been endorsed by the World Conservation Strategy (IUCN/UNEP/WWF, 1980) and by the Report of the World Commission on Environment and Development (1987) as part of their global initiatives for sustainable development. Both see conservation as part of development and, diplomatically, seem to promise something for everyone.

Those interested in wildlife see a chance to influence land use by redirecting tangible benefits of conservation to landowners, to stabilise markets for wildlife products, and to secure gene pools of species threatened in the wild. Those more interested in rural development are encouraged by opportunities to maintain productive landscapes, to reduce dependence on cultural energy, to tap new agricultural markets, and to maintain culturally consistent livelihoods for indigenous people.

But, others doubt not only whether wildlife production is a viable economic and environmental strategy, but also whether it runs counter to its claimed purpose. Some authorities believe that commercial game production simply won't work because of biological (Moss, 1975) or socioeconomic impediments (Parker & Graham, 1971). Some of the strongest opposition is philosophical; the essence of wildlife is that it is wild and a segment of the public responds to threats of it being any other way (Livingston, 1981; Evernden, 1986). Even pragmatic conservationists have expressed doubts about the wisdom of the strategy, enumerating a variety of biological and social threats of commercialisation (Geist, 1985).

This book evaluates this controversy by chronicling the current status of the industry and reflecting on the environmental and agricultural implications of these trends. The chapters in this section define the principal production modes, and document current international trade. But first, the purported role of wildlife production in the context of a broader conservation strategy is outlined.

Instruments of conservation

Conservation has moved broadly on four fronts; namely, regulating harvests, establishing sanctuaries, educating the public, and providing tangible incentives. Each has its advantages and limitations.

Sanctions

Since ancient times, the response to declining game resources has been to regulate harvests with sanctions as subtle as peer pressure or as absolute as edicts from powerful rulers. Today, game laws serve to establish the respective roles of landowners, the general public, and the state; and to establish the rules of harvest – what can be hunted by whom, where, when, and for what purpose.

Field enforcement by a central authority has always been a formidable challenge since wild animals tend to live in vast remote landscapes. The problem may be more than simply one of logistics since rural sympathies often run strong, especially when agencies pursue seemingly contradictory policies of game cropping or catering to foreign hunters. Enforcement can be caught up in an escalating war with poachers with calls for increasingly draconian measures.

More imaginative solutions are required than simply increasing fire-power. The motivation to poach is complex, involving any combination of hunger, desperation, anger, defiance, irresponsibility, or greed. For several species, the most serious factor is the high price of skins, horns, tusks, or organs on international markets for which a measure of control is possible through the Convention for International Trade in Endangered Species (CITES). For most others, the threat of habitat loss is more profound.

Sanctuaries

The greatest conservation achievement in modern times has been the protection of representative ecosystems in parks and reserves. Since Yellowstone (USA), the world's first modern national park, was established in 1872, the global network has grown to over 2000 reserves totalling almost 400 million hectares (IUCN, 1985). In addition to protecting genetic resources and maintaining life support systems, protected ecosystems play an important part in cultivating a conservation ethic, and may even pay their way by earning foreign exchange and generating employment.

Of course, the idea is not new. Reserves, largely to provide hunting, were established in the ancient world (Anderson, 1985) spreading with feudalism in the middle ages (Wolfe, 1970). Protected areas were not necessarily planned as an adjunct to legal sanctions; they often reflected the inability to enforce laws more widely. With the decline of the monarchies over the past several centuries, many of these sanctuaries were plundered but nevertheless survived to join the international network of protected areas.

Although protecting representative ecosystems clearly is a frontline activity, there remain questions about the long-term security and viability of such faunal enclaves. Despite proclamations to the contrary, parks and reserves are not always inviolate and do not always win local support. In the developed world, the popularity of parks rests in their amenity and recreational value. In the developing world, these values often are marketed to foreign tourists. This seems to justify protecting fairly large land bases but the economic benefits of tourism often fall into the hands of a few and create the impression that parks were established for rich foreigners (Myers, 1972). Although receipts from international tourism have been relatively stable, the costs of travel and threats of international terrorism and local insurgency contrive to interrupt the flow of benefits. Unless these benefits continue to balance the pressing needs of a growing population for pasturage and croplands, arguments of a global heritage cannot be expected to protect anything more than a few critical landscapes.

Without economic motivation, few nations apparently will sequester more than about 5% of their land area as sanctuaries. Even if these enclaves could be defended, the remaining 95% should not be willingly relinquished. After all, the survival of stocks in parks often depends on dispersal to surrounding areas, and the outlook for long-term maintenance of species in islands of protected habitat is not good for a variety of ecological, demographic, and genetic reasons (Soule, 1986). The maintenance of life support systems obviously depends largely upon what happens elsewhere.

Education

A great deal of faith has been placed in conservation education. In former times, it was as blunt as a royal proclamation of new laws with a warning of the grim consequences of infractions. Today, conservation education purports to enlighten people about the life support systems upon which they depend. However, there are several other agendas aimed at imposing moral sanctions, winning political support, soliciting funds, or more generally creating a climate for social change. Generally, campaigns are designed as appeals to sentiment, conscience, reason, and/or self interest.

Sentiment

An important handle on public interest is their sense of beauty, intricacy, cuteness, or adventure. Nature has always been an important subject of literature, painting, photography, and cinematography. Through art (form), it is possible to capture interest in the processes of nature (function). Ecological research has contributed more to providing interpretive depth than managing resources. Linnaeus's binomial classification has done a great deal to instil an appreciation of diversity and to add respectability to nature study. Comparative study stimulated by Darwin's theory of natural selection creates concern for seemingly insignificant taxa by turning them into pieces of a much larger puzzle. Pop-conservation relies on anthropomorphic portrayals of animals as cute, funny, or foolish. For large mammals, adventure is a recurrent theme. Although such appeals may sometimes seem undignified, they target specific publics of various ages and backgrounds.

Conscience

A more coercive appeal is based on conscience. At the turn of the century, vitriolic attacks were directed toward 'game hogs', 'market hunters', 'game butchers', or more diffusely against ethnic minorities. Today, it is more fashionable to speak of moral obligations, a world heritage, and future generations. All of this may be perfectly true, but making people feel stupid, selfish, or immoral seems to work only when detrimental activities are inadvertent or where non-compliance brings economic or political sanctions. The psychological effect is likely to be greater on followers of a movement (and hence its momentum) than on those whose behaviour is to be modified.

Reason

Conservation education can be a plea to behave responsibly: if we can't be good, we can at least be prudent (Livingston, 1981). Conservation rhetoric exhorts that 'our ability to save endangered species is a measure of our ability to ultimately save ourselves'. Since credulity is stretched by claiming that the loss of a link in an ecological web could lead to the collapse of an ecosystem and perhaps all of mankind, the more reasonable argument is posed that another species lost is another rivet popped on the earth's life support systems (Erlich & Erlich, 1981). It also is common to point out that wild species may offer new medical or agricultural contingencies in dealing with an uncertain future (Myers, 1983; National Research Council, 1983) or to enumerate current benefits of wildlife conservation (Prescott-Allen & Prescott-Allen, 1986).

Such appeals to reason can be powerful tools for gaining political or financial support and thus influencing long-term policies. However, they probably do little to change the behaviour of rural people that live in the midst of animals and often have the greatest impact on their future.

Self interest

Sometimes it may be enough to show that with no additional cost or inconvenience, landowners can enjoy the amenity values of wildlife or take personal pride in exemplary behaviour. However, especially in tough economic times, this argument is difficult to sell because conservation of wildlife, particularly large species, can be a liability for individual landowners. Talk of utilitarian values does little unless landowners can realise such values in a way and a time frame that matters. For this potentially important argument to work, institutional arrangements need to provide appropriate incentives. The benefits of conservation must be made more tangible, immediate, and personal.

Incentives

Vested interest already plays an important role in other aspects of wildlife conservation. Economic benefits of tourism as well as the amenity and recreational value this captures encourage the maintenance of parks and reserves throughout the world. Sport hunting maintains interest in wildlife habitat, ensuring priority on political agendas (Reiger, 1975). Whereas tourists and hunters may justify wildlife conservation in the public domain, they generally have less impact on land which is privately or communally owned. Landowners may even be disinclined to maintain wildlife habitat when it encourages trespass and vandalism.

The first step towards providing effective incentives is to fully cover the costs of wildlife damage to crops or livestock. Compensation programmes have never been entirely satisfactory; governments have been slow to provide full compensation for depredation and to streamline procedures for making claims, fearing that this might encourage false claims and careless management practices. Another approach is land easement in which the landowner is paid either directly or indirectly through tax concessions to relinquish the economic opportunities afforded by clearing, draining, or cropping (Ryder & Boag, 1981). Such programmes are expensive and have the disadvantage of paying people to do nothing, as if holding the public at ransom.

The problem with both compensation and land easement is that they may stem the tide but usually are not powerful enough to reverse it. What is needed is a way to integrate wildlife firmly into day to day economic decisions. An obvious solution is to allow landowners to profit from selling a harvestible surplus of game. Institutional arrangements can vary from complete transfer of ownership of certain game species, through various leasing arrangements, to access fees or charging for various services (Tomlinson, 1985). As logical as this may seem, there is considerable ambivalence about directing material or pecuniary benefits to individuals.

Elements of an effective strategy

Contrasting views about commercialisation of wildlife stem from the way people perceive relevant trade-offs, the appropriate relation between man and nature, human motivation, and social justice.

What choices?

Most people perceive the environmental challenge as serious but apparently differ in their faith in current systems of management to deal with it. The greatest perceptual gap relates to the nature of the trade-offs. With respect to commercial production, it is not always clear if the choices are whether wild animals will be fenced or free, public or private, present or absent. In terms of its environmental effect, recreational hunting of wild populations may be better than intensive game farming. But on good agricultural land, the only choice may be game farming or cultivation.

What's right?

Nature can be viewed either as resources rightfully subverted to serve human needs or as phenomena of intrinsic value. Management programmes make implicit assumptions about these values and are, therefore, constrained by what managers consider right in a philosophical sense.

Despite reverence and often mystical or spiritual associations, nature has been treated largely as a *resource*. Ancient Greeks and Romans professed a profound love of nature but were business-like in their dealings with it (Hughes, 1975). These philosophies are the basis of Western religions (Judaism, Christianity, and Islam) and presumably Western societies. It is not that nature is exploited simply for profit. Rather, it serves various humanistic goals such as providing nutritional sustenance, elevating socioeconomic status, or offering recreational or aesthetic experiences.

A contrasting view considers wild species to be elements of the cosmos no less important than man. Historically, civilisations have broadened identification and extended rights beyond the family, to the community, nation, race, and the entire human species. The environmental movement has proposed a further extension to higher animals, other living things, and even to landscapes. This serves as a statement of the human condition, measured either as a step away from animalness or a step closer to godliness. Consequently, conservation is transformed to a moral or spiritual duty.

These contrasting views of *intrinsic* versus *instrumental* value, of duty and worth, are usefully blended in the concept of stewardship in which wildlife is treated essentially as an *inheritance*. Mankind acts as 'God's resident manager' (Livingston, 1981) with a privilege to use resources but with two caveats; namely, to manage responsibly (pass on resources to future generations intact) and respectfully (ensure humane treatment).

What's effective?

There also is the parallel dichotomy of how conservation is best encouraged. What is the relative effectiveness of management by sanction *versus* by incentive? It would be nice to think that the future of wildlife could be defended simply as a matter of moral principle rather than the recreational and economic opportunities they present. But for this to happen, rural people may have to share our socioeconomic status. After all, popular support for the conservation movement comes largely from those whose survival needs are fully met and whose contacts with nature are mainly recreational. The problem is that this standard of living is only achieved by consuming perhaps 40 times the resources which are available to the rural poor (IUCN/UNEP/WWF, 1980).

What's fair?

A question that probably is too seldom asked is whether conservation as an end justifies any means. For example, conservation might be achieved by limiting the aspirations and political influence of the rural poor. It might be expedient to usurp traditional hunting rights of native people or limit the privileges of landowners. Recreational opportunities in parks and reserves might well be expanded by imposing the will of urban over rural people, rich over poor, or developed over underdeveloped countries. Such arguments probably were used by mediaeval kings to justify the harsh Forest Laws.

A pressing problem is to trace the flow of costs and benefits of conservation. There are many analyses of the collective economic benefits of conservation but fewer on to whom costs and benefits accrue. Social justice will be an important factor in the long-term future of environmental conservation.

Expedience or gratification?

The greatest dichotomy is whether to conserve wildlife primarily on the basis of intrinsic or instrumental value. To some people, this amounts to the choice of being guided by dignity or expedience – by what's right or by what works. Others see it as choosing between gratification derived from reaching easily quantifiable goals, and more diffuse but far-reaching achievement. Although there would seem to be ample opportunity to move on both fronts, this raises certain contradictions that have sometimes stalemated conservation programmes.

Experience has shown two problems with treating wildlife strictly as a resource. First, it would seem to leave little hope for species that have little immediate amenity, recreational, or economic value (Ehrenfeld, 1978). We can only hope they will be incidentally protected by conserving habitats for more charismatic species. Secondly, without caveats on utilisation, short-term optimisation may lead to the ultimate depletion of stocks (Clark, 1973). Also, resources are managed whereas the essence of wildlife is that it is wild (Ehrenfeld, 1978; Livingston, 1981; Evernden, 1986). Therefore, amenity values may be denigrated by developing recreational or economic values.

Basing conservation on intrinsic values and moral appeals overcomes these problems but introduces several others. First, it is not clear whether individual human rights can be correctly projected to populations of animals or species let alone lower lifeforms and inanimate objects (Norton, 1986). Secondly, there is the question of whether moral sanctions can ever work in anything but small integrated communities unless backed by strong legal sanctions (Shaw, 1987). A more subtle problem is that moral appeals tend to divide the world into man and nature and, more seriously, into good people and bad people. This has seen subsistence hunting redefined as poaching, poaching escalated to treason, the rural poor accused of greed, and the exclusion of aboriginal people from anthropogenic landscapes.

Why should rural people preoccupied with economic or even personal survival be greatly concerned for the welfare of wildlife particularly when faced with depredation of crops or livestock? Is it reasonable to expect those who share their lives with animals to show any more concern for our fulfillment than we show for their livelihood? Conservation programmes may even strengthen the resolve of local people who may see it as an attempt by those that benefit from wildlife conservation to take even more from those who pay. At best, many rural people simply cannot afford to conserve large mammals. At worst, those who benefit are trying to control those that pay. Conservation may become perceived as just another form of ideological terrorism.

This is the challenge which will carry us into the next century. Presently, we have little influence on land-use patterns and must develop structures to do so. As a first step in this direction, international agencies are beginning to talk of integration of conservation and development which have so long been at odds. The concept of 'sustainable utilisation of ecosystems' will go a long way towards ensuring a place for natural fauna and flora. However, large mammals present a particular problem since they often conflict with human enterprise. Unlike smaller innocuous species, they may not be incidentally protected by working toward the goal of sustainable development. For large, destructive, and even dangerous game, there may be little hope of survival except in protected areas. But the future of the majority of ungulates on private lands may depend on developing their instrumental value.

References

Anderson, J. K. (1985). *Hunting in the Ancient World*. Berkeley: University of California Press.

Clark, C. W. (1973). Profit maximization and the extinction of animal species. *Journal of Political Economy*, 81, 950–61.

Ehrenfeld, D. W. (1978). *The Arrogance of Humanism*. New York: Oxford University Press.

Ehrlich, P & Ehrlich, A. (1981). *Extinction – the Causes and Consequences of the Disappearance of Species*. New York: Random House.

Evernden, N. (1986). *The Natural Alien: Humankind and Environment*. Buffalo: University of Toronto Press.

Fennessy, P. & Drew, K. R. (eds.) (1985). *Biology of Deer Production*, Bulletin 22. Wellington: Royal Society of New Zealand.

Geist, V. (1985). Game ranching: threat to wildlife conservation in North America. *Wildlife Society Bulletin*, 13, 594–8.

Harding, R. S. O. & Teleki, G. (eds.) (1981). *Omnivorous Primates: Gathering and Hunting in Human Evolution*. New York: Columbia University Press.

Hughes, J. D. (1975). *Ecology in Ancient Civilizations*. Albuquerque: University of New Mexico Press.

IUCN. (1985). *United Nations List of National Parks and Protected Areas*. Cambridge: IUCN Conservation Monitoring Centre.

IUCN/UNEP/WWF. (1980). *World Conservation Strategy: Living Resource Conservation for Sustainable Development*. Gland: IUCN.

Krostitz, W. (1979). The new international market for game meat. *Unasylva*, 31, 32–6.

Kyle, R. (1987). *A Feast in the Wild*. Oxford: Kudu Publishing.

Livingston, J. (1981). *The Fallacy of Wildlife Conservation*. Toronto: McClelland & Stewart.

Martin, G. H. G. (1983). Bushmeat in Nigeria as a natural resource with environmental implications. *Environmental Conservation*, 10, 125–32.

Moss, R. (1975). Different roles of nutrition in domestic and wild game birds and other animals. *Proceedings of the Nutrition Society*, 34, 95–100.

Myers, N. (1972). National parks in savannah Africa. *Science*, 178, 1255–63.

Myers, N. (1983). *A Wealth of Wild Species: Storehouse for Human Welfare*. Boulder (Colorado): Westview Press.

National Research Council. (1983). *Little-Known Asian Animals with a Promising Economic Future*. Washington: National Academy Press.

Norton, B. G. (ed.) (1986). *The Preservation of Species: The Value of Biological Diversity*. Princeton: Princeton University Press

Parker, I. S. C. & Graham, A. D. (1971). The ecological and economic basis for game ranching in Africa. In *The Scientific Management of Plant and Animal Communities for Conservation*, ed. E. Duffey & A. S. Watt, pp. 393–404. Oxford: Blackwell Scientific Publishers.

Prescott-Allen, C. & Prescott-Allen, R. (1986). *The First Resource: Wild Species in the North American Economy*. New Haven: Yale University Press.

Prescott-Allen, R. & Prescott-Allen, C. (1982). *What's Wildlife Worth?: Economic Contributions of Wild Plants and Animals to Developing Countries*. London: Earthscan.

Reiger, J. F. (1975). *American Sportsmen and the Origins of Conservation*. New York: Winchester Press.

Ryder, J. P. & Boag, D. A. (1981). A Canadian paradox – private land, public wildlife: Can it be resolved? *Canadian Field Naturalist*, 95, 35–8.

Shaw, J. H. (1987). Assessing the progress toward Leopold's land ethic. *Wildlife Society Bulletin*, 15, 470–2.

Skinner, J. D. (1985). Wildlife management in practice: conservation of ungulates through protection or utilization. *Symposium of the Zoological Society of London*, 54, 25–46.

Soule, M.E. (ed.) (1986). *Conservation Biology: The Science of Scarcity and Diversity*. Sunderland (Massachusetts): Sinauer Associates.

Tomlinson, B. H. (1985). Economic values of wildlife: opportunities and pitfalls. *Transactions of the North American Wildlife and Natural Resources Conference*, 50, 262–70.

White, R. J. (1987). *Big Game Ranching in the United States*. Mesilla (New Mexico): Wild Sheep & Goat International.

Wolfe, M. L. (1970). The history of German game administration. *Forest History*, 14, 6–16.

World Commission on Environment and Development. (1987). *Our Common Future (The Brundtland Report)*. Oxford: Oxford University Press.

1

History and technology

ROBERT J. HUDSON

Abstract
A variety of terms have been used to describe emerging systems of commercial ungulate management. A classification is offered which identifies four main configurations (hunting, herding, ranching, and farming) on the basis of control of animal distributions and intensity of management. In hunting systems, distributions are unrestricted and technological innovation is limited largely to harvesting. Herding involves control of distributions by behavioural means and usually requires some degree of habituation. Ranching and farming are containment systems in which distributions are controlled by fences. Ranches are lightly stocked and animals are harvested in the field. Farms are more intensively managed with at least seasonal supplemental feeding and delivery of stock to central slaughter facilities for *ante-mortem* inspection. Although hunting systems have the earliest origin, they were not simply sequentially replaced by herding, ranching, and farming. Most contemporary herding systems appear to have developed secondarily from containment systems. Game ranching and farming have origins in the ancient world but have become practical in modern times only with improved and economical fencing products.

Introduction
The classic dichotomy of domestic and wild seemingly denotes an important change in man's economic relationship with animals. The distinction seems clear enough: domestic animals are husbanded rather than hunted; produced rather than procured. But this lends little insight into the diversity of ways animals have been exploited throughout human history and perhaps even less about future options. The distinction has been blurred with growing understanding of archaic man–animal relationships (Higgs, 1972, 1975; Price & Brown, 1985) and particularly the rising popularity of various forms of wildlife husbandry (Darling, 1960; Yerex, 1982).

Domestication denotes a symbiotic relationship between man and certain plants and animals (Brothwell, 1975). Although the term is usually reserved for human endeavours, man is not unique in forming interdependent relationships; symbiosis is a particularly widespread phenomenon in the insect world (Weber, 1966; Reed, 1984; Rindos, 1984). For example, various ants have 'domesticated' aphids, fungi, and plants. Neither is it clear that domestication was accomplished entirely by design. There is even the question of reciprocity; in response to altered selective pressures, human characters have changed as much as those of domestic animals.

Whereas it is generally conceded that *domestic* and *wild* simply denote polar conditions, many attempts have been made to dichotomise the gradient by zoological and cultural criteria (Jarman & Wilkinson, 1972; Jarman, 1976). The most common definition involves morphological changes arising from altered selection pressures whether inadvertent or planned. Despite the risks of misinterpretation, age or sex ratios of the kill also have been used as evidence of incipient domestication of conventional farm animals but with greater caution or not at all to other species (Collier & White, 1976; Wilkinson, 1976). Wilkinson (1972) notes this apparent double standard and urges consideration of a wider range of economic relationships. Nevertheless, it probably is better to find new terms than to continually broaden an older definition.

The term *domestication* should be reserved for the process which results in genetic adaptation of animals controlled by man. There must be detectable divergence from the wild genotype (the term semi-domestic usually is applied where cumulative changes are not great) whether or not this arises in populations displayed in zoos or raised on farms for meat. Although rates of genetic change differ according to management practices, domestication does not directly describe how animals are managed. Wildlife is not domesticated simply by being raised like conventional farm animals although that may be the ultimate outcome. Therefore, we are left with the problem of classifying the growing diversity of production systems.

Several terms have been proposed. *Husbandry* (Darling, 1960) and *ranching* (Dasmann, 1964) were forwarded as general terms to describe commercial exploitation of wild ungulates, generally for meat production. *Game ranching* attained most widespread use (Field, 1979; Walker, 1979) but subsequently acquired the more specific meaning of extensively managed but fenced systems. Eltringham (1984) applies *cropping* to the harvest of free-ranging animals and *culling* to reduction of over-population although another widespread interpretation of culling is removal of aged or genetically defective individuals. More intensive systems requiring a degree of habituation of animals have been called *game farming* (Yerex, 1982). However, to the urban public, game farms are private zoos lacking many of the amenities offered by large publicly supported zoological gardens.

These terms mean different things to different people. However, there are several more basic problems than simply gaining consensus. First, current conventions align production systems along a single gradient which is only a marginal improvement over the domestic/wild dichotomy. Second, they lack clear distinguishing criteria. This chapter offers a redefinition of the main modes of wildlife production and provides a brief history of their origin and emergence.

Classification

Any attempt to place a continuum of exploitive strategies into boxes is certain to meet with difficulty. But a step in the right direction is to identify the most important gradients. Ingold (1980) proposed three oppositions which frame recent transformations of the Scandinavian reindeer industry: namely, (1) shared versus divided access to the means of subsistence, (2) subsistence versus market production, and (3) predatory versus protective relationships. This system has the advantage of recognising both social and biotechnical criteria. However, it is not general enough nor was it intended to encompass the global spectrum of production systems.

The convention proposed here bases the primary classification solely on technical criteria; specifically, the style and intensity of management (Fig. 1.1). The fundamental tripartite division of the style axis describes how animal distributions are controlled: little or no control exists in hunting systems, control is achieved by behavioural means in herding systems, and by physical means in containment systems. The secondary opposition is procurement versus production, separating those based simply on harvest from those based on material or energetic subsidies. It is essentially a gradient of management intensity. Ordination of management systems between these two axes gives four clusters: namely, *hunting, herding, ranching*, and *farming*. Subsequent classification within each cluster can be based on social and other biotechnical criteria (Table 1.1).

Hunting

Hunting describes the harvest of essentially wild populations. Management involves little more than correctly estimating maximum sustained yields whereas technological innovation revolves around improving methods for harvesting, handling carcases, and distributing products. Hunting is practised for subsistence, commercial, and recreational purposes.

Subsistence hunting is practised by only a few hunter–gatherer societies but considerable quantities of bush meat still enter informal rural markets (Usher, 1976; Asibey, 1977; de Vos, 1973; Sale, 1983; Caldecott, 1986). Harvests are often unregulated and undocumented but, in a growing number of cases, quotas are allocated to communities and efforts are made to involve subsistence hunters in policy formulation and administration (Usher, 1986). Whether this form of hunting is best considered a cultural anachronism, a Stone-Age curiosity worth preserving, or an under-estimated appropriate technology for rural development is still debated.

Commercial hunting is distinguished from subsistence hunting in that it serves formal markets. It is perhaps most practical in centrally planned economies where uncertainties of public/private ownership do not complicate matters and where harvests can be closely regulated. Generally, *reduction cropping* in which over-population is periodically corrected by commercial harvest is more profitable than *sustained cropping* in which lower animal densities and behavioural modification from repeated harvests reduce procurement efficiency.

Sport hunting is practised mainly for recreational reasons but nevertheless

Fig. 1.1. Ordination of wildlife production systems.

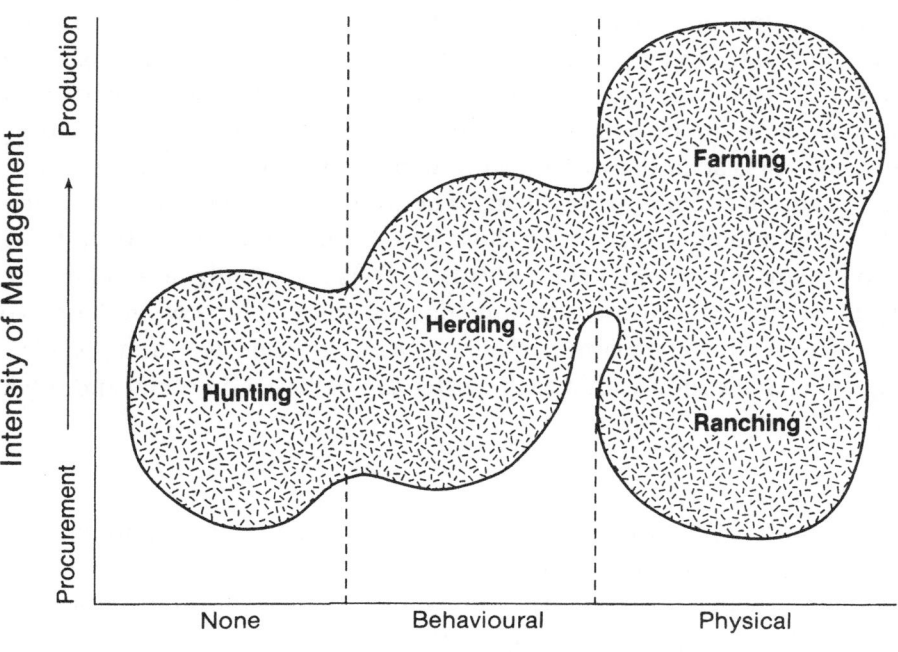

Control of Distributions

contributes significant quantities of meat. Sport hunting is organised in several ways. Simplest is licensed hunting in which harvests are regulated by issuing licences specifying species, sex, age, time, and/or place. Usually, unlimited numbers of licences are sold but, where allowable offtakes are low relative to expected demand, draws may be conducted. Safari hunting is based on provision of guiding and outfitting services by concessionaires to (usually) non-resident hunters. On private land, hunters may be required to pay various fees for access, services, leases, or trophies. The responsibilities of landowners and hunters are most carefully integrated in the revier system of continental Europe.

Herding

Herding denotes systems in which animal distributions are critically controlled by behavioural means (luring, herding, habituation, taming, etc.). The definition is a bit broader than applied to conventional domestic

Table 1.1. *Classification of wildlife production systems.*

HUNTING
   Subsistence hunting
     Traditional hunting
     Rural market hunting
   Commercial hunting
     Reduction cropping (culling)
     Sustained cropping
   Sport hunting
     Licence system
     Safari hunting (guiding/outfitting)
     Fee hunting
     Revier hunting

HERDING
   Seasonal mustering
   Free-camp system
   Loose herding
   Close herding
   Boma system

CONTAINMENT
   Ranching
   Farming
     Pasture systems
     Feedlot systems
     Confinement systems

animals, ranging from simple reinforcement of range-use traditions to close herding with periodic (seasonal, daily) confinement.

Seasonal mustering is one of the least intensive systems. Where animals are predictable in their seasonal movements, they can be captured at strategic times and locations for such things as shearing, velvet cutting, or selection of slaughter stock. Very little habituation is required. With the availability of aircraft and off-road vehicles, reindeer husbandry, particularly in Alaska and northern Canada, has become less intensive. Experimental systems at Pampa Galeras, Peru, in which vicuna are collected for shearing is another example (Rabinovitch, Hernandez & Cajal, 1985).

Free-camp systems are based on reinforcing home range traditions by baiting with managed habitats, salt, supplementary feed, or by periodic herding. The best examples are from reindeer husbandry in parts of the Soviet taiga (Chapter 10). Experimental systems of moose husbandry developed in the Soviet Union (Chapter 20) also are based on this principle.

Loose and *close herding* both imply rotation among seasonal ranges, differing mainly in levels of herd supervision. In loose herding, animals are not tended continuously except in certain seasons as in tundra and mountain forms of reindeer husbandry. Close herding differs by degree, involving almost constant tending by herders. *Boma systems* are those in which animals are penned at night to prevent predation or stock theft. Although they are used primarily for domestic livestock, at least one experiment in Kenya has evaluated traditional boma systems for eland, buffalo, and oryx in an effort to improve conditions for local pastoral peoples (King & Heath, 1975; King, Heath & Hill, 1977).

Containment

Containment systems are those in which animal distributions are critically controlled by physical barriers. Containment has become necessary with increasing costs of labour, and more practical with the availability of better and less expensive fencing. Containment offers opportunities for either more extensive or more intensive husbandry. The terms *ranching* and *farming* conventionally define this gradient.

Ranching is used in various contexts, but here it refers to populations which are fenced but otherwise managed as wild animals. It is distinguished from hunting in that animals are operationally owned (sometimes hedged by legal definition) by the enterprise and isolated by fencing from wild animals held in trust by the state. Usually, an assemblage of wild species is managed on natural vegetation and they are harvested in the field. Management focuses on range management with soft technologies such as range burning and selective harvesting.

Farming denotes intensive husbandry on fenced properties in which supplemental feeding, controlled breeding, and veterinary treatment are employed. Typically, supplementary feeding is practised seasonally and stock is transported to slaughter facilities for *ante-mortem* veterinary inspection. Farming is rapidly becoming the modal production strategy of the eminently successful world deer industry. *Pasture systems* are most practical and widespread. Industry stratification has occurred in some places and slaughter stock is finished in *feedlot systems*. The most intensive system, *confinement rearing*, is widely practised in zoos but it is rather uncommon as a commercial production system because of heavy requirements for capital and labour. Generally, it is viable only for highly valued special products such as musk (Zhang, 1983) or velvet antlers (Zhang, 1982).

Origins and emergence of game husbandry
To be of value, a classification system should lend insight into the emergence of animal production systems as well as simply segregate modern manifestations into approximately equal intuitive parts. As a backdrop to understanding future options, this section reviews the forces and events that may have forged present-day diversity of management systems.

From the classical Greeks to 19th-century historians, it was assumed that human cultural evolution passed through three distinct, sequential, and progressive stages: hunting–gathering, pastoral–nomadic, and agricultural–settled (Khazanov, 1984). Today, a more flexible view is taken. It is no longer clear that intensification of animal agriculture has been simply the result of technological innovation in a quest for ever increasing efficiency. It is better viewed in terms of input substitution see Fig. 1.2.

What is considered efficient depends on what resources (principally land, labour, capital, and cultural energy) are considered limiting. *Productivity* is a deceptive measure of efficiency in which land is considered the relevant resource. It follows that the viability of alternative production systems varies with the relative availability of resources. Options are constrained as much by available labour and technology as by environment and biology. What is practical in one setting is not in another. This section considers factors that may have favoured various transformations in animal agriculture.

From hunting to herding
A principal goal of primitive hunters was to improve harvest efficiency by developing new technologies and a better understanding of animal behaviour; insights which have long since been lost. Sadly, we are left with few evidences of the latter except inferences from artifacts left in the

archaeological record. Although conventional dogma holds that primitive hunters lived in equilibrium with the resources upon which they depended, the coincidence of megafaunal extinctions in the late Pleistocene and the path of human migration suggests that Palaeolithic hunters were able to influence populations even of large dangerous game (Martin, 1973). Although this proposition has been debated (Meltzer, 1986), historical and even contemporary evidence points to the effectiveness of human predation with or without modern weapons.

Once hunting began to have a significant impact on animal populations, attempts would be made to rationalise harvests particularly where human mobility was limited by technology or dependence on certain focal resources such as wood, stone, salt, or food/medicinal plants. Anthropologists point to a variety of taboos and other sanctions which served to limit harvests according to sex, age, and season. But this required some form of communal ownership (tribal territoriality) since it is difficult to manage open-access common-property resources.

Territories of specialised hunters tended to encompass the natural seasonal ranges of the animals on which they depended. The earliest evidence comes

Fig. 1.2. Substitution of land, labour, capital and cultural energy in ungulate production systems.

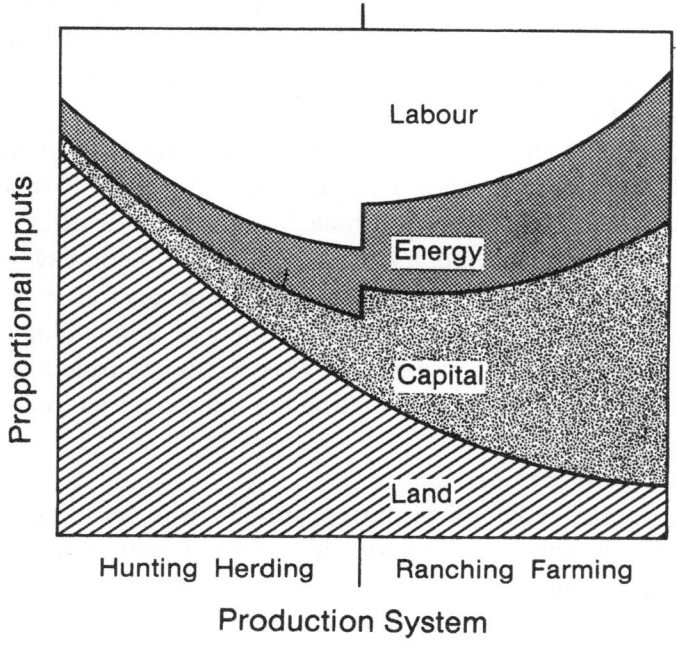

from Palaeolithic reindeer economies of western Europe (Sturdy, 1975). The pattern is clearer in the specialised bison economies of western North America because they were still intact at the time of European contact. Gordon (1979) identified two major bison herds in western Canada and provided evidence they were managed by different cultural groups. Where tribal territories could not encompass seasonal movements, regulation of harvests by other groups required control of animal distributions. This could be achieved most easily by habitat manipulation and/or artificial feeding to sharpen range-use traditions or perhaps by shortstopping extreme movements.

Fire has long been used in this way and even Palaeolithic man may have been an important agent of environmental change (Lewis, 1972). Indians of western North America used fire to control distributions of bison, particularly when commercial opportunities developed to serve the fur trade (Ray, 1974). The objective was to hold bison on tribal hunting grounds distant from trading posts thus increasing dependence of European traders and improving market demands. It also seems that winter feeding was practised in the European Mesolithic and early Neolithic, perhaps for 15,000 years preceding domestication of conventional farm animals (Simmons & Dimbleby, 1974).

More active and absolute control is achieved by herding. In its simplest form, it involves shortstopping seasonally extreme movements, a practice which has persisted to recent times. During the 1880s, the seasonal northward migration of bison was shortstopped by a cordon of soldiers to disadvantage Indians fleeing to Canada following an altercation with the US Army. Even today, Masai pastoralists sometimes shortstop wildebeest (*Connochaetes taurinus*) migrations to prevent transmission of malignant catarrhal fever to their livestock.

This change from hunting to incipient herding would occur only where it was practical and necessary. One such setting would be mountainous habitats where animal distributions were restricted seasonally and hence extremely vulnerable to over-hunting. Natural landscapes would channel seasonal movements making such species potentially manageable. Suitable species would be moderately gregarious with well-developed range-use traditions.

Incipient herding probably was a feature of at least some Mesolithic economies of Pleistocene Europe. Although it is difficult to reconstruct utilisation patterns, these economies probably were much more than simple hunting systems and therefore warrant distinction (Jarman, 1976). Reindeer (Sturdy, 1975), red deer (Jarman, 1972), fallow deer (Poplin, 1979), and gazelles (Legge, 1972) apparently were husbanded in some manner for millenia before the advent of the Neolithic.

Certain pastoral systems such as reindeer herding seem to be a natural development from specialised hunting economies. The main difference was simply that more absolute control was exerted over the daily movements of animals. This change would be necessary in the face of increasing human populations and sharper conflicts over local resources. Even today, the main reason for close herding is to minimise stock theft. Although productivity probably was sacrificed, risk was minimised, and a daily supply of animal products was ensured.

Classical domestication

Although classical domestication could be envisaged as a natural progression from hunting and incipient herding, it does not entirely explain dramatic changes at the advent of the Neolithic when familiar farm animals were domesticated. Although herding was adopted by hunting societies (Zvelebil, 1986), available evidence suggests that domestication of most ungulates was initially achieved by settled agrarian cultures rather than by nomadic hunters (Khazanov, 1984). In the Near East, gazelles and fallow deer were abruptly replaced by sheep and goats at a time when harvesting and later cultivating cereals was adopted by Natufian cultures (Henry, 1985).

It is difficult to re-create the conditions under which domestication took place. Indeed, it is not altogether clear that it was first practised to secure food supplies. Necessarily large herd sizes, long stockup periods, and low productivity of early domesticates would have been serious limitations. Where the option to hunt remained, it is unlikely to have been a viable alternative (Carr, 1977).

In some cases, the first step in the developing symbiosis may have been taken by animals rather than man. Conditions would have been right in early settled agrarian communities where land-use changes and reduced dependence on hunting would allow habituation to occur and the relationship to develop. The ease with which this occurs is evident in the habituation of animals in parks and the often serious depredation experienced by small farmers in areas still supporting wild ungulates.

At first, since contact probably was seasonal, animals might have been captured and stockaded to store meat on the hoof rather than to establish breeding populations. Further development would require the nutritional security of settled agriculture to allow experimentation but, perhaps more importantly, to escalate the value of animals beyond that of simply food. The time and resources required to rear animals for slaughter would be repaid only once animals offered sufficient prestige or value as items of exchange.

This principle is illustrated by Simoons & Simoons (1968) who proposed that loose herding of mithan, the domestic gaur (*Bibos gaurus*), may be a useful model of bovine domestication. Mithan usually roam freely in the densely forested hill country of Assam often interbreeding with wild gaur. Control is exercised simply by baiting with salt but the animals are docile and often wander through the villages. Hill tribes use the animal for sacrifices, settling debts, and other cultural purposes. Although the meat is eaten, there is little evidence that food production is a primary goal. Most men own only several animals, certainly fewer than would meet their family's nutritional subsistence were this important. But ownership confers prestige and this value alone apparently led to domestication. Until relatively recently, Bali cattle (*Bibos javanicus*) were managed in a similar manner (National Research Council, 1983; Mason, 1984). However, an earlier entry to a cash economy led to rationalisation and commercialisation of production.

Physical barriers to control animal movements became important when alternate demands for labour, and crop damage from habituated animals became more acute. Given the frailty of primitive structures and the physical prowess of wild ungulates, it is not surprising that such systems were slow to develop. Nevertheless, many of the traditional farm animals probably were tethered or stockaded and hand-fed during the early phases of their domestication. Size reduction and appearance of anomalies suggests rearing under less than ideal conditions. Increased variability suggests that animals were protected from predation and kept in genetically isolated and probably inbred populations.

Specialised pastoralism

Once the market value of animal products was established, animal production may have been specialised on marginal lands surrounding agricultural settlements. Large herd sizes would have required grazing increasingly distant pastures (Khazanov, 1984). Genetic changes accumulated during initial phases of domestication undoubtedly facilitated this transformation.

Given the modest reproductive and growth rates of large ungulates, it would have been rather impractical for herders to subsist on a harvestible surplus of slaughter stock unless exchanging goods with agriculturalists were possible. Exploitation of even more marginal lands isolated from opportunites for trade with agricultural settlements required specialisation of bioeconomic relationships. If a pastoral family were to make a complete commitment to their livelihood as herders, they needed animals which could be parasitised rather than consumed as prey; i.e., used as *interest* rather than *capital* resources. In addition to relying more heavily on interest resources such as milk or blood, there was an advantage in diversifying herds to include both small and large stock (camels, cattle, sheep, and goats) which provided the necessary nutritional contingencies to cope with harsh, unpredictable environments (Dahl & Hjort, 1976).

Containment systems

Menageries were a curious feature of affluent societies in the ancient world. Best known are the wild species (gazelles, oryx, and addax) apparently husbanded in the Old Kingdom of Dynastic Egypt (*c*. 2686–2181 BC). Pictographs provide only a hint of how they were kept and the purpose they served (Smith, 1969).

Ancient Assyrian and Persian nobles maintained hunting parks or *paradises* (Anderson, 1985). The Greeks shared this practice, although perhaps not on such a grand scale, primarily to provide sport and military training for young men defending the city states. Evidence that wild animals were husbanded more intensively for other purposes comes from the Greek term *theriotrophium*, literally mammal feeding areas (Anderson, 1985).

In urban centres of the Roman Empire, wild animals were used in the Sacred Games (Toynbee, 1973). Through time, increasingly spectacular displays of menageries were transformed into orgies of slaughter in which animals were goaded into combat with other animals or men. For example, 9000 animals were sacrificed during Titus's year-long dedication of the Colosseum in AD 80. Beyond the cruelty, the Roman games contributed to large-scale depletion of wildlife throughout mediterranean Europe and North Africa.

In the Roman countryside, native and exotic animals were maintained for sport and meat production in brick or wooden lattice enclosures called warrens or *leporia*. Since these enclosures were necessarily small, supplemental feeding was practised. From Varro's dialogue on country life quoted by Anderson (1985):

> Boars, too, can be kept in a warren with very little trouble, both captives and tame animals born on the premises, and, as you know, Axius, it is customary to fatten them. On the estate at Tibur that Varro here bought from Piso, a trumpet was sounded at regular hours and you saw boars and wild goats come for their food, which was thrown down from the exercise ground above, mast for boars and for the goats vetch or something else. Indeed, when I was at Hortensius's place at Laurentum, I saw the thing done more in the manner of the Thracian bard. There was a wood of more than 50 acres, so our host told us, surrounded by a park wall, which he called not a warren but a chase. There on high ground dinner was laid, and

as we were banqueting Hortensius ordered Orpheus to be summoned. He came complete with robe and lute, and was bidden to sing. Thereupon he blew a trumpet, and such a multitude of stags and boars and other four-footed beasts came flooding round us that the sight seemed as beautiful to me as the hunts staged by the aediles in the Circus Maximus – at least, the ones without African beasts.

Another Roman agricultural writer, Lucius Junius Moderatus Columella, provided greater detail on husbandry practices (Forster & Heffner 1968):

Wild creatures such as roebucks, chamois and also various kinds of antelopes, deer and wild boars sometimes serve to enhance the splendour and pleasure of their owners, and sometimes to bring profit and revenue ... if the cheapness of stone and labour make it advisable, certainly a wall of stone and lime is put round it; otherwise it is made of unburnt brick and clay. When neither of these methods serves the purpose of the master of the house, reason requires that they should be shut up with a post fence ... In this manner you can even close very wide regions and tracts of mountains ... the careful head of a household ought not to be content with the foods which the earth produces by its own nature, but, at the seasons of the year when the woods do not provide food, he ought to come to the help of the animals which he has confined with the fruits of the harvest which he has stored up, and feed them on barley or wheat-meal or beans, and especially, too, on grape-husks; in a word, he should give them whatever costs the least. Also in order that the wild creatures may understand that provision is being made for them one or two animals which have been tamed at home, and which, roaming through the whole park, may direct the hesitating creatures to the fare offered to them. It is advisable that this should be done not only during the scarce season of winter but also when those which were with young have brought them forth, so they rear them better ... But neither the antelope nor the wild boar nor any other wild creature should be allowed to live to a greater age than four years. For up to that time they advance in growth, after it they grow old and lean; and so they should be turned into cash while a vigorous time of life preserves their bodily comeliness. The deer, however, may be kept for many years, ... because it has been allotted a life of longer duration.

Enclosed deer parks became quite common in feudal Europe. By the year 1350, the number in Britain alone is placed at several thousand (Fletcher,

1982). Deer were contained by ditches with the spoil pile topped by thorn hedges or wooden structures so maintenance requirements were high. Popularity declined with the dismantling of feudalism. Deer farming became rather labour and capital intensive for all but the more wealthy estate-owners.

In North America, game farming attracted interest following the decline of wild populations in the 1800s (Moodie & Kaye, 1976). Perhaps because of their similarity to cattle, bison were considered earliest. But interest in white-tailed deer (*Odocoileus virginianus*) and especially wapiti (*Cervus elaphus*) was high. At first, fenced game parks were operated for hunting by clubs or individuals as in Europe (Tober, 1981). Later, the agricultural opportunity was promulgated (Lantz, 1910). Although considered to have a positive effect on conservation at the time, development was interrupted by attempts to curb market hunting and except for the effective domestication of the bison, game farming was all but forgotten until its recent rejuvenation.

From the time of Columella, fencing has been the critical obstacle to widespread adoption of game ranching and farming. Modern game fencing represents an important technological innovation which has opened a whole new set of opportunities. Indeed, this century has introduced more new agricultural animals than ever before in human history. No longer must commercial populations be slowly built from hand-reared orphans or the progeny of stockaded wild stock – means available at the advent of the Neolithic. Game production is now a practical alternative for the average farmer.

The future

Man–animal relationships will continue to evolve as they always have. With changing circumstances, different species and management systems gain prominence. Changes often are quite abrupt; for example, the Natufian gazelle economy was suddenly replaced by sheep and goat husbandry in the Levant, red deer and wild boar were replaced by cattle in Europe, and onagers were replaced by horses in the Middle East.

Domestication of new plants and animals has continued at an almost constant rate since the Neolithic, punctuated with periods of vigorous experimentation. Generally, these have been post-expansion societies with sufficient pressure on resources for a need but sufficient affluence to afford venture capital and to support specialty markets for meat or hunting. The current wave of interest is consistent with this historical pattern.

There is the question of how history will perceive these developments. Will this century be considered a Neo-Neolithic in which the spectrum of food-producing animals was broadened? In the frame of technical and cultural evolution, will it be considered a fleeting fancy similar to that of the ancient Egyptians and Romans? Or will it be seen as another tragic chapter in the enslavement of animals and appropriation of the earth's resources to serve mankind?

References

Anderson, J. K. (1985). *Hunting in the Ancient World.* Berkeley: University of California Press.

Asibey, E. O. A. (1977). Wildlife as a source of protein in Africa south of the Sahara. *Biological Conservation*, 6, 32–9.

Brothwell, D. R. (1975). Salvaging the term 'domestication' for certain types of man–animal relationships: the possible value of an eight point scoring system. *Journal of Archeological Science*, 2, 397–400.

Caldecott, J. O. (1986). *Hunting and Wildlife Management in Sarawak.* Kuching: World Wildlife Fund Malaysia.

Carr, C. (1977). Why didn't the American Indians domesticate sheep? In *Origins of Agriculture*, ed. C. A. Reed, pp. 637–93. Hague, Paris: Mouton.

Cohen, M. N. (1977). *The Food Crisis in Prehistory.* New Haven: Yale University Press.

Collier, S. & White, P. (1976). Get them young? Age and sex inferences on animal domestication in archaeology. *American Antiquity*, 41, 96–102.

Dahl, G. & Hjort, A. (1976). *Having Herds: Pastoral Herd Growth and Household Economy.* Stockholm Studies in Social Anthropology, Vol. 2. Stockholm: University of Stockholm.

Darling, F. F. (1960). Wildlife husbandry in Africa. *Scientific American*, 203, 123–34.

Dasmann, R. F. (1964). *African Game Ranching.* London: Pergamon.

de Vos, A. (1977). Game as food. A report on its significance in Africa and Latin America. *Unasylva*, 29, 2–12.

Eltringham, S. K. (1984). *Wildlife Resources and Economic Development.* New York: Wiley & Sons.

Field, C. R. (1979). Game ranching in Africa. *Applied Biology*, 4, 63–101.

Fletcher, J. (1982). United Kingdom. In *The Farming of Deer: World Trends and Modern Techniques*, ed. D. Yerex, pp. 45–54. Wellington: Agricultural Promotion Associates.

Forster, E. S. & Heffner, E. H. (trans.) (1968). *Lucius Junius Moderatus Columella on Agriculture.* Vol. II. Loeb Classical Library No. 407. London: Heinemann.

Gordon, B. H. C. (1979). *Of Men and Herds in Plains Prehistory.* National Museum of Man Mercury Series, Archaeological Survey of Canada Paper 84. Ottawa: National Museums of Canada.

Henry, D. O. (1985). Preagricultural sedentism: The Natufian Example. In *Prehistoric Hunter–Gatherers: The Emergence of Cultural Complexity*, ed. T. D. Price & J. A. Brown, pp. 365–84. Orlando: Academic Press.

Higgs, E. S. (ed.) (1972). *Papers in Economic Prehistory.* Cambridge: Cambridge University Press.

Higgs, E. S. (ed.) (1975). *Paleoeconomy.* Cambridge: Cambridge University Press.

Ingold, T. (1980). *Hunters, Pastoralists and Ranchers: Reindeer Economies and their Transformations.* Cambridge: Cambridge University Press.

Jarman, M. R. (1972). European deer economies and the advent of the Neolithic. In *Papers in Economic Prehistory*, ed. E. S. Higgs, pp. 125–47. Cambridge: Cambridge University Press.

Jarman, M. R. (1976). Early animal husbandry. *Philosophical Transactions of the Royal Society of London*, B 276, 85–97.

Jarman, M. R. & Wilkinson, P. F. (1972). Criteria of animal domestication. In *Papers in Economic Prehistory*, ed. E. S. Higgs, pp. 83–97. Cambridge: Cambridge University Press.

Khazanov, A. M. (1984). *Nomads and the Outside World*. Cambridge: Cambridge University Press.

King, J. M. & Heath, B. R. (1975). Game domestication for animal production in Africa. *World Animal Review*, 16, 23–30.

King, J. M., Heath, B. R. & Hill, R. E. (1977). Game domestication for animal production in Kenya: theory and practice. *Journal of Agricultural Science (Cambridge)*, 89, 445–57.

Lantz, D. E. (1910). *Raising Deer and Other Large Game Animals in the United States*, Bulletin Number 36. Washington: US Department of Agriculture, Biological Survey.

Legge, A. J. (1972). Prehistoric exploitation of the gazelle in Palestine. In *Papers in Economic Prehistory*, ed. E. S. Higgs, pp. 119–24. Cambridge: Cambridge University Press.

Lewis, H. T. (1972). The role of fire in the domestication of plants and animals in Southwest Asia: a hypothesis. *Man*, 7, 195–222.

Martin, P. S. (1973). The discovery of America. *Science*, 179, 969–74.

Mason, I. L. (ed.) (1984). *Evolution of Domesticated Animals*. London & New York: Longman.

Meltzer, D. J. (1986). Pleistocene overkill and the associational critique. *Journal of Archaeological Science*, 13, 51–60.

Moodie, D. W. & Kaye, B. (1976). Taming and domesticating the native animals of Rupert's Land. *Beaver*, 307 (winter), 10–19.

National Research Council. (1983). *Little-Known Asian Animals with a Promising Economic Future*. Washington: National Academy Press.

Poplin, F. (1979). Origin of the Corsican mouflon in a new palaeontological perspective: by feralizing. *Annales Genetique Selection Animale*, 11, 133–43.

Price, T. D. & Brown, J. A. (eds.) (1985). *Prehistoric Hunter–Gatherers: The Emergence of Cultural Complexity*. Orlando: Academic Press.

Rabinovich, J. E., Hernandez, M. J. & Cajal, J. L. (1985). A simulation model for the management of vicuna populations. *Ecological Modelling*, 30, 275–95.

Ray, A. J. (1974). *Indians in the Fur Trade: Their Role as Hunters, Trappers and Middlemen in the Land Southwest of Hudson Bay, 1660–1870*. Toronto: University of Toronto Press.

Reed, C. A. (1984). The beginnings of animal domestication. In *Evolution of Domesticated Animals*, ed. I. L. Mason, pp. 1–6. London: Longman.

Rindos, D. (1984). *The Origins of Agriculture: An Evolutionary Perspective*. New York: Academic Press.

Sale, J. B. (1983). *The Importance and Value of Wild Plants and Animals in Africa*. Gland: IUCN.

Simmons, I. G. & Dimbleby, G. W. (1974). The possible role of ivy (*Hedera helix* L.) in the mesolithic economy of western Europe. *Journal of Archaeological Science*, 1, 291–6.

Simoons, F. J. & Simoons, E. S. (1968). *A Ceremonial Ox of India. The Mithan in Nature, Culture, and History*. Madison: University Wisconsin Press.

Smith, H. S. (1969). Animal domestication and animal cult in Dynastic Egypt. In *The Domestication and Exploitation of Plants and Animals*, ed. P. J. Ucko & G. W. Dimbleby, pp. 307–16. London: Duckworth.

Sturdy, D. A. (1975). Some reindeer economies in prehistoric Europe. In *Palaeoeconomy*, ed E. S. Higgs, pp. 55–95. Cambridge: Cambridge University Press.

Tober, J. A. (1981). *Who Owns the Wildlife: The Political Economy of Conservation in Nineteenth Century America*. Westport (Connecticut): Greenwood Press.

Toynbee, J. M. C. (1973). *Animals in Roman Life and Art*. London: Thames & Hudson.

Usher, P. J. (1976). Evaluating country food in the northern native economy. *Arctic*, 29, 105–20.

Usher, P. J. (1986). *The Devolution of Wildlife Management and the Prospects for Wildlife Conservation in the Northwest Territories*. Ottawa: Canadian Arctic Resources Committee.

Walker, B. H. (1979). Game ranching in Africa. In *Management of Semi-Arid Ecosystems*, ed. B. H. Walker, pp. 55–81. Amsterdam: Elsevier.

Weber, N. A. (1966). Fungus-growing ants. *Science*, 153, 587–604.

Wilkinson, P. F. (1972). Current experimental domestication and its relevance to prehistory. In *Papers in Economic Prehistory*, ed. E. S. Higgs, pp. 107–18. Cambridge: Cambridge University Press.

Wilkinson, P. F. (1976). 'Random' hunting and the composition of faunal samples from archaeological excavations: a modern example from New Zealand. *Journal of Archaeological Science*, 3, 321–8.

Yerex, D. (ed.) (1982). *The Farming of Deer. World Trends and Modern Techniques*. Wellington: Agricultural Promotion Associates.

Zhang, B. (1983). Musk deer, their capture, domestication and care according to Chinese experience and methods. *Unasylva*, 35, 16–24.

Zhang, P. (1982). China. In *The Farming of Deer: World Trends and Modern Techniques*, ed. D. Yerex, pp. 11–14. Wellington: Agricultural Promotion Associates.

Zvelebil, M. (ed.) (1986). *Hunters in Transition: Mesolithic Societies of Temperate Eurasia and their Transition to Farming*. Cambridge: Cambridge University Press.

2

International trade

RICHARD A. LUXMOORE

Abstract

Monitoring the international trade in wildlife products is one way of investigating the importance of wildlife production systems around the world. In many countries, subsistence hunting or domestic trade may be of far greater importance in terms of volume, but export trade has particular economic significance in that it generates foreign exchange. International trade in game meat, estimated from Customs statistics, amounts to at least 30,000 tonnes/year (worth approximately US$ 100 million), less than half of which is probably attributable to large ungulates. The skins of some ungulates are traded, but they tend not to be as valuable as those of furbearing carnivores and rodents. Of far greater significance is the trade in medicinal products, particularly velvet, the soft antlers of deer. The total volume of international trade in medicinal products is low, but they have a high unit value and probably account for over US$ 30 million a year. Foreign exchange also can be generated by wildlife-based tourism either through hunting or simply game viewing. This is difficult to quantify, but it may be of greater economic significance than trade in wildlife products. Although international trade in game meat is insignificant in comparison with trade in domestic livestock products, wildlife affords economic opportunities which may be very important in countries which have few other exports.

Introduction

All animal products once used by primitive man were derived from wildlife, but, with the development of domestic animal husbandry, the importance of wild animals, particularly as food, has declined. Developed societies tend to believe that virtually all animal protein is derived from agriculture, but this is to ignore the world's fishing industry which supplies about 70 million tonnes/year, the great majority from wild stocks. It is a simple matter to obtain from the UN Food and Agriculture Organisation (FAO) statistics dealing with the volume of the fish catch in different countries, but similar assessments of the contribution of terrestrial wild animals to the human diet are not readily available (Krostitz, 1979).

One approach is to estimate wildlife production from the viewpoint of the producer. The chapters in this book describe the different wildlife production systems and, where available, estimate the amounts of products derived from them. An alternative approach to estimating trade volumes is from the viewpoint of the consumer. In this chapter, I indicate the scale of international trade in game products as this is one area where some statistics are available. It must be appreciated that domestic markets and non-commercial (subsistence) uses are not quantified. In many cases, these categories may be far greater in volume, although the export trade has particular economic significance as it generates foreign exchange. Domestic production also can contribute to the balance of foreign trade by supplying what might otherwise have to be imported. It is hoped that by quantifying world markets for game products, the scale of the industries described in the ensuing chapters can be seen in context.

Consistent with the focus of this book, an attempt is made to isolate production from wild ungulates. Meat is given primary consideration, but often only the other products are traded internationally, the meat being consumed locally. Therefore, the economic significance of trade in non-meat products is assessed, although the data are less complete.

Game meat

Game meat traded internationally is recorded in published Customs statistics. These give the weight and value of each commodity traded, specifying the source country for all imports and the destination of all exports. There are two main international commodity classifications, both of which include game meat under much broader standard classifications: the Standard International Trade Classification (SITC) under commodity heading 011.89, 'meat and edible offal, not elsewhere specified'; and the Customs Co-operation Council Nomenclature (CCCN) under commodity heading 02.04 'other meat and edible offals'. Fortunately, many countries provide finer subdivision of these commodities to allow the distinction of game meat, or meat from wild animals. However, there are several major problems in compiling an analysis of international trade:

- Published Customs statistics are not available for many countries.
- Switzerland, the United States, the United Kingdom and others, do not specify game meat as a separate commodity but include it with other offals and meat products.
- Some countries use subtly different commodity classifications. For instance, the Scandinavian countries specify reindeer meat separ-

ately from game meat. Austria divides game meat into feathered
game, furred game, and rabbits/hares, the latter also including
some domestic rabbits.
   – Time periods may differ. European countries give statistics for
calendar years, whereas Australia and New Zealand report annual
statistics from July to June.

Fortunately, many of these problems can be overcome. It is possible to infer a
considerable amount about the trade balance of countries for which adequate
Customs data are not available from the records of their trading partners.

Sources and methods

A list of the Customs statistics consulted is included at the end of
this chapter, together with the commodity classifications. Where different
commodity classifications are shown for the same country, the values were
added together. Weights were converted to metric tonnes, and the financial
value to US$ from the local currencies, using the International Financial
Statistics published by the International Monetary Fund.

Annual trade from country *a* to country *b* was estimated from either the
volume of exports to country *b* reported by country *a* or of imports from
country *a* reported by country *b*, the larger of the two values being taken for
each pair of trading partners. The gross exports from each country were then
estimated from the sum of its reported export trade to all other countries, and
its gross import trade from the sum of its imports from all other countries.
Minimum net export and net import trades were then calculated as the
difference between gross exports and gross imports, negative values being
ignored in both cases. The minimum volume of international trade was then
estimated as either the sum of net imports or of net exports, both values being
equal. All calculations were performed separately on the weights and the
financial values of the transactions.

Volume of trade

Total annual volumes of international trade in game meat varied
from 30–35 kilotonnes between 1980–85, for a value of US$ 85–146 million
(Table 2.1). West Germany imported the largest quantity, accounting for
over half of the total trade. The next major importer was France, with about
25% of the total trade, followed by Belgium–Luxembourg, Italy, and
Switzerland, all importing similar quantities. The major exporter was Argentina, accounting for around a third of the total trade, followed by the United
Kingdom, Hungary, Poland, South Africa, Spain, China, Czechoslovakia,
New Zealand, the Soviet Union, Yugoslavia, and Austria (Table 2.2).

Table 2.1. *Estimated minimum net imports of game meat (tonnes).*

Net importers	1980	1981	1982	1983	1984	1985
Algeria	—	—	1	1	—	—
Angola	—	—	—	—	—	1
Andorra	—	—	13	14	17	1
Belgium-Luxemb.	2,135	1,615	1,550	1,625	1,688	1,965
Cameroon	—	—	9	—	—	10
Congo	—	—	—	—	—	3
Denmark	290	209	252	132	144	225
Djibouti	—	—	—	—	—	3
Faroe Is.	—	—	—	1	—	—
France	8,313	8,353	7,210	9,241	8,088	8,696
French Guiana	—	—	—	11	—	18
French Polynesia	—	—	—	—	—	10
Gabon	11	14	21	18	20	20
Germany, West	19,253	19,656	17,147	15,692	16,887	17,431
Guadeloupe	—	—	12	18	14	14
Hong Kong	6	5	7	3	—	14
Iraq	—	—	—	1	1	3
Italy	2,029	1,551	1,198	2,085	1,213	2,900
Ivory Coast	—	—	—	—	—	4
Japan	2	3	33	10	36	58
Kuwait	—	1	—	19	—	—
Lebanon	46	—	21	—	—	—
Martinique	—	21	21	25	29	22
Netherlands	1,728	*	*	*	*	*
New Caledonia	—	—	—	—	—	7
Nigeria	—	—	1	—	—	—
Norway	—	411	483	153	*	158
Oman	—	—	—	—	—	2
Portugal	—	—	2	—	—	—
Reunion	24	23	13	21	15	19
Saudi Arabia	52	—	7	—	—	13
Seychelles	—	—	—	—	—	1
Singapore	—	—	—	—	—	1
St Pierre	—	—	—	—	—	1
Sweden	*	519	739	120	903	1,153
Switzerland	1,383	1,604	1,301	1,467	1,658	1,541
Togo	—	—	—	—	—	1
UAE	20	28	17	—	12	—
USA	*	*	—	2	11	27
Zaire	—	—	—	—	1	—
Country unknown	*	30	43	*	*	*
Total tonnes	35,292	34,043	30,101	30,659	30,737	34,322
US$ million	146	122	97	85	87	122

*Net exporter in this year.

Table 2.2. *Estimated minimum net exports of game meat (tonnes).*

Net exporters	1980	1981	1982	1983	1984	1985
Algeria	2	—	—	—	—	—
Argentina	12,098	10,468	9,304	9,599	8,986	11,627
Australia	382	1,860	1,403	337	146	103
Austria	1,015	999	873	953	192	718
Brazil	—	—	10	22	20	31
Bulgaria	108	58	31	150	194	174
Canada	24	—	—	—	—	—
Chile	—	50	49	189	137	147
China	1,444	876	853	1,719	2,110	1,528
Czechoslovakia	977	918	1,180	1,184	1,488	1,316
Finland	24	53	21	31	95	59
Germany, DR	—	—	—	—	—	32
Greece	—	—	—	—	30	17
Greenland	39	51	51	41	56	38
Hungary	3,086	2,534	2,369	2,934	2,763	3,097
Iceland	—	—	—	—	—	16
Ireland	26	26	52	70	54	63
Israel	—	8	—	—	—	—
Mongolia	121	133	138	167	—	166
Morocco	—	—	—	—	14	15
Netherlands	*	627	746	710	350	313
New Zealand	1,017	1,641	1,197	929	831	1,291
Norway	—	*	*	*	6	*
Poland	1,729	2,217	1,858	3,449	3,119	2,850
Romania	678	429	523	291	293	89
South Africa	3,479	2,197	1,546	560	835	1,033
Spain	1,296	1,560	1,620	1,574	1,758	1,632
Sweden	270	*	*	*	*	*
Tunisia	—	26	—	28	34	10
Turkey	33	—	—	—	—	—
UK	5,041	4,761	3,857	4,061	4,788	4,695
Uruguay	358	237	181	218	280	495
USA	37	37	—	*	*	*
USSR	883	1,133	1,321	578	1,075	1,433
Yugoslavia	1,081	1,144	918	850	1,040	1,278
Country unknown	44	*	*	15	43	56
Total tonnes	35,292	34,043	30,101	30,659	30,737	34,322

*Net importer in this year.

The accuracy of these estimates is influenced by two major systematic sources of error: (1) Customs statistics were not available or suitable for many countries; and (2) levels of trade reported by one trading partner were seldom equal to those reported by the other.

Estimating the first source of error is difficult. If correlations between reported imports and exports are strong, Customs reports of only the importing countries could be used to estimate total trade. Of the major exporters, only Spain, Yugoslavia and Austria provided sufficiently detailed statistics for this analysis; whereas for importers, only Switzerland's Customs statistics could not be consulted. However, the Customs reports largely exclude non-European countries. For example, New Zealand Customs report venison exports of 1269, 1591, and 1379 tonnes annually from 1981/82 to 1983/84. Of this amount, 72%, 74%, and 55%, for the three periods, respectively, were reported as exported to the European countries forming the basis of the present analysis. The drop in the proportion in the final year was largely due to increased exports to the United States. A similar consideration of Australian Customs records of kangaroo meat exports (Dixon, 1985) shows that between 1980/81–82/83, 93%, 81%, and 79% of annual exports went to European countries.

The second source of error is easier to quantify. In a comparison of selected

Table 2.3. *Comparison of trade between trading partners whose Customs reports were analysed.*

Country	Exports reported by		Imports reported by	
	Country	Importers	Country	Exporters
Austria	9,393	10,178	698	21,227
Belgium—Luxemburg	541	389	6,419	7,118
Denmark	170	135	496	329
Finland	173	212	5	20
France	4,120	4,803	10,999	13,740
Germany,West	6,136	4,061	15,321	15,606
Italy	3,828	2,718	887	1,641
Netherlands	8,856	4,847	3,104	2,856
Norway	512	743	569	796
Spain	2,921	4,450	0	9
Sweden	1,877	1,439	1,179	708
Yugoslavia	6,661	5,702	0	17
Total (tonnes)	45,188	39,677	39,677	64,067

trading partners with suitable trade statistics, reported imports accounted for 88% of reported exports (Table 2.3). Two major discrepancies were apparent; the first concerns exports reported by the Netherlands, which were almost double the corresponding quantity of imports reported by other countries; and the second concerns imports reported by Austria, which were considerably less than reported by the exporters. This suggests that some systematic error may be responsible, such as a broader classification of game meat used by the Netherlands. Austria's main supplier, Yugoslavia, included a separate category for game offals which were not included in the Austrian reports. There was a surprisingly high correlation between the remaining transactions.

The method of calculating minimum net trade underestimates the true volume of international trade. Several countries, notably Austria and the Netherlands, both imported and exported large volumes of game meat. Thus, calculating net trade avoids the possibility of counting meat twice if it is imported and later re-exported. If, as is probably the case, most meat imported is consumed within the country, and different meat is exported, then the gross trade would give a better estimate of trade volume. The gross trade in game meat fluctuated between 36–43 kilotonnes between 1980–85.

Composition of game meat in international trade

Customs statistics give little indication of the species composition of the international game meat trade. Austrian statistics divide meat into rabbits/hares, feathered game, and furred game. Scandinavian statistics distinguish reindeer meat, and Norway additionally distinguishes feathered game, but beyond that there is no information. Therefore, it is necessary to turn to the exporting countries to ascertain the species in trade.

Argentina, the main supplier, exports almost exclusively meat of the brown hare (*Lepus capensis*) (Mares & Ojeda, 1984), a species which has proliferated throughout the Pampas since its introduction at the turn of the century (Jackson, 1986). Jackson reported that meat exports from Argentina from 1975–83 varied between 10,000–14,200 tonnes/year, figures comparable with those in Table 2.2. Corroboration comes from Austrian import statistics, which record only 'rabbit and hare' meat from Argentina.

The second major exporter, the United Kingdom, published no separate statistics, but the total amount of venison (mainly the meat of red deer [*Cervus elaphus*] and roe deer [*Capreolus capreolus*]) approaches 2000 tonnes/ year (Anon., 1983a). Even if this were all exported, it would account for less than half of inferred exports (about 4400 tonnes) so the rest must therefore comprise small game (hares, rabbits, and birds).

Game bags in other European exporting countries similarly comprise a mixture of small game and ungulates, albeit with a higher percentage of large game than the United Kingdom. Austrian imports from other European nations comprised 87% 'furred game' and only 13% 'feathered game, hares and rabbits'.

Australian game exports, largely kangaroo meat, declined from 1671 tonnes in 1980/81 to 225 tonnes in 1983/84 (Dixon, 1985; cf. Table 2.2). Considerable quantities of meat from feral pigs also were exported; 2120 tonnes in 1984 of which 75% went to West Germany (Tisdell, 1987). It is uncertain whether Customs officials recorded feral pigs as swine or wild game. Exports from New Zealand are primarily venison (especially red deer), totalling 914 tonnes in 1981/82 and 1175 tonnes in 1982/83 to European countries whose import figures formed the basis of the analysis in Table 2.2.

South Africa exported 1446 tonnes of game meat in 1980/81 (Jooste, 1983), mostly of springbok (*Antidorcas marsupialis*) and blesbok (*Damaliscus dorcas phillipsi*) (Luxmoore, 1985), although about 1000 tonnes of ostrich meat are also exported annually (Conroy & Gaigher, 1982). Most is exported to Switzerland and therefore will not be reflected in Table 2.2.

To conclude, if all known exports of birds and small game are subtracted from the 195,154 tonnes traded between 1980–85, and if the remainder is said to comprise a minimum of 13% small game, then less than half the game meat on international markets appears to have derived from ungulates.

Patterns of trade

Many factors determine patterns of international trade. Common factors include the available supply and the location of suitable markets. Supply is affected not only by the presence of suitable game species, but by the ability to harvest them economically and by the surplus of supply over domestic demand in the source country. This may be primarily an economic surplus, realised when export prices exceed local prices. International market demand is similarly determined by a surplus of consumer demand over domestic production, and by the availability of foreign exchange to pay for it. It is also strongly affected by consumer preference for the type of game meat sold.

Vagaries of consumer preference can generate marked fluctuations. For example, in 1980, South Africa was the third largest source of game meat, but exports declined sharply to only 560 tonnes in 1983 (Table 2.2). Its exports of game meat had increased from low levels in the 1970s in response to West German demands for the small carcases of blesbok and springbok. German consumers were accustomed to roe deer and fallow deer (*Dama dama*), and were prepared to substitute similar-sized antelopes. Butchers and wholesalers saw no reason to clarify the confusion and sold the meat merely as 'venison'. The preference for small species was apparent in South Africa where premium prices were paid for them in comparison with the larger species, such as eland (*Taurotragus oryx*) and kudu (*Tragelaphus strepsiceros*) (see Luxmoore, 1985). Australia then entered the market, supplying kangaroo meat from the growing culling programme. Exports to West Germany alone exceeded 1000 tonnes by 1980/81 and 1981/82 (Dixon, 1985; cf. Table 2.2). Much of this kangaroo meat was also retailed as 'venison' until this practice was revealed by the German press in 1983. Consequently, Australian exports to West Germany dropped to 80 tonnes in 1983/84 (Dixon, 1985). Imports from South Africa similarly plunged. Consumer resistance to unidentified 'venison' was appeased by labelling the South African meat correctly as 'antelope', and sales of this 'new' product slumped. Repercussions were felt in South Africa, where the price paid to farmers for carcases fell from R 2.30/kg in 1982 to R 0.80/kg in 1983 (Luxmoore, 1985). There is some evidence that the imports from South Africa started to recover in 1984 as a result of improved marketing and advertising.

The absence of Canada and the United States from the list of major importers of game meat results from public health restrictions on imports as well as restrictions on venison sales in many states and provinces. In common with many countries, North American regulations require that all meat for human consumption must receive *ante-mortem* veterinary inspection as well as satisfy stringent requirements on treatment and storage. These constraints preclude the import of game meat shot in the wild, and this was a major factor causing New Zealand to adopt legislation which requires all farmed deer to be slaughtered in Deer Slaughter Premises which cannot be used for traditional livestock (Chapter 16).

Veterinary restrictions impose additional controls on international trade. European countries will not import meat of ungulates from countries having diseases such as endemic foot-and-mouth disease. This effectively rules out most of Africa, except those areas in South Africa where extreme measures have delimited the disease. Lagomorphs (rabbits and hares) are not affected by this regulation, and this, coupled with the general longstanding culinary acceptance of the European hare, may explain the strong demand. The meat of wild boar also is very popular in Europe, but boars are affected by several diseases transmissible to domestic swine, notably African swine fever, so international trade is severely restricted.

It is apparent that international trade in game meat is not a simple matter. Even if game can be successfully harvested, many bureaucratic impediments must be overcome before export markets can be accessed. Transport is also difficult, as the meat must be preserved, usually by refrigeration. When all these factors are considered, it is perhaps surprising that so much game meat is exported when other wildlife products are far simpler to trade.

Other products

Often, products other than meat dominate exports, the meat all being consumed locally. These commodities include furskins, leather, fibre, decorative items (such as horns, mounted heads, etc.), medicinal products (especially velvet antler and various internal organs), and live animals. Though notoriously difficult to quantify, the value of wildlife in attracting international tourists is financially comparable to exporting meat in terms of foreign exchange earnings. It includes both game viewing and recreational hunting.

Skins and furs

The most valuable furs are those of the carnivores: mustelids, felids, and canids. However the most numerous furs, in terms of trade in wild-caught skins, are from herbivorous rodents, especially muskrat (*Ondatra zibethicus*), beaver (*Castor canadensis*), and nutria (*Myocaster coypus*). All are outside the scope of this book. Few ungulates are exploited primarily for their fur or hair, and any such trade is usually a by-product of the meat industry. Some skins of deer and antelope are tanned with the hair on and sold as decorative rugs, while others and those of peccaries (Tayassuidae) are processed for leather.

Few international trade figures are available. New Zealand publishes international trade statistics detailing deer skins (CCCN classification 211.99.01), which showed that 48,305 skins were exported in 1981/82, 33,424 in 1982/83, and 29,554 in 1983/84, with respective values of US$ 439,049, US$ 369,309, and US$ 358,227, accounting for 8.7%, 6.4%, and 6.4% of the value of venison exports. The majority were purchased by West Germany and Japan.

Imports of 'deer, buck and doe skins' to the United States averaged US$ 1.2 million a year from 1976–80; 42% came from Sweden, 15% from West Germany, and 13% from Canada (Prescott-Allen & Prescott-Allen, 1986). Japanese Customs statistics show that imports of 'deer skins and elk skins' (SITC Category 41.01-260) averaged 622 tonnes/annum between 1980 and 1985 at an average annual value of US$ 2.52 million. The chief sources were China, USA, Canada, and New Zealand. Most deer carcases exported from the United Kingdom are sold with the skin still on; this is termed 'German dressed' and is common practice in Europe. The value of the skin is therefore included with that of the meat.

Grimwood (1968) estimated that the number of skins of the South American brocket deer (*Mazama americana* and *M. gouazoubira*) exported annually from Peru averaged 29,663 (7070–54,852) between 1946–66, individual skins being worth £ 0.27 each. Trade in animal products from the Amazonian region of Peru has been banned since 1973, although *M. americana* was included on a list of common species which could be hunted for subsistence purposes. In 1981, it was feared that uncontrolled hunting of skins for the export trade could jeopardise the species as a food resource; so regional export quotas were imposed. The total export quota for the whole country was 26,500 skins, but it is not known whether this was reached. Total prohibition of animal trade was enacted in Brazil in 1967, but prior to that, export of pelts of brockets from Amazonia totalled 222,859 from 1950–65 (Carvalho, 1967).

African antelope skins are generally more decorative than those of cervids, and some are exported tanned with the hair on. In Namibia in 1980, export of game hides earned R 303,652 (US$ 390,320), equivalent to about 5% of the meat revenue (Jooste, 1983). In South Africa, most meat processing and export is conducted by large companies in central abattoirs. Unlike in Europe, the carcases are usually butchered and packaged prior to export and prepared skins sold separately (Freudenberger, 1982). One of the largest processing companies, Kovisco, derives about 10% of its income from skins (Visser, 1983). Extrapolated to the export trade of about R 8 million (US$ 10.3 million) annually (Jooste, 1983), skin exports might be worth around US$ 1 million/year. The value of hides varies greatly among species. With the exception of elephant (*Loxodonta africana*), zebra (*Equus (Hippotigris) burchelli*) are by far the most valuable large herbivore, and Eltringham (1984) pointed out that several game cropping enterprises in Tanzania relied heavily on revenue from zebra skins. He showed that from 1970–72 at Loliondo Ranch, a total of TSh 507,250 was earned from zebra skins, TSh 26,286 from meat, and TSh 11,260 from other skins.

The same was true in Zimbabwe in 1983: tanned zebra skins were worth some Z$ 450 (US$ 445), whereas impala skins received only Z$ 20. Tannery purchases of antelope hides from 1978–83 averaged 13,800 a year, at a retail value of approximately Z$ 276,000 (assuming Z$ 20 a skin), and 697 zebra hides, worth about Z$ 317,000 (B. Child, personal communication). In the six months ending December 1974, the gross income of the tanning industry in Zimbabwe (then Rhodesia) was estimated to be US$ 236,132, of which 52% was from elephants, 8% from zebra, and 36% from antelope (Mossman & Mossman, 1976). It is not known what percentage of skins are exported, but if the majority were, exports would approach US$ 500,000.

The skins of South American peccaries (mostly *Tayassu tajacu* and *T. peccari*) make fine leather, used especially in glove manufacture, and they have long been exported to Europe, Japan, and North America for processing. Between 1946–66, nearly three million skins were exported from Iquitos alone (Grimwood, 1968). Figures supplied by the Peruvian Ministry of Agriculture show that between 1966–85 the exports of skins of *T. tajacu* fell from 184,201 to 26,046, averaging 100,181 annually; whereas exports of *T. pecari* skins fluctuated between 5285 and 86,261, with a mean annual export of 40,676 over the same period. In the three years prior to 1967, when all wildlife exports from Brazil were banned, 1,932,469 peccary skins were exported (Smith, 1977) and peccary skins are still the most commonly confiscated wild mammal product. Peruvian exports declined in 1982 with the imposition of export controls, but totalled 101,000 for the two species of peccary. Paraguay is another important source, exporting up to 96,000 skins a year (Broad, 1984). From 1976–79, an annual average of 43,093 skins of peccaries were exported from Argentina, at a declared export value of US$ 256,250 (Mares & Ojeda, 1984). Prices paid to hunters are low: US$ 5 for *T. tajacu* and US$ 2.50 for *T. peccari* in Paraguay in 1979 (Sowls, 1984), and there is evidence that skins are only a by-product of what is primarily meat hunting (Broad, 1984).

The South American Camelidae serve an important fibre industry. The vicuna (*Vicugna vicugna*) is perhaps best known and it has long been exploited for its extremely fine wool. It was so seriously depleted that it was effectively banned from international trade by being placed on Appendix I of the Convention of International Trade in Endangered Species of Wild Fauna and Flora (CITES). Recent conservation measures have been so effective that populations have now built up in Peru to allow a harvest of fibre by shearing herded animals, and this strictly limited trade was re-opened in 1987. The guanaco (*Lama guanicoe*) has less fine wool but it is nevertheless traded, both as spun cloth and also as whole pelts, the juveniles, or 'chulengos' being particularly favoured. Most of the surviving guanaco live in Argentina from where the majority of skins are exported (Franklin, 1982). Skins exported from 1976–79 averaged 55,902 a year, with a value of US$ 1.4 million (Mares & Ojeda, 1984). The guanaco is on Appendix II of CITES, which means that Party nations must report all trade in skins and wool products. An analysis of these annual reports (Broad *et al.*, 1988) showed that the annual international trade during the period between 1980–85 averaged 11,452 whole skins and 11,890 cloth items, almost all of which originated in Argentina.

Medicinal products

Western medicine uses many products derived from animals, mostly hormone extracts and antibodies but these are usually obtained from domestic animals. Traditional Oriental medicine places a far greater reliance on the products of a variety of wild animals, from pickled snakes to gall bladders of bears. Many of the most popular products are derived from deer; the most valuable being velvet antlers. Other products include hard antlers, tail, bones, penes, heart, liver, sinews, placenta, blood, and skin (Lee & Ch'ang, 1985) and musk (Green, 1986).

Some velvet antler is obtained from deer shot at the appropriate time of year, but most is removed from living animals. Occasionally free-living animals are herded for this purpose, as happens with reindeer, but the great majority derives from animals reared on farms. The use of antler is most prevalent in the Far East, although there is a substantial demand in expatriate Chinese communities in other parts of the world. Much of the velvet used in China, Korea, and Taiwan is produced locally but the remainder is imported. One of the main products of New Zealand's deer farming industry is velvet, almost all of which is exported. International trade in velvet is recorded separately in the overseas trade statistics of New Zealand, Republic of Korea, Taiwan, and Thailand, reflecting the economic significance to these countries. Other major trading countries, such as Hong Kong, record antler together with horns of various species and a variety of other deer products.

Annual imports of velvet (CCCN commodity classification 0509.0120 'Lu jung') to Taiwan have fluctuated between 2–11 tonnes since 1980 (Table 2.4). The major sources have been New Zealand (63% of the total), Canada (11%), the United States (7%), West Germany (6%), Australia (3%), Singapore (3%), Hong Kong (2%), and Singapore (2%). However, Taiwan records no direct imports from mainland China, for political reasons, although it is possible that some is imported via Hong Kong (Lee & Ch'ang, 1985). The total reported imports from Hong Kong amounted to only 705 kg over the six years.

Imports of velvet to the Republic of Korea rose from 3 tonnes in 1977 to 11.3 tonnes in 1979 (Anon., 1980). Several categories of antler are reported in the Customs import statistics; the two which correspond most closely with velvet are 'Antlers, in whole' (CCCN category 0509.0101) and 'Other antlers, including powder and waste' (CCCN category 0509.0199). Total imports of these, also shown in Table 2.5, exceeded 32 tonnes in 1982 before falling back to 17 tonnes in 1984. The major sources were New Zealand (27% of the total), the United States (25%), China (22%), West Germany (6%), and

Hong Kong, Japan, Netherlands and Denmark (all with 2%). The higher unit value of velvet imported to Korea (US$ 351/kg) compared with that imported to Taiwan (US$ 112/kg) is probably due to the fact that the former is usually dried and the latter frozen (Rennie, 1982).

Thailand specifies 'soft antler' (CCCN category 0509.02) in its Customs statistics, and annual imports of this commodity from 1980–84 were 160, 1708, 1890, 3140, and 1000 kg, respectively. Almost all derived from China, with insignificant amounts from the Soviet Union, New Zealand, and Burma.

New Zealand is the only producing country to separate velvet in its export

Table 2.4. *Imports of medicinal products of deer and antelope to Taiwan.*

	1980	1981	1982	1983	1984	1985
Velvet antler (kg)	6,753	9,835	8,226	6,361	10,892	2,555
(US$ 1,000)	1,246	1,213	717	600	1,006	216
Hard antler (kg)	72,565	143,321	47,541	61,928	82,229	86,487
(US$ 1,000)	267	378	194	160	235	262
Deer penes (kg)	3,219	12,255	13,611	7,189	4,646	6,115
(US$ 1,000)	153	697	480	236	216	252
Deer tendons (kg)	0	1,355	290	800	300	2,659
(US$ 1,000)	0	12	3	4	2	15
Antelope horn (kg)	590	258	571	—	1,141	110
(US$ 1,000)	81	54	71	2	177	14
Musk (kg)	2	4	18	9	2	34
(US$ 1,000)	85	42	98	77	51	108

Table 2.5. *Imports of medicinal products of deer to the Republic of Korea.*

	1980	1981	1982	1983	1984	1985
Velvet antler (kg)	18,411	17,373	32,122	22,525	17,553	20,420
(US$ 1,000)	6,885	8,052	8,853	7,432	6,571	7,295
Hard antler (kg)	253,268	364,091	413,356	422,434	310,848	415,584
(US$ 1,000)	2,219	2,624	3,152	3,092	2,308	2,906
Reindeer antler (kg)	0	663	11,562	3,257	12,553	0
(US$ 1,000)	0	35	99	24	100	0
Musk (kg)	56	37	33	11	111	132
(US$ 1,000)	1,277	850	638	779	1,305	1,253

statistics (CCCN commodity classifications 291.16.25 'Velvet, frozen' and 291.16.29 'Velvet, other'). A total of 9.7 tonnes were exported in 1981/82, valued at NZ$ 1.94 million; 15.1 tonnes in 1982/83, valued at NZ$ 3.98 million; and 19.5 tonnes in 1983/84, valued at NZ$ 3.83 million. Of the total, 40% went to Taiwan, 25% to Hong Kong, 21% to the Republic of Korea, 6% to the United States, and 6% to Japan, with lesser amounts to Australia, Singapore, and Guam. Although it is not possible to make direct comparison with the import statistics of Taiwan and the Republic of Korea because of the different reporting periods, there is a reasonable correlation between them.

Lee & Ch'ang (1985) summarised the world trade in velvet from 1977–82, and found that the main exporter was China, producing some 40–50 tonnes/year, followed by the Soviet Union, exporting 12–14 tonnes. The majority of this trade was routed through Singapore and Hong Kong to Taiwan, Thailand, the Koreas, and Hong Kong itself. Hong Kong's international trade in velvet cannot be estimated directly from Customs statistics, and this represents a serious gap in the analysis. However, China and the Soviet Union have been joined by New Zealand, the United States, West Germany, and Canada as major sources of velvet.

Velvet produced in China is mostly from *Cervus nippon*, with lesser amounts from *C. elaphus*, almost all farmed. New Zealand exports almost exclusively velvet of farmed *C. elaphus*. The United States and Canada have only small deer farming industries and the majority of their velvet probably derives from reindeer in Alaska and Canada. Most of the deer farmed in West Germany are *Dama dama*, the velvet of which is of lesser value (Fletcher, 1984), and so it is probable that some of the exports are of *C. elaphus* shot in the wild.

Other deer products are specified in Customs statistics in a variety of forms. Hard antler is traded in large quantities, but is usually included with a variety of other horns, including those of domestic animals. The Republic of Korea provides a category, 'Other horns of deer, including powder and waste' (CCCN category 0509.299), which probably represents mainly hard antlers. Imports of this from 1980 to 1985 were between 253–422 tonnes, with a mean value of US$ 7.47/kg (see Table 2.5). The major sources were the United States (21% of the total), West Germany (17%), India (15%), Denmark (12%), and Sweden (9%).

The corresponding categories in Taiwanese Customs statistics are 'Horns, antlers (whether with wool or not)' (CCCN category 0509.0111) and 'Horns, antlers or scrap' (CCCN category 0509.0112). Imports of these from 1980–85 varied from 47–147 tonnes (Table 2.4). The major sources for these were Singapore (41%), Hong Kong (17%), West Germany (17%), Indonesia

(6%), and New Zealand (4%). Imports from Hong Kong and Singapore almost certainly represent re-exports from other countries. Hard antlers can be obtained from wild animals, either after shooting them or by simply collecting cast antlers. Supplies are thus not confined to countries with farmed or herded populations of deer, although they fetch much lower prices than velvet antler.

Hong Kong is one of the major entrepôts for trade in eastern Asia (see Lee & Ch'ang, 1985). The importance of this trade is shown by the Hong Kong Customs statistics in Table 2.6, giving the reported imports and re-exports of 'Deer parts, fresh, chilled or preserved' (SITC category 291.988). This is separate from 'Horns, antlers, etc.' (SITC category 291.161), and therefore presumably excludes hard antler, although it may include velvet and various internal organs.

Taiwan, the destination of many of Hong Kong's re-exports, reports two further categories of deer products in its Customs statistics, deer penes (CCCN category 0514.0500, 'Lu pien') and deer tendons (CCCN category 0515.0490). Although quantities are low, these commodities have a high unit value and collectively represented the major imports in 1985 in financial terms (Table 2.4).

The horns of certain antelope are also believed to have medicinal properties. Martin (1983) described how saiga antelope (*Saiga tatarica*) horns from the Soviet Union may be used in Oriental medicine as a substitute for rhinoceros horn, primarily as a fever reducant. Taiwanese Customs statistics include a category for antelope horn (CCCN category 0509.0130 'Horns, antelope'), imports of which ranged from zero to over 1 tonne between 1980–85 (Table 2.4). Of the total, 56% was imported from Hong Kong and 31% from Singapore, with only 1% from South Africa. It is likely that the majority of the antelope horn was saiga.

The most valuable of all ungulate products traded internationally is musk, a secretion from the preputial gland of musk deer (*Moschus* spp.), which is

Table 2.6. *Imports and re-exports of 'Deer parts, fresh, chilled or preserved' (SITC category 291.988) reported in Hong Kong Customs statistics.*

	1980	1981	1982	1983	1984	1985
Imports (kg)	863,156	106,959	211,414	111,021	178,108	156,289
(US$ 1,000)	15,426	9,637	10,721	13,628	18,976	
Re-exports (kg)	55,818	19,004	22,393	21,526	28,493	25,186
(US$ 1,000)	7,209	7,302	3,398	3,907	4,579	

used in Oriental medicine and also in Western perfumery. Imports to Japan, which ranged from 109–727 kg annually between 1960–83, reached a peak value of US$ 24,031/kg in 1978 (Green, 1986). Imports to the Republic of Korea (CCCN category 0514.0101 'Musk') and Taiwan (CCCN categories 0514.0110 'Musk, of granules' and 0514.0120 'Musk, of powder in bottle') are shown in Tables 2.4 and 2.5. The unit values indicated by these data are variable owing to the small quantities reported and to the fact that musk is often adulterated and therefore of variable quality (Green, 1986).

Data on the international trade in medicinal products are even less complete than those on meat. However, Thailand, Taiwan, the Republic of Korea, and Hong Kong are probably the main importers of deer products. The exports of velvet and antler from New Zealand showed that from 1981–84 a total of 83% by value were exported to these four countries. The total value of the imports of these products to Taiwan and the Republic of Korea from Tables 2.4 and 2.5, plus imports of velvet to Thailand, the net imports to Hong Kong and the imports of musk to Japan amounted to US$ 29.4 million in 1984, the most recent year for which full data are available. Average annual exports of velvet and antlers from New Zealand to other countries were US$ 0.6 million. Thus, the true value of international trade in medicinal products of ungulates is probably well in excess of US$ 30 million a year. There is also a considerable illegal, and therefore unrecorded, trade. Martin (1983) pointed out that animal and herbal medicinal products were traded covertly, and that they constituted the most important category of goods confiscated by Korean Customs, worth a total of US$ 1.14 million in the first 11 months of 1980.

Tourism

The potential for foreign exchange earnings from tourism is clearly recognised, but, except for hunting, it is difficult to establish how much is attributable to wildlife. Game hunting is a major industry in Africa, attracting almost exclusively foreign hunters. Although variable, the usual practice is to charge a fixed daily fee for accommodation and the services of a professional hunter and camp staff, with additional trophy fees depending on the species of game shot. Figures published in 1983 in the American hunting magazine, *Game Coin*, showed that the average daily rates varied from US$ 333 in Namibia to US$ 1200 in Zaire. Trophy fees ranged from US$ 19 for a duiker to US$ 6783 for a white rhinoceros. Total fees for a week's hunting can exceed several thousand US$, particularly if prized species such as lion (*Panthera leo*), leopard (*Panthera pardus*), elephant, rhinoceros (*Ceratotherium simum*), and buffalo (*Syncerus caffer*) are included. Kettlitz

(1983) estimated that the average trophy hunt in South Africa cost R 9000, but that this could rise to R 20,000 if it included a lion or rhinoceros. Child (1984) underlined the importance of buffalo in the hunting economy of Zimbabwe, increasing the value of an average hunt from Z$ 3500 to Z$ 7500.

Hunting tourists often remain in the country after the safari and spend money on other goods and facilities, thereby increasing the revenue generated, and making it difficult to estimate the true value of sport hunting. Figures published by the Department of Agriculture and Nature Conservation in Namibia show that some R 4.4 million (US$ 5.6 million) were attributable to trophy hunting, representing 35% of all the income generated by wildlife in the country (see Jooste, 1983). The safari hunting industry in Zimbabwe earned over Z$ 6 million (US$ 5.9 million) in foreign exchange in 1983 (Child, 1984) and this is believed to have risen to nearer Z$ 10 million in 1985 (G. R. T. Child, personal communication). Eltringham (1984) quotes figures indicating that the total fees from hunting licences in Zambia rose to around £ 640,000 (US$ 1.2 million) in 1978.

Hunting ungulates in Europe also attracts overseas visitors. Jarvie (1978/79) found that 60% of the clients hunting deer in Scotland were foreigners, and estimated that the total income in foreign currency from hunting fees and associated expenditure was around £ 800,000 (US$ 1.5 million). This can be compared with a wholesale value of £ 2.1 million for venison sold in the same year. Other European countries have similar hunting industries; the contribution of foreign clients is not known, but it is thought to be substantial in Eastern Europe (Krostitz, 1979). Hunting in North America is mainly carried out by residents, although there is a small number of foreign visitors, particularly Americans to Canada. Many South American countries prohibit commercial hunting, and in any case have few large ungulates attractive to trophy hunters. The status of trophy hunting in Asia is not known, although some countries, such as Mongolia, are trying to develop an industry.

The economic value of game viewing is even harder to quantify. Few national parks, considered in isolation, achieve an operating profit, but the tourist industry as a whole can undoubtedly attract foreign exchange. The proportion of this which is attributable to wildlife, rather than other attractions in the country, is impossible to estimate. Eltringham (1984) concluded that Africa was visited chiefly for its wildife, and that Kenya was the country with the largest tourist industry, worth some £ 90 million in 1981. Thresher (1981) estimated the value of tourism in Amboseli National Park in Kenya. In 1976, 97,000 tourists visited the park, of which 60,000 were overseas visitors, contributing a net foreign exchange earning of US$ 3.4

million. Computer modelling of future trends predicted that by 1991 the net foreign exchange earnings would rise to US$ 10.6 million at 1976 prices. Bophuthatswana has a total of six national parks. The largest of these, Pilanesberg, mainly attracts visitors from neighbouring South Africa and caters for both viewing tourists and hunters. An economic review carried out in 1983 found that the total income from viewing was R 40,000 (US$ 35,960) and from hunting and game harvesting R 340,000, set against a total expenditure of R 1,550,000, and therefore running at a considerable loss. However, it was estimated that substantial development of game viewing was possible and could increase revenue to R 1,285,000, turning the park into a financial asset (Anon., 1983b).

Significance of wildlife production

Krostitz (1979) used the FAO computer databank to calculate the total volume of international trade in game meat. In 1978, it amounted to some 52 kilotonnes, having increased from about 25 kilotonnes a year in the late 1960s. The volume of trade indicated by the present analysis was only 35 kilotonnes in 1980 (Table 2.1). Although this is only a minimum estimate and the sources were slightly different to those used by Krostitz, it appears that the volume of trade may have fallen slightly since 1978. Table 2.1 further indicates that the decline has continued between 1980–84, but the trade increased again in 1985. This fall may reflect changes in supply and demand, or it may be due to the enforcement of stricter veterinary and medical controls on trade.

Krostitz (1979) estimated that the total production of game meat in 1978 was 750 kilotonnes, or 0.5% of the total production of meat from domestic livestock (130 million tonnes). The *1984 FAO Production Yearbook* indicates that meat production from domestic livestock had risen to about 250 million tonnes, and, although no estimate of wildlife production has been made on this occasion, it is unlikely that it yet makes a significant contribution to meat production on a global scale. The current analysis also indicates that wild ungulates contribute less than half of the game meat traded, small game supplying the remainder. However, this is not to underestimate the local importance, and there are some societies where it still represents the major item of animal protein.

Another way to assess the importance of game is in economic terms, and it is here that the international trade, representing foreign exchange, is particularly relevant. Although less than half of the total weight of game meat traded is attributable to ungulates, the higher unit value of their meat means that they may account for more than half of the value, perhaps US$ 50 million in

1984. International trade in wild ungulate skins is difficult to quantify, but it is probably worth over US\$ 5 million/year. Tanned hides are unaffected by the same veterinary restrictions as fresh meat, and may therefore be exported even from areas subject to foot-and-mouth disease. Skin exports can therefore be financially significant in areas where meat is not exported.

Trade in medicinal products is worth well over US\$ 30 million a year, and rivals the trade in meat in economic volume. Like the skin trade, it is less affected by import controls, and the small volume and high value of the commodities further facilitate trade. However, the markets are restricted to areas where Oriental medicine is practised. The sale of medicinal products can be run as an adjunct to meat production, as in New Zealand, or it can be the only commercial export, as in India and China.

The amount of foreign exchange generated by tourism is poorly quantified, but the few figures that are available suggest that it may be substantial. While tourism generated by hunting is important in Southern Africa, it is probable that the more passive forms of wildlife tourism are more important on a global scale.

The production of meat is traditionally seen as the major goal of wildlife production systems, but it can be seen that, in economic terms, other products are often more important. Finally, although wildlife may make only a small contribution to the world economy, its local significance should not be underestimated, particularly when it can be used to generate export trade from areas with few other cash crops.

References

Anon. (1980). *Korean Deer Farming*, 80(12), 61.

Anon. (1983a). *Countryside Sports: Their Economic Significance*. Survey by Cobham Resource Consultants. Reading: Standing Conference on Countryside Sports.

Anon. (1983b). Pilanesberg National Park: the next five years. *Tshomarelo News*, National Parks Board of Bophuthatswana, 14, 1–21.

Broad, S. (1984). The peccary skin trade. *Traffic Bulletin*, 6(2), 27–8.

Broad, S., Luxmoore, R. & Jenkins, M. (1988). *Significant Trade in Wildlife, A Review of Selected Species Listed in CITES Appendix II. Volume 1, Mammals*, pp. 158–67. Cambridge: IUCN.

Carvalho, J. C. (1967). A conservacao de natureza e recursos naturais na Amazonia Brasileira. *Atas Simp. Biota Amazonica*, 7, 1–47.

Child, G. R. T. (1984). *The Organization of Hunting within the Parks and Wildlife Estate of Zimbabwe*. Harare: Department of National Parks and Wildlife Management.

Conroy, A. M. & Gaigher, I. G. (1982). Venison, aquaculture and ostrich meat production: action 2003. *South African Journal of Animal Science*, 12, 219–33.

Dixon, A. M. (1985). The European trade in kangaroo products. *Traffic Bulletin*, 6(5), 73–82.

Eltringham, S. K. (1984). *Wildlife Resources and Economic Development*. New York: John Wiley.

Fletcher, T. J. (1984). Other deer. In *Evolution of Domesticated Animals*, ed. I. L. Mason, pp. 138–44. New York: Longman.

Franklin, W. L. (1982). Biology, ecology and relationship to man of the South American camelids. In *Mammalian Biology in South America*, ed. M. A. Mares & H. H. Genoways. Special Publication Series, Vol. 6, Pymatuning Laboratory of Ecology, University of Pittsburg.

Freudenberger, D. (1982). Southern Africa. In *The Farming of Deer: World Trends and Modern Techniques*, ed. D. Yerex, pp. 31–44. Wellington: Agricultural Promotion Associates.

Green, M. J. B. (1986). The distribution, status and conservation of the Himalayan musk deer *Moschus chrysogaster*. *Biological Conservation*, 35, 347–75.

Grimwood, I. R. (1968). *Appendix III to Recommendations on the Conservation of Wild Life and the Establishment of National Parks and Reserves in Peru*. London: Ministry of Overseas Development.

Jackson, J. E. (1986). The hare trade in Argentina. *Traffic Bulletin*, 7(5), 72.

Jarvie, E. (1978/79). *The Red Deer Industry*. Edinburgh: Scottish Landowner's Federation.

Jooste, J. F. (1983). Game farming as a supplementary farming activity in the Karoo. *Proceedings of the Grassland Society of Southern Africa*, 18, 46–9.

Kettlitz, W. K. (1983). Trophy hunting in the Transvaal. *Fauna and Flora*, 40, 26–9.

Krostitz, W. (1979). The new international market for game meat. *Unasylva*, 31(123), 32–6.

Lee, C. H. & Ch'ang, T. S. (1985). Marketing and utilisation of deer products in Asia. *Royal Society of New Zealand, Bulletin*, 22, 307–10.

Luxmoore, R. (1985). Game farming in South Africa as a force in conservation. *Oryx*, 19(4), 225–31.

Mares, M. A. & Ojeda, R. A. (1984). Faunal commercialization and conservation in South America. *BioScience*, 34(9), 580–4.

Martin, E. B. (1983). *Rhino Exploitation*. Hong Kong: World Wildlife Fund.

Mossman, S. L. & Mossman, A. S. (1976). *Wildlife Utilization and Game Ranching*. IUCN Occasional Paper No. 17.

Prescott-Allen, C. & Prescott-Allen, R. (1986). *The First Resource: Wild Species in the North American Economy*. New Haven: Yale University Press.

Rennie, N. (1982). Antler velvet. In *The Farming of Deer: World Trends and Modern Techniques*, ed. D. Yerex, pp. 141–4. Wellington: Agricultural Promotion Associates.

Smith, N. J. H. (1977). Human exploitation of terra firme fauna in Amazonia. *Ciencia e Cultura*, 30(1), 17–23.

Sowls, L. K. (1984). *The Peccaries*. Tuscon: University of Arizona Press.

Thresher, P. (1981). The present value of an Amboseli lion. *World Animal Review*, 40, 30–3.

Tisdell, C. (1987). The Australian feral pig and the economics of its management. In *Biologie des Suides*, ed. F. Spitz & R. Barrett. Paris: INRA, (in press).

Visser, G. S. (1983). *The Local and Overseas Market of Venison*. Paper presented at the Game Industry Conference, Kimberley.

Appendix–Customs reports consulted

Statistics used in game meat trade analysis

Austria: Der aussenhandel Osterreichs, Ser 1A, Spezialhandel nach Waren und Landern Gesamtubersichten. Osterreichs Statistiches Zentralamt. CCCN 02.04.11 Fleisch u geniessb. schlachtanfall von hasen und kaninchen; 02.04.19 Fleisch u geiessb. schlachtanfall von anderen haarwild; 02.04.20 Fleisch u geniessb. schlachtanfall von feder wild.

Belgium–Luxembourg: Bulletin Mensuel du Commerce Exterieur de l'Union Economique Belgo-Luxembourgeoise Ministere des Affaires Economiques, Institut National de Statistique. CCCN 02.04.300 Viandes et abat comestibles de gibier (1980–1983).

Denmark: External trade. Danmarks Statistik. CCCN 02.04.300 Kod og spiseligt slagteaff. af vildt.

Federal Republic of Germany: Aussenhandel nach Waren und Landern (Spezialhandel). Reihe 2 Herausgeber: Statistisches Bundesamt Wiesbaden. CCCN 02.04.300 Fleisch v wild.

Finland: Foreign trade. Board of Trade, Official Statistics of Finland. 02.04.100 Poronliha; 0204.309 Muu liha (1981–1984).

France: Statistiques du Commerce Extérieur de la France. Importations/Exportations Ministere du Budget, Direction Générale des Douanes et Droits Indirects. CCCN 02.04.300 Autres viandes et abat comestibles: de gibier.

Italy: Statistica Mensile del Commercio con l'Estero Instituto Centrale di Statistica, Roma. CCCN 02.04.300 Carni e frattaglie: di selvaggina.

Netherlands: Maandstatistiek van de Buitenlandse Handel per Goederensoort, Central Bureau voor de Statistiek. CCCN 02.04.300 Vlees van ander wild.

Norway: Utentikshandel. Central Bureau of Statistics, Norway. CCCN 02.04.100 Reindeer meat; 02.04.301 Meat of woodfowl; 02.04.305 Other meat. (1981–1984).

Spain: Estadistica del Commercio Exterior de Espana. Ministerio de Economia y Hacienda. CCCN 02.04.300 Carnes y despojos comestibles: de caza (1982–1984).

Sweden: Utrikeshandel. Sveriges Officiella Statistik. CCCN 02.04.101 Reindeer meat, fresh; 02.04.105 Reindeer meat, frozen; 02.04.109 Reindeer offals; 02.04.301 Game meat, fresh; 02.04.305 Game meat, frozen. (1981–1984).

Yugoslavia: Statistics of Foreign Trade of the SFR. of Yugoslavia. SITC 01.18.910 Hare meat; 01.18.950 Other game meat; 01.18.990 Edible offals of game. (1980–1981).

Other statistics used

Hong Kong: Hong Kong Trade Statistics, Census and Statistics Department, Hong Kong.

Japan: Japan Exports and Imports: Commodity by Country, Japan Tariff Association.

Korea, Republic of: Statistical Yearbook of Foreign Trade, Department of Customs Administration.

New Zealand: Overseas Trade Statistics, New Zealand.

Taiwan: The Trade of China (Taiwan District), Statistical Series No. 1, Statistics Department, Inspectorate General of Customs, Taipei.

Thailand: Foreign Trade Statistics of Thailand (B.E. 2519), Department of Customs, Bangkok.

SECTION B

Subsistence hunting

DAVID R. KLEIN

Subsistence relies on local production and distribution of goods and services. Emphasis is placed on maximising material and psychological security. Characteristics of capitalistic systems, such as production for profit, market structures, formal contracts, capital accumulation, and surplus value are generally absent in subsistence economies (Lonner, 1980). However, both subsistence and cash-based systems encompass trading, technological development, skill training, storage, transportation, and equipment production and maintenance.

Most of the world's human population lives in rural areas and engages in some form of subsistence activity (Reining & Lenkard, 1980). Today subsistence economies almost always interface with cash economies and the pressures for change imposed by the industrialised world tend to erode traditional patterns of existence. Nevertheless, it is at least theoretically possible for subsistence and cash economies to coexist while maintaining their distinctive characteristics. The success of subsistence hunting societies in maintaining the structure and character of their economic base while faced with external pressures for change has varied greatly throughout the world. Specific examples are discussed in the following three chapters which introduce the principles of non-agricultural management and review subsistence hunting in tropical and northern environments.

Many subsistence hunting economies exploit large mammals, but usually rely on a much broader resource base. Peoples living near the sea rely on marine mammals, sea birds, fish, and invertebrates, whereas small mammals, birds, fish, and in some areas, reptiles and amphibians are important to most subsistence peoples living inland, at least seasonally. Archaeological evidence suggests that where there has been a lack in diversity of potential subsistence resources, human populations dependent on hunting large mammals have been transitory. Whereas subsistence hunting societies were largely self-sufficient, today government-sponsored welfare programmes and opportunities for cash income through wages or sale of locally produced commodities make it possible for subsistence hunting to persist although transformed from traditional ways.

Employment opportunities have variable effects on subsistence activities and resources. Time devoted to employment limits time available for hunting and wage incomes earned decrease dependency on subsistence resources. Opportunities to learn hunting and other subsistence skills also are reduced. On the other hand, cash income generally leads to purchase of often more effective hunting equipment, such as rifles, boats and motors, and all-terrain vehicles. Since employment available to subsistence societies is usually sporadic and seasonal, the net effect may be greater hunting pressure through improved transportation and hunting weapons. Those who acquire newer hunting equipment as a result of their cash income also may increase their hunting activities as rationale for their equipment purchases and to obtain subsistence foods to be shared within the community.

The availability of markets encourages the sale of wildlife products, often to the detriment of wildlife populations. The classic example is the high market value of rhinoceros horn which has provided incentives for poaching, leading to the extirpation of both the black (*Diceros bicornus*) and white rhinoceros (*Ceratotherium simum*) throughout much of Africa. The killing of wildlife to market meat, hides, ivory, horn, trophies, and other products is often practised by peoples formerly dependent upon the same wildlife species for subsistence. To peoples in transition between cultures the cash income from such ventures may be a greater incentive than the traditional subsistence products. Unfortunately, illegal markets usually place pressure on wildlife populations inhabiting areas where poaching cannot be effectively controlled.

Aboriginal people who in the past lived in relative balance with local resources have been freed of this direct relationship through adoption of subsistence agriculture, integration with market economies, reliance on modern medical services and government welfare. The consequent breakdown of cultural constraints on population size has allowed populations to expand well beyond the capability of local subsistence resources to sustain them. Where wildlife habitat is being destroyed through land clearing for agriculture, overgrazing by domestic livestock, and clearing forests for fuel and wood, there appears to be little opportunity for preservation of wildlife and the subsistence hunting societies that depend upon them. The likely irreversible destruction of wildlife habitat and the cultural changes associated with exposure to market economies, adoption of foreign educational and religious systems, and changed aspirations also impose constraints on the return to subsistence life-styles once they have been abandoned.

The destruction and degradation of wildlife habitats continues at a rapid pace, especially in the wet tropics with the removal of rainforests, and in semi-arid regions through overgrazing by domestic livestock. In other parts of the world, particularly the northern regions of Eurasia and North America, wildlife habitats remain largely intact and local communities with long traditions of subsistence dependencies have been able to maintain healthy relationships with the land and its resources. Although human populations are increasing, the adoption of market economies and growing opportunities for wage employment are reducing dependence on subsistence resources. Therefore, it should be possible for subsistence resources to continue to provide an important, but not total, contribution to the economies of northern communities as they continue to evolve a balance between local self-sufficiency and participation in the world market economy. It is important, however, that harvests of subsistence resources are balanced with their productivity and habitats are protected from destruction or degradation.

References

Lonner, T. (1980). Subsistence as an economic system in Alaska: theoretical and policy implications. *Technical Paper*, 67, 1–36. Juneau: Alaska Department of Fish & Game.

Reining, P. C. & Lenkard, B. (1980). *Village Viability in Contemporary Society*. Boulder (Colorado): Westview Press.

3

Non-agricultural management of plants and animals: alternative burning strategies in Northern Australia

HENRY T. LEWIS

Abstract

A growing number of studies based on historical reconstruction or on more definitive field interviews and direct observations, demonstrate the environmental significance and technological sophistication of indigenous burning practices in parts of North America and Australia. One of the better studied areas is the Northern Territory of Australia where both traditional and introduced practices are widely used by two different groups: hunter–gatherers and cattle ranchers. The tool assemblages of each are fairly distinct, with cattlemen now employing a variety of sophisticated tools which include aerial ignitions from light aircraft. Aborigines, at least those living in more remote areas, now use butane lighters instead of fire sticks and cover distances in four-wheel drive vehicles rather than on foot. Despite the more impressive array of tools and techniques employed by cattlemen, the technological knowledge of Aboriginal hunters–gatherers involves a much more complex understanding of environmental networks of cause and effect. The differences of relative complexity relate directly to the different resource strategies involved: for stockmen it involves cattle, for hunter–gatherers it involves a broad spectrum of natural resources.

Introduction

A small but growing number of anthropologists have undertaken studies of a largely overlooked but highly important aspect of virtually all hunting and gathering technologies: the prescribed uses of habitat fires for influencing the numbers and distributions of preferred plants and animals. To date, detailed descriptions of uses of fire have been made for only a limited number of hunter–gatherer societies in North America and Australia, although similar practices have been widely noted for Africa, Latin America, New Guinea, and parts of South and Southeast Asia. Incentives to the anthropological study of hunter–gatherer burning practices derive from developments both inside and outside the discipline.

The motivation for examining the pyrotechnics of hunting–gathering adaptations relates to the enormous growth of studies on human–environmental relationships, a subfield of the discipline variously called *ecological anthropology* or *cultural ecology*. Actually, the more important stimulus has come from the biological sciences, specifically fire ecology and the recognition that controlled burning has enormous potential for managing natural resources. Itself a relatively new field, fire ecology has provided anthropologists with the outlines for understanding the significance that prescribed burning has for hunting–gathering adaptations.

This new-found concern with the *folk technology* and *folk ecology* of fire does not represent the first time that anthropologists have considered indigenous practices of burning. In the 1950s, anthropologist Omer C. Stewart (1951, etc.) provided examples and made interpretations of the overall importance of habitat burning by North American Indians. In attempting to convince others that Indian uses of fire had significant consequences for natural environments and human adaptations, he stated:

> The materials I have presented establish that Aboriginal man has had a tremendous and decisive influence on several aspects of his physical environment. Our knowledge of such effect is far from complete because anthropologists, to a large degree, have assumed that though natives adjusted to their environment, they did not change the physical world sufficiently to warrant careful investigation. It is time that anthropologists revised their thinking on this matter (Stewart, 1954:248)

The geographer, Carl O. Sauer, argued that habitat burning must have been employed by prehistoric foragers, pastoralists, and farmers (Sauer, 1947, 1956, 1970, 1975). Also, Gordon Day (1953), a biologist, compiled a list of historical references to Indian burning practices in the northeastern United States and from these examples suggested a range of environmental consequences and related advantages for hunters and gatherers. However, for reasons discussed elsewhere (Lewis, 1982a), anthropologists were slow to follow Stewart's advice or the example of others. It was not until the late 1960s that interest by anthropologists was renewed.

In most references to Aboriginal fires, researchers have tried to single out reasons for indigenous burning practices, the pre-eminent question expressed as *why*: 'Why do (or did) hunter–gatherers burn forests, brushlands, and prairies?'. The implication of such a question is that hunter–gatherers perceive a one-to-one relationship between setting fires and achieving specific goals. In contrast, field studies indicate that hunting–gathering peoples in

North America and Australia understand and manipulate networks of multiple causes and effects that are ecologically much more important and interesting than the one or two reasons that they may have given for setting a particular fire.

In most modern societies, public knowledge on the effects of fire is still presented from an over-simplified and one-sided perspective, i.e. that fires are inherently destructive and must be prevented. Given this popular view, plus the broadly accepted position of anthropologists and others that foragers (in contrast to farmers) did not significantly influence or intentionally manage environments, relatively few studies have done more than merely state that hunter–gatherers set fires to drive game or improve grazing.

Most of what we know about North American Indian uses of habitat fires is derived from secondary historical and anthropological sources, usually in brief notations made by explorers or settlers. Interpretations of what these practices might have meant for hunting–gathering adaptations have necessarily relied upon the conclusions of fire ecologists and their view of the multiple effects that fires have on natural settings. By the late 1960s, sufficient ecological information was available to indicate many of the specific advantages that prescribed burning would have had for hunting–gathering adaptations.

Using a variety of historical references, I was able to demonstrate how fire technologies of Indians paralleled models for prescribed burning proposed by Biswell (1967) for managing natural resources in California:

> Thus, it (is) possible to take isolated facts and give them new contextual meanings within a multivariate system of relationships because of what is now known ... about the human ecology of California's wildlands (Lewis, 1973:82)

This approach was followed by similar reconstructions (Bean & Lawton, 1973; Norton, 1979; Timbrook, Johnson & Earle, 1982; Boyd, 1986) which reconfirmed that aboriginal burning practices undoubtedly had great significance for a range of environmental zones in the American West. A similar historical reconstruction of Indian fire technology in the Great Plains region of Canada exists (Arthur, 1975).

A different approach was possible in northern Alberta where I (Lewis, 1977, 1982b, 1985a) interviewed older native people, some of them still active as hunters and trappers, about traditional understandings and the burning techniques they employed until the late 1940s. This and a similar study by Ferguson (1979) have shown that an older generation of indigenous peoples in the boreal forest region still have sophisticated understandings of the ecological dynamics that make up the complex patternings of secondary successions in fire-influenced and fire-maintained environments. By comparing these with earlier documented studies, Lewis was able to isolate general patterns which characterise hunter–gatherer uses of fire in widely separated and environmentally different regions. Lewis (1982a) demonstrated that hunter–gatherer fires differ from natural fires in terms of four related considerations: the seasons within which fires are set, the frequency with which a given area is reburned, the intensity of particular fires, and the purposeful selection of sites in terms of preferred resources.

Jones' (1969) brief article on Tasmania, itself motivated by the research of fire ecologists (especially Jackson, 1968), was an important impetus to other Australians to examine historical references to the burning practices of Aborigines (Blainey, 1976:67–83; Flood, 1983:200–215; Hallam, 1975, 1985; Horton, 1982; Jones, 1969; Nicholson, 1981). The Australian record has also included research based upon ethnographic interviews and actual field observations (Tindale, 1959; Gould, 1971; Jones, 1975, 1980; Latz & Griffen, 1978; Haynes, 1982, 1983; Kimber, 1983; Lewis, 1985b). Whereas not all writers agree on the significance of Aboriginal practices, Australia is the only country within the Western industrialised world where traditional uses of hunter–gatherer fires can still be examined *in situ*, albeit such practices are almost entirely restricted to remote or *outback* areas. Clark (1981, 1983) and Horton (1982) minimise the importance and sophistication of Australian Aboriginal fire technology on the basis of hypothetical argument though neither conducted field studies.

As with all aspects of human behaviour, anthropologists assume that the pyrotechnics and related knowledge systems are understandable and coherent in their own contexts, both environmental and cultural. It is not assumed that they will be immediately or necessarily rational in our cultural terms nor involve the same environmental goals. However, given the fact that such practices relate directly to adaptation and survival, the logic of such practices and understandings is more readily appreciated cross-culturally than are the more esoteric and symbolic aspects of culture. The following example comparing hunter–gatherers and cattle ranchers in northern Australia provides a contrast of two somewhat rationally different systems but with areas of great overlap.

Evidence for and related interpretations of the prehistoric influences of hunter–gatherers have been suggested for a number of regions in the world, but some of the most concerted work and debated issues have been carried on in Australia. The most recent statement, one that argues strongly for the significance of Aboriginal practice, is that by Hallam (1985), and in this she provides an overview of the arguments and literature on the subject. Whether or not it will ever be possible to do more than infer the precise significance of prehistoric man's impact upon natural environments, on the basis of historic and ethnographic evidence, it seems reasonable to assume that the adaptations of hunting and gathering societies were not merely passive responses to environmental factors. Most assuredly, fire, one of nature's four fundamental elements, has been truly Promethean.

Alternative burning strategies in Northern Australia

Fire, grass, kangaroos, and human inhabitants seem all dependent on each other for existence in Australia; for any one of these being wanting, the others could no longer continue. Fire is necessary to burn the grass, and from those open forests, in which we find the large forest-kangaroo, the native applied that fire to the grass at certain seasons, in order that a young green crop may subsequently spring up, and so attract and enable him to kill or take the kangaroo with nets. In summer, the burning of long grass also discloses vermin, birds' nests, etc., on which the females and children, who chiefly burn the grass, feed. But for this simple process, the Australian woods had probably contained as thick a jungle as those of New Zealand or America, instead of the open forests in which the white men now find grass for their cattle, to the exclusion of the kangaroo (Mitchell, 1848).

Across much of the northern half of Australia two groups employ technologies of fire, practices that are deceptively simple to use but which require sophisticated understandings of natural systems to be adaptive over time. Prescribed burning is applied in a variety of environmental zones and, within them, a range of habitat types as an important feature, a 'tool', of both hunting–gathering and pastoral adaptations. Despite significant mechanical changes in making and setting fires – from the use of fire drills to butane lighters, and from cowboys throwing matches from horseback to dropping chemical incendiaries from airplanes – the requisite understandings of how particular fires can behave at given times in a range of habitats, how they can be controlled and limited in intensity and extent, how it is possible to predict immediate short-term benefits, and how to judge long-term environmental consequences, are the subject matter of these two 'folk sciences' and the still relatively new field of fire ecology.

Environmental indications are reasonably clear that since the arrival of humans in Australia more than 30,000 years ago, fires have been used to influence communities of plants and animals. Though prehistoric evidence of indigenous uses of fire is indeterminate and limited to inference (cf. Clark, 1981, 1983; Singh, Kershaw & Clark, 1981; Horton, 1982), historical references are quite clear. Two interpretive studies, both based on historical descriptions of Aboriginal uses of fire, provide excellent evidence of habitat management by prescribed burning at the time of European contact (Hallam, 1975, 1985). Much more detailed information on indigenous burning technologies is found in a number of studies that outline contemporary Aboriginal adaptations in remote areas where habitat burning is still employed to affect the distribution and relative abundance of preferred plants and animals (Jones, 1980; Haynes, 1983; Kimber, 1983). This combination of historic and ethnographic accounts adds strength to the argument that Aboriginal burning practices were significant in prehistoric times (Hallam, 1985).

Throughout the history of Australia, cattle and sheep pastoralists set fires to manage pastures. Euro-Australian pastoralists also frequently observed the effects of Aboriginal burning and perhaps most important, employed large numbers of Aborigines as cowboys.

For stockmen in northern and central Australia, fire remains their most important management tool, for without the prescribed use of fire, tropical pastures have little grazing value. On a large scale, pastoralists employ fires to induce more uniform growth of grasses and, on a much more limited basis, to convert local paddocks to the growth of introduced grasses and legumes. As Johnson & Purdie (1981) have summarised, the use of fire in combination with heavy grazing ultimately results in the reduction of weedy or undesirable species and an increase of grasses over trees and shrubs.

Burning practices of both stockmen and hunter–gatherers are heavily influenced, though not specifically determined, by climatic patterns. Natural fires would normally occur at the end of the dry season, about mid December, with the build-up of cumulus cloud which heralds the onset of the wet season. Thunderstorms, sometimes accompanied by rain but more often not, precede the monsoon rains that persist until early or even late April with an annual rainfall of 1200–1500 mm. Except for brief dry periods that can occur in January and February, it is difficult to burn during the wet season. From mid April to mid May intense storms, sometimes cyclones, occur and the 2-metre-high stands of sorghum grass (*Sorghum intrans*), spear grass (*Heteropogon triticeus*), and other subtropical species are flattened. Humidity remains high until early or mid May when burning is initiated by both pastoralists and foragers. The dry season begins with cooler and dryer weather, which becomes increasingly hot by mid August, and the 'dry' persists until the return of monsoon rains in December.

The following discussion of pastoral burning practices is based upon a general consensus of 18 stockmen interviewed during 1983 in an area that extended east to the Arnhemland border, west to the Ord River, south to Wave Hill, and east again to an area south of the town of Katherine. Four individuals were Aborigines who were equally aware of the different aims and practices of stockmen and hunter–gatherers.

The outline of hunting–gathering burning practices is largely based upon the work and personal collaboration of C. D. Haynes (1982, 1983) plus my own observations and interviews with Aborigines in Kakadu National Park. For those Aborigines in outback areas, where hunting and gathering still constitute an important subsistence base, the use of habitat fires remains an integral and significant feature of subsistence technology. Much of Aboriginal life in the more remote areas of northern and central Australia is today related to the *outstation movement* whereby Aborigines have returned to traditional tribal lands after years of residence in and dependence on missionary or government settlements and pastoral stations. Knowledge about traditional subsistence activities varies geographically, though the dynamics of earlier human–environmental relationships are remembered by many elders and still applied in northern and central regions of the country (Meehan, 1982; Bell, 1983).

Whereas rifles have largely replaced spears and boomerangs, matches and butane lighters have supplanted fire sticks, and the movement from area to area is now in four-wheel drive vehicles rather than on foot, many aspects of traditional hunting and gathering are still important features of Aboriginal life in more remote areas and fire continues to be a significant technological feature of bush life. It is through the uses of selected burning that these populations of part-time hunter–gatherers still make a pronounced impact on local environments.

Though we can only estimate the impact of hunter–gatherer uses of fire in the prehistoric or historic past, to ignore the effects of Aborigines on local resources is simply naive. For reasons discussed elsewhere (Lewis, 1972, 1982b), this important feature of hunting–gathering adaptations has been ignored when reconstructing human prehistory. Given our emerging understanding, it is difficult to imagine that some 300,000 to 500,000 nomadic foragers would have made an impact any less significant that what is today made by a smaller number of pastoralists in Australia.

Hunter–gatherer fires

If asked why they burn, answers of Aborigines are usually site-specific: improving the relative abundance of preferred plants, altering or maintaining the habitats, cleaning a campsite to rid it of insects and snakes, encircling game during a hunt, reducing accumulations of fuel, establishing fire guards around patches of rainforest, and even for the sheer joy that marks the beginning of the dry season. Yet, even the broadest inventory of reasons for burning only touches upon a people's more comprehensive knowledge of fire.

Inland from the coastal fringes of sand dunes and mangrove forests, the coastal region between Darwin in the west and Gove Peninsula on the eastern tip of Arnhemland, are eucalypt dominated open forests (about 30%) and woodlands (about 40%). The trees of the open forest, of which stringybark (*Eucalyptus tetrodonta*) and woollybutt (*E. miniata*) are most common, are tall (15–19 m) with a patchy understorey of shrubs, palms, and grasses. Woodlands on poorer shallow soils are composed of smaller and more widely dispersed trees (10–12 m) especially boxwood (*E. tectifica*) and bloodwood (*E. latifolia*), and have a uniform understory of grasses. In both habitats, open spaces are dominated by tall (1–2 m) sorghum grasses which are highly flammable in the dry season, and when mature, are poor forages.

The remaining areas are composed of freshwater floodplains, paperbark swamps, and isolated stands of rainforest. Inland the coastal plain is bordered by a sandstone escarpment. Each biome is exploited to varying degrees, and managed by fire in distinctive ways. The burning strategy depends upon their perceived value, accessibility, and characteristics of the community of plants and animals.

A number of seasonal calendars have been produced by researchers for Aborigines of the north coast. For example, the calendar for the South Alligator River area, represented by the Gundjeidmi language, has six major seasons. Though roughly equal to the Gregorian calendar, specific parts of a given year are derived from climatic and biological events. It is the combinations of seasonal incidents (swarming of dragonflies, final storms of the year, blooming of various fruit trees, etc.) that indicate the appropriate seasonal changes for setting fires in particular habitats.

Though most common in the dry months, a few fires may be set during short breaks (*gularr gaimigo*, or 'fine hot spells') in the monsoon season, the *gudjewg*. These are usually limited to clearing local campsites and settlement areas, with ground cover being a problem at this time of year when snakes move to high ground, a concern for people living in an area which has the world's five most venomous species.

The first dry season fires are lit after the monsoon rains have stopped but before the last convective storms have passed – the *banggereng* (approximately April) – in which the sorghum grasses are levelled. Burning begins along the margins of the floodplains where water levels have dropped and the exposed grasses and sedges are dried before those in other habitats. In addition to the animals that will be attracted to new grass regrowth, adjacent stands of rainforest and paperbark swamps are fireguarded against the larger and hotter fires that are set a few weeks later on the floodplains and in adjacent stands of tall forest and woodland.

At the start of the dry season, floodplain fires are set in stages, weeks apart, and burn from the previously burned strips towards the wetter, lower lying areas where they simply extinguish as they reach the damper green growth. Consequently, the central parts of a floodplain can be safely fired months after burning was initiated on the margins. With a fire-induced production of green pick, the larger macropods (kangaroo and wallaby) frequent these sites well into the dry season. In some instances, where continued soil moisture allows for complete recovery of grasses, portions of the floodplain will be fired a second time in the same season.

Waterfowl are an important floodplain resource, especially magpie geese which are exploited for both eggs and meat, and burning is considered by the Aborigines to be significant to the birds' nesting and feeding requirements. Geese must be able to feed around the nest, but where unburned detritus has accumulated, it is difficult for them to obtain new roots and sand. Without calcium from sand, eggshells are soft and the number of goslings is reduced. Finally, as the dry season progresses, many snakes withdraw into the tall grass and become increasingly hazardous for hunters. Burning reduces this danger and the last fires are set well before waterfowl nest.

By early May, coinciding with the flowering of the woollybutt trees, fires are begun well beyond campsites and settlement areas. This is the start of the *wurrgeng*, the 'cold weather season', marked by the arrival of cool southeasterly winds and lower night temperatures.

Three broad considerations are effectively the same for burning open forests and woodlands, though, overall, the two areas are valued and managed in somewhat different ways. First, informants emphasise that they should be set during windy midday periods. Fires can therefore be more easily directed and flames and convection columns are bent forward with the result that the height of scorch damage is reduced to less than 3 m and often less than 1.5 m. This protects the flowers of fruiting trees, which provide a component of the diet for both Aborigines and some of the animals they hunt, such as flying fox and possum.

Secondly, fires at this time are irregular and substantial tracts of intervening vegetation are left unburnt (40–50%), and within burnt sections the effects are patchy. By contrast, larger, late season fires leave much less of a mixed habitat and individual sites are burned more severely and evenly.

Thirdly, early season fires burn out during the late afternoon or early evening. Whereas at the beginning of the season each fire lasts for little more than an hour, by mid July fires go out with the early morning dew. The mosaic of burnt and unburnt areas makes it possible to control fires in the late part of the season because it is then possible to 'aim' them towards burnt sites.

Relative humidity is lowest in mid August with the arrival of the *gurrung* ('hot weather time') by which time burning should be completed within the open forest. The only fires still set there at this time are for encircling kangaroos and wallabies during hunting. As a part of the mosaic, the unburned sites are places where the animals hide during the day. In addition to the larger macropods, smaller animals such as bandicoots, marsupial rats, and goannas can be taken by hunters. Human foragers are not alone in exploiting fire drives: forked-tailed kites or 'fire hawks' and other opportunistic, predatory birds are attracted by the first signs of smoke.

In the immediate hours and days following a fire, considerably before the emergence of new growth, kangaroos, wallabies, and goannas congregate on newly burned sites to dig for the roots of grasses and wild yams. Aborigines are aware of the behavioural responses of animals to fire, and desired species are hunted accordingly. Their understanding of these relationships and their selective employment of fire allows them to manage and predict what are otherwise natural events.

Whereas the general patterns of woodland fires are similar to those of open forests, the characteristics of woodland stands result in different burning practices. Fires in eucalypt woodlands are generally larger and hotter because, with reduced shade and with winds less modified by the canopy of trees, the uniform understory of sorghum grasses burns with greater intensity. In addition, many woodland fires are set much later, some as late as mid November with the onset of the pre-monsoon storm season, the *gunemeleng*. Resources are more limited in the eucalypt woodland, with hunting largely focused on kangaroos and wallabies that feed upon the green pick. The major difference between eucalypt woodland and the open forest concern the scale, intensity, and length of burning periods.

Patches of rainforest and paperbark swamps are normally not burned. The concerns Aborigines have for monsoon forests relate to the plant community (which is much less fire tolerant), the animals which use rainforest for concealment, and the ritual and totemic significance of these areas. However, comments by informants indicate a degree of variation regarding fires in rainforest stands. Whereas some Aborigines maintain that fires are not set, others have indicated that they are sometimes, but only after they have been fireguarded against late season fires from surrounding vegetation types, and at a time when the fires merely burnt along the floor of the rainforest stand, with flames less than 10 cm. These occasional fires remove leaf litter which can inhibit and conceal the growth of yams. Paperbark swamps are burnt every 5–10 years. As with rainforest, the accumulation of litter is the main factor involved. These areas can only be burned after water levels recede and fuels dry. They are burned before fuels are fully cured so as to limit damage to paperbark trees and other less fire-tolerant plants.

Difficulty in traversing the rough, deeply fissured escarpment country has discouraged resource exploitation on the plateau. Spinifex and the other wiry grasses and shrubs of the plateau country are highly flammable, and fires that occur in pockets of eucalypt woodland and stands of heath and scrub can be very intense.

Hunting fires atop the escarpment are frequently set from the base of a cliff by igniting strands of spinifex at places where a stream or river has cut deep into the plateau. These are timed to take advantage of the prevailing midday winds from the southeast so that when the fire crests it will be blown back towards the face of the escarpment. The animals which are caught in its path are forced up game trails where waiting hunters are able to shoot or snare them. Additional species pursued here include the black wallaroo and the rock wallaby. Though pockets of woodland may be reached by trail, the sandstone plateau is less important to coastal plains people.

Though practical considerations are important, aesthetic concerns are involved. The themes of 'cleaning' and 'taking care' of one's country (Jones, 1980) are strong motivations for using fire. Areas dominated by rank grass, thick leaf litter, or a tangle of undergrowth are considered not to have been cared for. Corrective burning is instituted irrespective of the time of year since, in the view of informants, further delays can only make a bad situation worse. Aborigines sometimes give the impression of having an almost manic compulsion about re-establishing fires in neglected environments.

The fires that result from their concern to clean an environment, especially those torched late in the dry season, appear to be destructive conflagrations. Aborigines are well aware that corrective burning demonstrates the problems which develop when excessive fuels accumulate. The offense is in having allowed such an unkept condition to develop, not in the seemingly draconian means necessary to remedy the situation. Aborigines have very strong emotive and ethical concerns regarding the uses of fire, and their actions cannot be explained by our overly simplified view of fire as being essentially bad or inherently dangerous.

Traditional burning practices thus involve the selective setting of fires (in some cases withholding of fires), at various times of the year (or even day), across a variety of habitats (burned with varying frequencies and intensities), in order to influence the relative productivity and spatial distribution of a broad spectrum of plant and animal resources. The apparent casualness with which all of this is done gives little or no indication of the hunter–gatherer's understanding of the wide range and variable effects that fires have for local habitats. At the same time, much of how the fire technology of hunter–gatherers differs from that of cattle pastoralists derives from the fact that the former exploit a broad spectrum of resources and the latter a much narrower one. Much more is involved than the fact that the one hunts kangaroos and the other 'hunts' cattle.

Pastoral fires

The burning technology described to me by cattlemen differs not only from that of Aboriginal hunter–gatherers but also from the prescribed burning practices of the Bush Fires Council of the Northern Territory, the government agency primarily concerned with the prevention and control of bush fires. As noted previously, the cattleman's burning practices are different from those of the hunter–gatherer in terms of the focus on cattle. It differs from the activities of the Bush Fires Council in its focus upon the distribution and quality of forage and not, as in the case of the Bush Fires Council, primarily the reduction or mitigation of fire hazards.

The practices of the Bush Fires Council are rationalised and coordinated through five regional offices in Darwin, Batchelor, Katherine, Tennant Creek, and Alice Springs. The most dramatic service provided by the Council is its *Protective Aerial Controlled Burning Programme* which involves aerial ignitions along the several hundred kilometres that make up the boundaries of a typical cattle station – all at no cost to owners. Set early in the dry season when fires burn out during the night, these wide and irregularly shaped firebreaks then provide a protective line against the spread of fire from or into adjacent properties.

Within the boundaries of a station some firing will already have begun well before the perimeter firebreaks are in place. Traditionally carried out by Aboriginal cowboys from horseback, most burning is now done from helicopters or four-wheel drive vehicles. Except for the advantage of speed, the older cattlemen interviewed have a negative view of 'copter cowboys'. Complaints include a belief that helicopters are 'too removed' from what happens on the ground, with a consequent loss of contact with and understanding of plant and animal conditions in remote areas.

Like hunter–gatherers, cattlemen schedule burning in terms of biological factors that are close at hand. Rather than being based on regional weather forecasts and the availability of aircraft, the traditional stockman pattern for burning is based on biological characteristics: the swarming of dragonflies, the flowering of selected trees and bushes, the seasonal arrival of certain migratory birds, and whether grass stems are still tinged with green indicating that there is sufficient moisture in the soil to effect a growth of green pick. Older stockmen maintained that these local conditions simply cannot be adequately judged from a plane or helicopter, and that aerial ignitions are simply too alienated from conditions on the ground to assure that fires are set at appropriate times and places.

Burning understory grasses and shrubs of eucalypt woodlands and open forests normally occurs during the first two months of the dry season. The timing differs slightly from the Aboriginal pattern in that it begins and ends a few weeks earlier, and this relates directly to the sole concern of cattlemen for livestock. The very first fires are set adjacent to homesteads and corrals as fireguards. Nearby fenced paddocks of native grasses are also fired and the subsequent emergence of green pick acts as a lure for cattle from surrounding areas. Depending upon the duration and intensity of an extended wet season, these fires can be ignited as early as mid March or, in exceptionally wet years, as late as early May. During this time cowboys are posted to outlying paddocks and corrals to set similar guard fires and initiate mustering. By clearing detritus (making cross-country travel easier and safer) and initiating new growth (concentrating cattle in selected areas), mustering is made possible while palatable forage is available for stock until the onset of the following wet season.

Because the understory fuels of eucalypt woodlands become dry two or three weeks earlier than those in stands of open forest, fires are initiated there first. These burnt-out corridors limit the spread of fires which are set later in the lower, intervening areas. Open forests are torched in essentially the same way as this habitat type, for the stockman does not consider the open forest as a different resource, as do Aboriginal hunter–gatherers. The delay in burning open forests occurs only because of damper fuels.

An important consideration of the somewhat earlier pattern of dry season burning by cattlemen is that the fired areas become a mix of green pick, partially burned, and unburned grasses. Though mature native grasses provide low levels of nutrients, cattlemen stated that old growth supplements regrowth, which by itself is inadequate. The pattern of burning within open forest also differs from that of hunter–gatherers. Individual fires are larger (20–1000 versus 0.5–25 hectares), the aim being to create a maximum effect for cattle rather than a mosaic of burned and unburned patches, set over a longer period (4–6 weeks longer), which supports a broad range of plant and animal resources for hunter–gatherers. Pastoral burning within both eucalypt woodlands and open forests is usually completed by mid June, after which soil moisture is normally insufficient to initiate further green pick.

The amount of eucalypt and open forest burned during the early dry season was variously stated as being 40–60%. A major difference from that of hunter–gatherer burning is that the remaining unburned areas, and even some of the partially burned areas, are fired at the end of the dry season, usually after one or two mid-December rains. The result is that as much as 90% of the total area is burned annually. In addition to providing still more new growth at the end of the year, cattlemen claim that these year-end fires reduce the amounts of sorghum and increase the productivity of preferred native types such as kangaroo grass (*Themeda australis*).

The estimated carrying capacity of cattle for eucalypt woodlands (6/km^2) is more than twice that of open forests (Perry, 1960), and cattlemen claim that with careful management (limiting over-grazing, and burning to increase preferred grasses) considerably larger numbers of cattle can be maintained in both areas. Advantages of eucalypt woodlands to pastoralists include greater uniformity of grasses, whereas hunter–gatherers prefer the diversity of plant and animal species in open forests.

Although limited in area and restricted to coastal regions, floodplains are important areas for the northernmost cattle stations. When waters recede, the reeds and grasses provide excellent grazing. Informants were less precise about burning floodplains, this in part depending upon the amount of grazing pressure involved, with those floodplains most heavily grazed not fired. The main reason for burning floodplains is to reduce ticks. All informants qualified their comments about burning floodplains by emphasising particular local conditions. Because soils remain moist longer, floodplains are fired much later than surrounding areas, with the lowest portions being burned as late as November and early December.

Similar variable practices also apply to swamps and the grass fringes of rivers. Smaller more interior flood basins or 'flats' were also burned as conditions permit and, like floodplains and swamps, later into the dry season than nearby stands of eucalypt woodlands or open forest. Unlike the subcoastal floodplains, interior flats support 'sour grass' or 'blady grass' (probably *Imperata*) which are only palatable during early stages of growth.

The small intermittent stands of monsoon rainforest or 'jungles' found within tall, open forests and nearby floodplains were not mentioned as being important and informants stated that the early season fires that they set within tall, open forests seldom carry into these stands. However, where cattle might use rainforest for shelter during dry periods, stands of monsoon forest might be burned to drive the animals into the open, thus causing damage. Unlike the Aborigines, cattlemen do not specifically fireguard rainforest, though they argue that the decrease in such habitats are a consequence of developers and water buffalo hunters rather than their own practices of firing.

With the exception of fireguarding the margins of coastal floodplains, northern cattlemen normally confine fireguards to stations and cultivated paddocks, man-made structures, and fenced pastures of introduced grasses. Further south, in the Mitchell grass country which ranges from the Barkly Tablelands intermittently across to the Victoria and Ord rivers, fires are purposefully excluded, though corridors of spinnifex and other fire responsive grasses and shrubs are fired. Mitchell grass (*Astrebla pectinata*), which is adversely affected by regular burning, is the most nutritious forage in northern Australia and is found on treeless plains south of the area considered here. However, just under half (8) of the informants had worked in the region at various times and were familiar with the need for excluding fire from stands of Mitchell grass while at the same time burning other forage types.

Both Euro-Australians and Aborigines were asked whether Aboriginal hunter–gatherers ever set hunting fires in Mitchell grass areas. All answers were negative, with the Aborigines interviewed stating that burning was neither necessary nor desirable for purposes of hunting–gathering. If Aborigines did regularly attempt to exclude fire from these regions (approximately 1 million km^2), this would be a most impressive example of fire exclusion similar to what they do on a much smaller scale with rainforests in the north.

Contrasts

Essential differences between cattlemen and hunter–gatherers are related to the nature of resources exploited. Within a stand of open forest cattlemen are not concerned with the varieties of plants and animals taken by hunter–gatherers nor with the food chains involved. They are not concerned that fires might damage wild yams within a rainforest stand, or that the flowers of fruit trees might be damaged in the absence of wind, or that large fires reduce the necessary mix of areas needed by macropods and smaller ground-dwelling species.

The fundamental difference is that fire technology of hunter–gatherers is more elaborate and complexly structured than is that of cattlemen. I do not argue that one system is better. Given the resource base of each, the two fire technologies are reasonable and coherent systems of knowledge in their own right. In terms of one group understanding the fire technology and ecology of the other,

Aboriginal hunter–gatherers have been in a better position to appreciate the needs of a pastoral economy than the cattlemen have been for understanding the needs of a hunting–gathering economy. Large numbers of Aborigines became cowboys and few, if any, Europeans became hunter–gatherers.

Most Euro-Australians possess little understanding of, much less appreciation for, traditional Aboriginal technology within which burning was an integral part. Most 'Territorians' merely denigrate traditional burning practices, even those who have been in a position to observe Aboriginal practice for a long period of time. As put to me by one station owner:

> You're not one of those anthropologists who's going to try and tell me that the Black Fellas knew what they were about ... that they were some kind of conservationist? What a lot of bloody nonsense!

Even those who are reasonably sympathetic or prejudiced toward Aborigines usually lack any real understanding of what habitat maintenance and modification by fire involves. As one such individual, a missionary, commented: it was one of the 'unfortunate customs' that Aborigines still practice.

However, two of the Euro-Australian stockmen that I interviewed demonstrated a fairly objective and receptive understanding of indigenous practices and how they related to pastoral burning technology. One of these individuals stated:

> Aboriginal stockmen didn't have to be taught to burn; they grew up with it ... it was important to their hunting way of life. All we had to do was give them a few boxes of matches and they went off and burned the places that they were supposed to and left the rest alone ... and that's the way we learned about burning this country.

The other individual had come to the north in the early 1920s and, though he knew about burning for sheep and cattle pasture in South Australia, the needs for pastoral burning in the Top End were entirely new. His education about burning, he said, came from Aborigines, many of whom were alternately cowboys and hunters during different times of the year:

> For Abos, the change from hunting and burning for roos to hunting and burning for cattle was easy. It just meant that they had to deal with one animal instead of a hundred-and-one other damn things.

Knowledge as technology

Aboriginal technologies, like those of hunting–gathering peoples elsewhere, have frequently been described as simple or primitive compared to those of more advanced or modern societies. Perspectives such as this are commonly couched in evolutionary assumptions and supported by the superficial equation that tools equal technology. However useful this may be for organising museum displays, it is much too simplistic for characterising the technologies of modern hunter–gatherers as represented over the past 20–30 thousand years. Robin Riddington has argued for a much more meaningful approach to understanding hunting–gathering technologies:

> Perhaps because our own culture is obsessed with the production, exchange, and possession of artifacts, we inadvertently overlook the artiface behind technology in favour of the artifacts that it produces ... I suggest that technology should be seen as a system of knowledge rather than an inventory of objects ... The essence of hunting and gathering adaptive strategy is to retain, and be able to act upon, information about the possible relationships between people and the natural environment (Riddington, 1982:471)

In this respect, an increasing number of studies by anthropologists and others are showing the considerable complexity of hunter–gatherer technologies when seen as systems of information which affect human–environmental relationships, and not merely as catalogues of tools and traits (e.g., Gladwin, 1970; Nietschmann, 1973; Blurton-Jones & Konner, 1976; Feit, 1978, 1983; Freeman, 1979, 1982, 1985; Jones, 1980; Johannes, 1981; Lewis, 1982b). As the above outline of Aboriginal burning practices has shown, knowledge of how plants and animals in a variety of habitats are influenced by variable uses of fire is infinitely more complex than merely the knowledge that underlies the tools and techniques for making fire. Given the diversity of environments inhabited and the variety of resources exploited by post-Pleistocene hunters and gatherers, it is difficult to imagine that they could have successfully adapted with only simple or primitive conceptions of environmental relationships. These adaptations would, of necessity, have required complex and sophisticated understandings of local ecosystems.

As some studies have recently shown (Johannes, 1981; Freeman, 1985), the knowledge that indigenous peoples have regarding environmental phenomena can provide important insights for and guidelines to scientific research and habitat management policy. In the Northern Territory, a few scientists, most of them working with the Commonwealth Scientific and Industrial Research Organization's Division of Wildlife and Rangelands Research in Alice Springs, have been looking at questions relating to the remaining numbers and distributions of endangered small animal species in the central deserts. Aboriginal knowledge of plants, animals and fire has played an important role in this research (Latz & Griffen, 1978; Burbidge, 1985). At the same time, both the Federal and Northern Territory governments have involved Aborigines in the management and operation of National Parks, most notably Kakadu, Uluru (Ayer's Rock), and Cobourg Peninsula. Though their involvement in the direct management of these parks appears to be constrained and overly institutionalised, they continue to carry out part-time hunting and gathering activites and, as an integral part of that, habitat burning.

In parts of Asia, Africa, Australia, the subarctic and arctic, and South and Central America, where remnant populations of hunters and gatherers are still found, it is possible for conservationists and environmentalists to benefit from indigenous systems of knowledge and perspectives on wildlife management. If national governments recognise the potential wealth of information that foraging peoples have, hunter–gatherers could continue to make an important contribution to the preservation of at least small portions of the once enormous areas that their forebearers – and ours – managed over past millennia.

Acknowledgements

This is a rewritten and expanded version of a paper included as part of a symposium, *Fire Ecology and Management in Western Australian Ecosystems* in Perth in May 1985 (Lewis, 1985b). The research, carried out in the Northern Territory of Australia between April and August, 1983, was supported by a grant from the Social Sciences and Humanities Research Council of Canada. I am also thankful for support provided by the North Australian Research Unit, Australian National University, Darwin, and the Australian National Parks and Wildlife Service personnel at Kakadu National Park and at their offices in Darwin.

References

Arthur, G. W. (1975). *An Introduction to the Ecology of Early Historic Communal Bison Hunting Among the Northern Plains Indians*. Archaeological Survey of Canada Paper 37. Ottawa: National Museum of Man.

Bean, L. J. & Lawton, H. W. (1973). Some explanations for the rise of cultural complexity in native California with comments on proto-agriculture and agriculture. In *Patterns of Indian Burning in California: Ecology and Ethnohistory*, ed. H. T. Lewis, pp. v–xvii. Ramona (California): Ballena Press.

Bell, D. (1983). *Daughters of the Dreaming*. Melbourne: McPhee Gribble/George Allen & Unwin.

Biswell, H. H. (1967). The use of fire in wildland management in California. In *Natural Resources: Quality and Quantity*, ed. S. V. Circiacy-Wantrap & J. J. Parsons, pp. 71–86. Berkeley: University of California Press.

Blainey, G. (1976). *Triumph of the Nomads*. Woodstock (NY): Overlook.

Blurton-Jones, N. & Konner, M. J. (1976). !Kung knowledge of animal behavior

(or: the proper study of mankind is animals). In *Kalahari Hunter–Gatherers,* ed.
R. B. Lee & I. de Vore, pp. 325–48. Cambridge (Massachusetts): Harvard University
Press.

Boyd, R. (1986). Strategies of Indian burning in the Willamette Valley. *Canadian Journal of Anthropology,* 5, 65–86.

Burbidge, A. (1985). Fire and mammals in hummock grasslands of the arid zone. In *Fire Ecology and Management of Western Australian Ecosystems,* ed. J. Ford, p. 91. Perth: Western Australian Institute of Technology.

Clark, R. L. (1981). The prehistory of bushfires. In *Bushfires: Their Effect on Australian Life and Landscape,* ed. P. Stanbury, pp. 61–73. Sydney: Sydney University.

Clark, R. L. (1983). Pollen and charcoal evidence for the effects of Aboriginal burning on the vegetation of Australia. *Archaeology of Oceania,* 18, 32–7.

Day, G. M. (1953). The Indian as an ecological factor in the northeastern forest. *Ecology,* 34, 329–46.

Feit, H. A. (1978). *Waswanipi Realities and Adaptations: Resource Management and Cognitive Structure.* Ph.D. Thesis. Montreal: McGill University.

Feit, H. A. (1983). Decision-making and the management of wildlife resources: contemporary and historical perspectives on Waswanipi Cree hunting. Paper presented at XI International Congress of Anthropological and Ethnological Sciences, Quebec City.

Ferguson, T. A. (1979). *Productivity and Predictability of Resource Yield: Aboriginal Controlled Burning in the Boreal Forest,* MA Thesis. Edmonton: University of Alberta.

Flood, J. (1983). *Archaeology of the Dreamtime.* Sydney: Collins.

Freeman, M. M. R. (1979). Traditional land users as a legitimate source of environmental expertise. In *The Canadian National Parks: Today and Tomorrow – Conference 11: Ten Years Later,* ed. J. G. Nelson, R. D. Needham, S. H. Nelson & R. C. Scace, Studies in Land Use History and Landscape Change. No. 7, pp. 345–69. Waterloo (Ontario).

Freeman, M. M. R. (1982). An ecological perspective on man-environment research in the Hudson and James Bay region. *Naturaliste canadien,* 109, 955–63.

Freeman, M. M. R. (1985). Appeal to tradition: different perspectives on Arctic wildlife management. In *Native Power: The Quest for Autonomy and Nationhood of Indigenous Peoples,* ed. J. Brosted, J. Dahl, A. Gray, H. C. Gullov, G. Henriksen, J. B. Jorgensen & I. Kleivan, pp. 265–81. Bergen: Universitetsforlaget As.

Gladwin, T. (1970). *East is a Big Bird: Navigation and Logic on Puluwat Atoll.* Cambridge (Massachusetts): Harvard University Press.

Gould, R. A. (1971). Uses and effects of fire among the Western Desert Aborigines of Australia. *Mankind,* 8, 14–24.

Hallam, S. J. (1975). *Fire and Hearth.* Canberra: Australian Institute of Aboriginal Studies.

Hallam, S. J. (1985). The history of Aboriginal firing. In *Fire Ecology and Management in Western Australian Ecosystems,* ed. J. Ford, pp. 7–20. Perth: Western Australian Institute of Technology.

Haynes, C. D. (1982). *Man's firestick and God's lightning: bushfire in Arnhemland.* Paper presented to the ANZAAS 52nd Congress. Sydney, NSW.

Haynes, C. D. (1983). *The pattern and ecology of Munwag: traditional Aboriginal fire regimes in north-central Arnhemland.* Paper presented at Wet–Dry Tropics Symposium. Darwin, Northern Territory.

Horton, D. R. (1982). The burning question: Aborigines, fire and Australian ecosytems. *Mankind,* 13, 237–51.

Jackson, W. D. (1968). Fire, air, water and earth: an elemental ecology of Tasmania. *Proceedings of the Ecological Society of Australia,* 3, 9–16.

Johannes, R. E. (1981). *Words of the Lagoon: Fishing and Marine Lore in the Palau District of Micronesia.* Berkeley: University of California Press.

Johnson, R. W. & Purdie, R. W. (1981). The role of fire in the establishment and management of agricultural systems. In *Fire in the Australian Biota*, ed. A. M. Gill, R. H. Groves & I. R. Noble, pp. 497–528. Canberra: Australian Academy of Science.

Jones, R. (1969). Fire-stick farming. *Australian Natural History*, 16, 224–8.

Jones, R. (1975). The neolithic, paleolithic and the hunting gardeners: man and land in the Antipodes. In *Quaternary Studies*, ed. R. P. Suggate & M. Cresswell, pp. 21–34. Wellington: Royal Society of New Zealand.

Jones, R. (1980). Hunters in the Australian coastal savanna. In *Human Ecology in Savanna Environments*, ed. D. R. Harris, pp. 107–46. London: Academic Press.

Kimber, R. G. (1983). Black lightning: Aborigines and fire in central Australia and the Western Desert. *Archaeology in Oceania*, 18, 38–45.

Latz, P. K. & Griffen, G. H. (1978). Changes in Aboriginal land management in relation to fire and to food plants in central Australia. In *The Nutrition of Aborigines in Relation to the Ecosystem of Central Australia*, ed. B. S. Hetzel & H. J. Frith, pp. 77–85. Melbourne: CSIRO.

Lewis, H. T. (1972). The role of fire in the domestication of plants and animals in Southwest Asia: a hypothesis. *Man*, 7, 195–222.

Lewis, H. T. (1973). *Patterns of Indian Burning in California: Ecology and Ethnohistory*. Ramona (California): Ballena Press.

Lewis, H. T. (1977). Muskata: the ecology of Indian fires in northern Alberta. *Western Canadian Journal of Anthropology*, 7, 15–52.

Lewis, H. T. (1982a). Fire technology and resource management in aboriginal North America and Australia. In *Resource Managers: North American and Australian Hunter-Gatherers*, ed. N. M. Williams & E. S. Hunn, pp. 46–7. AAAS Selected Symposium No. 67. Boulder (Colorado): Westview Press.

Lewis, H. T. (1982b). *A Time for Burning*. Occasional Publ. No. 17. Edmonton (Alberta): Boreal Institute for Northern Studies, University of Alberta.

Lewis, H. T. (1985a). Why Indians burned: specific versus general reasons. In *Proceedings: Symposium and Workshop on Wilderness Fire*, ed. J. E. Lotan, B. M. Kilgore, W. C. Fischer & W. R. Mutch, pp. 75–80. Ogden (Utah): Intermountain Forest and Range Experiment Station, USDA Forest Service.

Lewis, H. T. (1985b). Burning the 'Top End': kangaroos and cattle. In *Fire Ecology and Management in Western Australian Ecosystems*, ed. J. Ford, pp. 21–31. WAIT Environmental Studies Group Report No. 14. Perth: Western Australian Institute of Technology.

Meehan, B. (1982). *Shell Bed to Shell Midden*. Canberra: Australian Institute of Aboriginal Studies.

Mitchell, Maj. T. J. (1848). *Journal of an Expedition into the Interior of Tropical Australia*. London: Longman, Brown, Green & Longmans.

Nicholson, P. H. (1981). Fire and the Australian Aborigine: an enigma. In *Fire and the Australian Biota*, ed. A. M. Gill, R. H. Groves & I. R. Noble, pp. 55–76. Canberra: Australian Academy of Science.

Nietschmann, B. (1973). *Between Land and Water*. New York: Seminar Press.

Norton, H. H. (1979). The association between anthropogenic prairies and important food plants in western Washington. *Northwest Anthropological Research Notes*, 13, 175–200.

Perry, R. A. (1960). *Pasture Lands of the Northern Territory, Australia*, Land Research Series No. 5. Melbourne: CSIRO.

Riddington, R. (1982). Technology, world view, and adaptive strategy in a northern hunting society. *Canadian Review of Sociology and Anthropology*, 19, 469–81.

Sauer, C. O. (1947). Early relations of man to plants. *Geographic Review*, 37, 1–25.

Sauer, C. O. (1956). The agency of man on the earth. In *Man's Role in Changing the Face of the Earth*, ed. W. L. Thomas, pp. 49–69. Chicago: University of Chicago Press.

Sauer, C. O. (1970). Plants, animals and man. In *Man and His Habitat*, ed. R. H. Buchanan, E. Jones & D. McCourt, pp. 34–61. London: Routledge.

Sauer, C. O. (1975). Man's dominance by use of fire. *Geoscience and Man*, 10, 1–13.

Singh, G., Kershaw, A. P. & Clark, M. (1981). Quaternary vegetation and fire history in Australia. In *Fire and the Australian Biota*, ed. A. M. Gill, R. H. Groves & I. R. Noble, pp. 23–54. Canberra: Australian Academy of Science.

Stewart, O. C. (1951). Burning and natural vegetation in the United States. *Geographical Review*, 41, 317–20.

Stewart, O. C. (1954). The forgotten side of ethnogeography. In *Method and Perspective in Anthropology*, ed. R. F. Spencer, pp. 221–48. Minneapolis: University of Minnesota Press.

Stewart, O. C. (1955a). Fire as the first great force employed by man. In *Man's Role in Changing the Face of the Earth*, ed. W. L. Thomas, pp. 115–33. Chicago: University of Chicago Press.

Stewart, O. C. (1955b). Forest fires with a purpose. *Southwestern Lore*, 20, 42–6.

Stewart, O. C. (1955c). Why were the prairies treeless? *Southwestern Lore*, 20, 59–64.

Stewart, O. C. (1955d). Forest and grass burning in the mountain west. *Southwestern Lore*, 21, 5–9.

Stewart, O. C. (1963). Barriers to understanding the influence of use of fire by Aborigines on vegetation. *Annual Proceedings of the Tall Timbers Fire Ecology Conference*, 2, 117–26.

Timbrook, J., Johnson, J. R. & Earle, D. D. (1982). Vegetation burning by the Chumash. *Journal of California and Great Basin Anthropology*, 4, 163–86.

Tindale, N. B. (1959). Ecology of primitive Aboriginal man in Australia. *Monographiae Biologicae*, 8, 36–51.

4

Small-scale hunting economies in the tropics

STUART A. MARKS

Abstract

This paper is about the linkages between a society and its resources wherein developments in the former are inevitably reflected in the latter. Whether for subsistence, market or sport, hunting manifests both place-based and non-place-based phenomena. This paper claims the ubiquity of larger scale political–economic influences in all societies, that inequalities among and between societies are the main forces transforming social variables, that history is important for understanding these processes, and that ungulates serve purposes other than food. A review of hunting dynamics in tropical societies supports these claims. A model describes the cause and effect relationships of the main variables affecting human predatory behaviour and their effects on wild African buffalo (*Syncerus caffer*) in the Luangwa Valley of Zambia. Fluctuations in buffalo populations close to developing human communities have been a function of disease, range quality, and hunting pressure. Transformations in society including hunting tactics, technologies, and political–economic relationships compared to fluctuations in buffalo numbers suggest the dynamic relationships between hunters and their quarries.

Introduction

Wild ungulates and other animals are generally acknowledged as valuable sources of meat and other commodities in many small-scale tropical societies. Yet existing information on hunting within these societies tends to be fragmented, often obtained tangentially in the course of other investigations, and of short duration, lacking a sense of process or history. Attempts to synthesise this disparate information prior to the 1970s are fraught with epistemological and interpretive difficulties. Recognising these difficulties, this chapter briefly reviews the literature on hunting within marginal tropical economies, discusses some conceptual problems relative to the published information, proposes the integration of social and environmental aspects of hunting to create a processual model, and illustrates the model's utility with a case study from Central Africa.

Hunting economies in the tropics today
Africa

The !Kung of Botswana and the Mbuti of Zaire began this century on the farthest perimeter of the developing world system. For both groups, hunting was a basic and time-consuming activity. Despite their different landscapes of bush and forest, neither society was a closed economic system nor functioned autonomously. Both have become radically transformed and marginalised from their original states and drawn into the economic and class conflicts of the modern world.

Southern Africa

Long-term studies on the G/wi of the central Kalahari (Silberbauer, 1981) and the !Kung San along the Botswana/Namibia border (Lee, 1979) provide baselines for the hunting bands in Botswana. These groups depended upon wild products, gathered mainly by women, as the mainstays of nourishment. Hunting, by men, provided fewer calories but held a central role in the ritual life of the bands.

Silberbauer (1981) estimated that a band of 80 members (16 active hunters) consumed an average of 93 kg of meat per capita annually during the 1958–1966 period. Most of the larger mammals taken were ungulates. Observations on the same band by Tanaka (1976) showed a smaller figure (81.9 kg per capita) for 1967–68. Weapons employed by the G/wi included bow and poisoned arrows, snares, barbed probes, spears, and clubs. This band also ran prey down and scavenged meat from predator kills. These hunters killed game in the proximity of their camps and only occasionally ventured further afield.

Among the !Kung San, band size averaged 30 people and included 8–10 hunters (Lee, 1979). For this group, the basic weapon kit was the bow and arrow supplemented by snares. Snares were used to take small ungulates such as the steenbok (*Raphicerus campestris*) during the dry season. Dogs assisted in the pursuits of warthogs (*Phacochoerus aethiopicus*) and gemsbok (*Oryx gazella*) (Lee, 1979). Unfortunately, Lee (1979: 230) does not provide data on the numbers of game killed or for meat consumed during a year. Yet, he estimates that offtakes from herds were comparatively small and that the average hunter might take 2–3 large ungulates a year, or up to 120 in a lifetime. Elsewhere (p. 234), he mentions that these figures probably overestimated kills by individuals. The bulk of meat came from kills of giraffe (*Giraffa camelopardalis*), warthog, gemsbok, kudu (*Tragelaphus strepsiceros*), wildebeest (*Connochaetes taurinus*), eland (*Taurotragus oryx*), roan (*Hippotragus equinus*), and hartebeest (*Alcelaphus buselaphus*).

Hunting was the lifetime occupation for men. Kills of significant mammals punctuated the ritual stages within a hunter's career. The relative successes recorded for 127 men, together with the numbers of kudu (a major prey species) killed, show a wide range of individual achievements with younger hunters accounting for the most kills (Lee, 1979).

Whereas killing game and distributing meat among band members brought prestige and status to individuals, these attributes were tempered by the band's cultural themes of modesty, reciprocity, and egalitarianism. These cultural norms helped balance individual and band needs until events external to the band forced drastic changes during the late 1970s.

Historical and archaeological records on these Kalahari societies reveal repeated overlaps in the time-honoured categories of hunter and herder and of forager and farmer. Such mixing of subsistence means suggests that bands have survived through tactics of flexible opportunism (Schrire, 1980). Its current people continue to show similar adaptations in their affiliations with other groups. Yet, the level and intensity of recent involvements between band members and more powerful groups are unprecedented. By the 1970s, the !Kung San and other bands were experiencing rapid transformations to farming and herding, formal schooling for their children, the potential loss of their land, and environmental impoverishment. All these changes were exacerbated by drought. In addition, the !Kung San were caught in the middle of an armed conflict between whites and blacks in southern Africa. Increasingly, men were recruited by the South African Police who sought to use their hunting and tracking skills to halt the movements of freedom fighters across the Kalahari (Lee, 1979).

Central Africa

Bands of Mbuti pygmies in the Ituri forest of Zaire employ three hunting strategies. Each of these strategies involves different sets of weapons, a distinct labour organisation, and is focused on different prey (Harako, 1976, 1981; Hart, 1978). Spear hunting for elephant (*Loxodonta africana*), buffalo, okapi (*Okapia johnstoni*), and giant forest hogs (*Hylochoerus meinertzhageni*) involves ambushing prey at close quarters in dense forest. Such endeavours are rather risky and exclusively masculine. In contrast, net hunting is a cooperative strategy in which both sexes drive medium sized and small prey toward set nets. Net hunting is used primarily to capture duikers (*Cephalo-*

phus spp.) and the water chevrotain (*Hyemoschus aquaticus*) and usually produces large quantities of meat (Hart, 1978). Calculated on a per human body weight basis, yields from net hunting in the central Ituri forest are among the highest hunting returns reported (Hill, 1982). The most generalised strategy practised by the Mbuti is bow hunting. Used in conjunction with stalking and driving, bow hunting enables hunters to kill arboreal species, thus increasing the spectrum of species harvested (Harako, 1981).

The precariousness of resources within the Ituri forest of Zaire suggests the Mbuti may have always been peripheral to agricultural societies (Hart & Hart, 1986; for Pygmies elsewhere, see Bahuchet & Guillaume, 1982). Today, Mbuti bands depend on trading meat for agricultural products which, in some bands, account for over 60% of their calorific intake. Mbuti, in contact with commercial traders, increase their frequencies of net hunts, time spent hunting, numbers of casts, and proportion of nets used each day. This intensification permits them to capture more game for sale, but less is used for home consumption. Furthermore, these heavy offtakes decrease the availability of local antelopes (Hart, 1978).

Western Africa

Wildlife of all sorts, locally called *bushmeat*, is widely consumed in many West African countries (Asibey, 1974; de Vos, 1977; Jeffrey, 1977). Most forms exploited today are the smaller species, particularly rodents, bats, primates, and occasionally smaller antelopes. The decline of the korrigum (*Damaliscus lunatus*), one of West Africa's commonest grassland ungulates, in competition with cattle and human developments suggests what has happened to most large ungulates (Sayer, 1982). As yet, there is no comprehensive account for forest wildlife, either documenting its use in the region or its contributions to the food supply of local human communities (Ajayi, 1979).

An economic survey of bushmeat use in Brendel State, Nigeria, suggests the turn wildlife exploitation takes once rural areas are integrated into the regional and national economies. Martin's (1983) surveys show that over 50% of the people ate bushmeat regularly. Since bushmeat commanded such a high price, rural people exported it to town rather than consuming it themselves. In this way, rural people were able to garner products which they lacked or no longer produced.

Central and South America

Along the Atlantic coast of Nicaragua, the subsistence system of the Miskito Indians has changed as its people have responded to commercial and state interests (Nietschmann, 1973). Although most of their food came from agricultural cultivation, in the late 1960s the Tasbapauni group still relied to a substantial extent on turtling, fishing, and hunting. Their location along the coast directed most non-agricultural activities towards the sea. Of the 169 adult men at Tasbapauni, 124 did some fishing and hunting. Of these, 65% turtled, 20% hunted and fished, and 15% hunted. Men hunted as individuals or as small groups, and used shotguns. Of the approximately 50,270 kg of meat consumed by this village of 1000 people in 1968/69, 70% came from harvests of green turtles (*Chelonia mydas*), 7% came from white-lipped peccary (*Tayassu peccari*), and 5% came from white-tailed deer (*Odocoileus virginiana*). The rest came from a mélange of other oceanic and terrestrial species.

In Panama, Bennett (1968) chronicled the demise of the larger ungulates. Increases in the human population linked with deforestation, logging, farming, and hunting has impoverished the habitat and its large fauna throughout the isthmus.

With its expansive rainforest and diversity of animals, Amazonia is home for many previously isolated hunter–gatherers. Extensive clearing, road building, and other developments link these people today with larger economies causing changes in their weapons, social organisation, demographics, and expectations from their environments. In a study of hunting efficiencies in northern Brazil, Saffirio & Scaglion (1982) compared villages situated on a highway with villages further removed from easy contact with outsiders. Along the highway, villagers depleted some wildlife species, yet these villages showed higher yields because they had modified their hunting techniques and changed their tactics. Acculturated villagers used primarily bow and arrows and participated more in group hunts. Although these authors questioned the sustainability of these yields, Smith (1976) described highway settlements elsewhere in which individuals continued to hunt many of the same ungulates after 15 years of settlement.

Elsewhere in the Amazonia, Hames (1979) compared the hunting efficiencies of the shotgun and the bow and arrow for the Ye'Kwana and Yanomamao Indians in southern Venezuela. An immediate effect of using the shotgun by the Ye'Kwana was to decrease time spent hunting. However, the demonstrable superiority of the shotgun over aboriginal weapons led to a noticeable decline in the proximity of larger prey and to economic changes brought about by the Ye'Kwanas' need to spend more time growing crops for market to afford the new guns and ammunitions.

The targets reported for most Amazonian hunters included an array of birds, amphibians, reptiles, fish, and insects in addition to a few ungulates (de

Vos, 1977). Yet, the take of these few ungulate species added substantially and significantly to the group's diet.

Asia

I am less familiar with ethnographies of hunter–gatherers in Asia. Today, human groups subsisting on populations of wild ungulates are relatively rare. Returns from hunters in Papua New Guinea show that most species obtained are monotremes, marsupials, and rodents rather than ungulates (Dwyer, 1974, 1983). In the Philippines, Agta hunters frequently exchanged wild pig and deer meat with Palanan farmers (Peterson, 1981).

Fox (1969) suggested that the economic specialisation of several hunter–gatherer groups in India is a rather recent development, the result of contact with other larger and dominant groups. The smaller hunter–gatherer groups exploited marginal environments for deer meat and other forest products desired by members of the larger encompassing society. Fox suggested that when forest products become no longer rewarding economically, the smaller groups revert to less specialised activities such as agriculture and herding.

The problem

To define subsistence hunting as a field of study is to make assumptions about variables perceived to operate in nature and in society, together with their forms and interactions. To define any field of investigation is to draw boundaries, including some variables while excluding others. Given the academic compartmentalisation of knowledge, cultural aspects of human subsistence in tropical countries became the province of social scientists who provided descriptive accounts of technologies, lists of exploited species, the social organisation of labour, and the distribution of proceeds, while assuming a stability of environmental forms. For their part, biologists and ecologists contributed through their studies of wild mammalian populations and productivity while either assuming a homogeneity of human influences or denying their pertinence. Since I perceive subsistence hunting as an interdisciplinary study incorporating variables from both the social and biological sciences, I begin with some assumptions about variables and contexts, together with their linkages and relationships.

Assumption one

All small-scale societies today are influenced by the world political–economic systems; that is, by the industrialised economies, by the political developments of colonial empires, and by the independent or quasi-indepen-

dent nations that replaced these colonial empires. The processes of incorporating the natural resources, labour, and markets of lesser technologically
and politically developed societies into larger frames dominated by northern
industrial countries were greatly accelerated during the latter half of the 19th
century. These processes continue today at different rates and stages in
various areas of the tropics (Cohen, 1973; Chirot, 1977; Schrire, 1984).

The political–economic entanglements between the smaller, reciprocal, and
kin-based societies and the larger, bureaucratic, market-mediated, and
heterogeneous societies have occurred sporadically and at different intensities. Many of the small-scale societies currently persisting with some measure
of autonomy are located on the frontiers between states or in marginal zones.
In these localities, topographic and cultural barriers may prevent their ready
incorporation within the larger units. Some scholars believe that the incorporation of these smaller social and cultural groups into larger ones is an
essential process in the modernisation of a state and therefore constitutes an
essential objective of economic development. Others are not so certain about
the inevitability of this model nor the sanctity of current boundaries between
tropical states.

Despite this contention, most development schemes sponsored and economically supported by states have involved the increasing removal of
cultural, political, and economic control from the local to the national and
international levels (Leonard, 1985). As a consequence, small-scale societies
are everywhere in transition, their destinies mediated by their various linkages
within the political economies of states. A corollary is that with the social
structures of marginal societies indeterminate in form and function, their
biological environments which previously provided for most of their wants
and needs are likewise in flux.

These assertions imply that researchers undertaking a field study need to
acquire knowledge of how activities of individuals and groups at the local
level articulate and are influenced by the broader context of regional and
state-level events. The researcher must cope simultaneously with place-based
ecological relationships and with remotely acting political–economic processes, integrating and synthesising the previously separate paradigms of
resource ecology with those from the social sciences (Blaikie, 1985).

Assumption two

For individuals within small-scale societies, the main focus of
activities is to secure food and other resources to meet essential needs. An
unwarranted assumption about many of these societies was that they existed
in harmony and balance with their resources and that they carefully managed

their numbers and needs so that degradation of their essential resources was inconceivable. It is doubtful if any society has lived in such harmony for long. Some local populations may show strategies which restrain access to critical resources, but such exclusionary strategies may not necessarily have long-term management consequences. The efficacy of such control-access strategies is contingent upon population densities remaining below carrying capacity under existing modes of production. In societies where records have been kept, technologies, cropping patterns, sustenance tactics, and per capita food consumption show qualitative and quantitative variation with time.

In small-scale societies, food is not just to acquire, it is to distribute. To distribute food is to participate in an economy embedded within a social context. Within these societies, the reciprocal giving and receiving of food is an important characteristic of how the social group defines itself in contradistinction to other groups. Domestic production is mainly for use with households producing to sustain their biological, social, and cultural needs and responsibilities. If misfortune befalls a household, its needs may be sustained through its webs of kinship with other local or more distant households, who expect reciprocity if and when they fall on hard times. In this way, the distribution of food is a means to express social identities while allowing for the recognition of family and community cohesion.

Given this rather static model for small-scale societies, there appear few incentives for its members to produce beyond immediate needs or to accumulate a surplus. The production for surplus and the maximisation of production, together with the frequent erosion of resource units, are assumed to come from outside these small-scale social systems and from larger political–economic systems possessing differing organisational principles. This assertion is based upon the observation that in small-scale societies, food has more social and use value than it has exchange value. Therefore, as the autonomy of local economic systems is eroded, individuals, formerly within them, find that their market transactions involving food come into conflict with the reciprocal exchange patterns which had previously motivated them. Consequently, as economic organisation changes from reciprocal community-based exchanges to more money-based market transactions, its social complement becomes increasingly individualised accompanied by the fragmentation of community endeavours and identities. A consequence of development by differentiation and by fragmentation is that the fortunate few do not share responsibility for the unfortunate many (Sahlins, 1965).

The Tragedy of the Commons was popularised by Hardin & Bladen (1977) as the problem of regulating private resources obtained from common held lands. I perceive that an important driving force behind over-use of resources from common lands is the inequalities existing within and between societies (Marks, 1984; Blaikie, 1985).

Assumption three

Sustenance within small-scale societies may be studied as a system of functionally related components of technologies, ideas about resources, and activities. Many of these societies possess diversified strategies which normally lower the probability of failure by combining a wide spectrum of food-getting pursuits with different means, within different locales, and at different seasons. Theoretically, such diverse and wide-ranging strategies keep exploitation within the time and spatial confines of their resources' capacity to reproduce.

Hunting or trapping of wild ungulates are not autonomous activities. Rather, they were embedded within a matrix of other activities which include cultivating or trading for crops, gathering wild plants, fishing, and participating in the labour markets of larger dominant societies. Whether a group's technology is locally produced or acquired through trade is a relevant question. Consequently, the interrelationships among all food-acquiring tactics and modes of production are important if one aspires to understand the use, seasonality, and sustainability of resource exploitation within any society. The range of culturally accepted and utilised resources, and their temporal and spatial differences, may show kaleidoscopic changes for any society. Such a range of resources and their uses are not readily assessed during a short field study.

Assumption four

Wild ungulates are important beyond their contributions to the meat supply in small-scale societies. Wild mammals are the source of powerful metaphors and symbols, which provide vivid images and plans of behaviour for society members. The meanings for such associations are found in folk etymologies and expressed through their uses in magic, medicines, devination, and as cultural symbols. For example, among the Bisa of Zambia, abundant mammals, such as the zebra (*Equus burchelli*) and hippopotamus (*Hippopotamus amphibius*) were not exploited for many years. Rather than sources of meat, these mammals were important cultural markers between insiders and outsiders (Marks, 1976). As with many cultural phenomena, such associations may be as ephemeral or as long lasting as the circumstances that initially invested them with meaning.

A case study from Central Africa

Both the place-based human and environmental relationships and the remotely acting political–economic processes for the central Luangwa Valley in Zambia are described elsewhere (Marks, 1976, 1984). I review this information below to illustrate the processes of recent societal transformations and their impacts on hunting of a major prey species. My abbreviated account here makes no claim for impartial coverage of all social and ecological complexities or of issues covered in detail elsewhere. It is intended to illustrate some of the forces by which human communities throughout rural central Africa were marginalised, their cultures transformed, and resource processes within their immediate environments impaired (Palmer & Parsons, 1977).

This particular study is based upon 100 years of documented interactions between a human community of cultivators, traders, warrior–hunters, and subsequently cultivators and migrant labourers known as the Bisa, and the density of African buffalo, a major meat resource. During this time, Bisa society and culture have become increasingly subordinated within a larger politico–economic framework now known as the Government of the Republic of Zambia. This domination has profound effects upon the lives and lifestyles of the Bisa as well as upon the resources within the Luangwa Valley. The relative abundance of buffalo within the environs of Bisa villages reflects an amalgam of local and more distant selective forces acting on this species. Buffalo were not the sole targets for Bisa and other hunters throughout this time period. Yet, the importance of buffalo meat within the local community, together with its large size and reputation as dangerous prey, enables me to trace its relative abundance through oral and written records.

In the early years of this century, the prospects for minerals, particularly zinc, silver, and copper, brought British capital to central Africa (Roberts, 1976). The development of mines in what is now the nation of Zambia generated the need to recruit labourers for the mines. Labourers stationed at the mines required an inexpensive source of food. The colonial government provided labour for the mines through the imposition of a tax levied on adults payable only in cash, and through the eviction of Africans from lands reserved for European settlers. Subsequently, European settlers established large commercial farms along the line of rail which connected the mines to the distant seaports of South Africa.

Capitalistic relations of production in which foreigners controlled the methods of production while indigenous people supplied the labour became dominant in the mines, in the developing commercial and administrative towns, and on the settlers' farms. Elsewhere, Africans were left partly in control of their means of production provided they paid taxes and provided labour.

In Zambia, where population densities are relatively low and scattered, such unprecedented demands by the colonial administration for labour, for tax revenue, and for marketable crops led to widespread environmental degradation in some areas (Vail, 1977). Yet, not all areas succumbed to colonial pressures within the same time frame. Some groups, such as the Bisa whose hunting in the Luangwa Valley is chronicled below, were able to moderate outside influences to some extent.

While a number of significant changes occurred when Zambia became politically independent in 1964, government policies continued to place high priority on provisioning cheap and reliable food supplies to the cities and on urban developments. Today urban centres contain almost half of the Zambian population. Zambia's economy is characterised by sharp regional differences between its provinces which are located along the railways and the others that have remained largely rural. There is a continued emphasis on commercial crops and research to benefit large farms while crops for local production and consumption receive little or no research. The accelerated migration to urban areas by rural people seems to reflect the deterioration of economic conditions in the hinterlands (Marter & Honeybone, 1976; Marks & Robbins, 1984).

The model

The main variables affecting the population of buffalo within the central Luangwa Valley including the types of human predation can be depicted in a graphic model (Fig. 4.1). In this model, buffalo herd size is a function of the availability of water and quality of pasture during the dry season, the presence or absence of virulent epizootics (such as rinderpest), and the nature and extent of carnivore and human predation (Sinclair, 1977; Mloszewski, 1983). The numbers of male buffalo and herds is related to population densities. Grassland and savannas along the rivers are the preferred habitats for buffalo in the central Luangwa. These habitats are also where Bisa cultivate their crops and situate villages. Some human activities such as clearing and abandoning cultivation sites, cutting trees for house and other constructions, and setting fires during the dry season create additional grassland at the expense of woodland. These human-created environments are less favourable for the tsetse fly, the main vector for trypanosomiasis, a potentially fatal disease in humans and domesticated livestock. The presence of the tsetse has precluded raising domesticated livestock, and contributed in a positive way toward the persistence of subsistence and other types of hunting in the central Luangwa.

The spatial distribution of the Bisa population and their villages are determined by the decisions of elders within the matrilineages, the main organising principle of its social structure, and by cultural needs. Most lineage members live in the same or adjacent villages and cultivate land together. Lineage elders allocate roles, control socioeconomic activities, enforce norms, and oversee the distribution of assets. Prior to the establishment of British domination, lineages were largely autonomous, subsisting on local resources and trading for some items not locally available. At this time, hostilities with other ethnic and local groups were intermittent and occasionally devastating.

In the past 100 years, Bisa society has undergone several transformations as its status has shifted from a largely autonomous one to one more marginal. In the latter state, their politics and economy have become dictated increas-

Fig. 4.1. Model for human–buffalo processes in the central Luangwa Valley of Zambia. Human and environmental variables affecting the buffalo population are linked with arrows pointing in the direction of assumed causality. A minus sign suggests that higher values of the preceding variable result in lower values for the succeeding one, while a plus sign produces higher values.

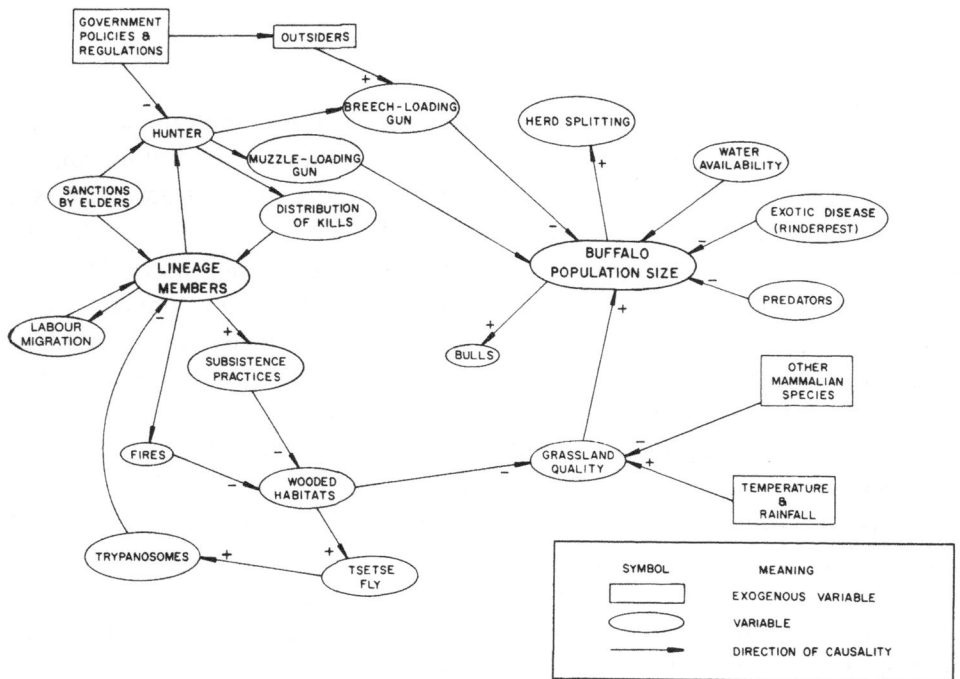

ingly by other groups, mediated through the bureaucracy of the state. These changes have affected Bisa culture, including their technology, their modes of subsistence and production, and their social structures. Furthermore, these changes have affected important components in the population ecology of the buffalo.

Transformations in Bisa society

During the 19th century, the relative prosperity and self-assertiveness of the Bisa impressed explorers as they passed through the Luangwa Valley. Bisa villages were strategically located along the major trading routes and controlled access into the hinterland. Elsewhere, Bisa men were known as proficient traders throughout central Africa (Gamitto, 1960; Roberts, 1976). Besides long-distance trading, men farmed. If they hunted large game, they belonged to professional guilds organised around the tools and techniques of the chase. At this time, the driving ambition for most men was to become a leader within their matrilineages. These positions became open upon the death of their incumbents and were filled by elders acting under strong factional interests. Elders evaluated candidates upon their reputations for political astuteness and upon their achievements in trade, war, and hunting.

The closing decades of the 19th century were turbulent, marked by increased trading, warfare, and outbreaks of new diseases affecting both people and wildlife. People died from the many new diseases transmitted by the slave and ivory caravans, and an introduced epizootic, rinderpest, decimated many ungulates, particularly buffalo. The establishment of British colonial control had its own far-reaching effects. British termination of the slave and ivory trades transformed the indigenous political–economic system by subordinating it to colonial whims. Taxes on all indigenous men, imposed by the British, directed Bisa men to look for employment wherever they could find work. Consequently, Bisa men became scattered throughout southern Africa for various lengths of time. The conscription of labour was facilated by disruptions in the previous economy, by disasters in food supplies and of diseases, and by indigenous desires for manufactured goods to replace locally made items. These widespread changes were to have long-term consequences for the Bisa homeland. For with the energies and talents of young and middle-aged men siphoned elsewhere, the Bisa homeland began to stagnate and progressively lose its social dynamism.

Transformations within the environment

Colonial policies adversely affected the environment as well (Vail, 1977). To dampen any opposition to its control, the colonial administration acted quickly to limit numbers of modern, breech-loading guns and ammunitions in African hands. The government allowed limited numbers of muzzle-loading guns, by then obsolete for insurgency, to remain in use. At the same time, the British made it illegal to take wildlife with earlier techniques which included the use of weighted spears, downfalls, poisons, and bows and arrows. These latter prohibitions effectively disbanded the guilds, thereby removing an indigenous means for acquiring animal protein without providing for a replacement. Furthermore, these injunctions undercut the major means by which wildlife populations were kept in check around villages. When enforced in conjunction with the new labour laws, these colonial policies initiated a series of trends which produced major environmental and societal reversals in the central Luangwa Valley. Prohibitions on the taking of wildlife by traditional means together with the long-term absences of men set the stage for increasing wildlife populations. Increases among some wildlife species enabled the tsetse fly, the main vector, to spread typanosomiasis throughout the Luangwa Valley (Ford, 1971). A plague of sleeping sickness (trypanosomiasis) forced the closing of administrative posts on the valley floor and frequent closure of the valley to outside travellers. Furthermore, the virulence of the sickness contributed to the valley's reputation, at least among foreigners, as an unhealthy place, subsequently isolating the valley from further government improvements.

From the 1920s through the 1970s, increasing wildlife around villages forced lineage residents to maintain two residences if they expected to produce and consume their own food. One house was a link in a circle of houses surrounding the large field which members of a village cultivated. Each household resided in these field houses from the time of planting until the harvest. At harvest, households returned to the village where their houses encircled the granaries. In 1940, the government designated large tracts of land in the Luangwa as game reserves and resettled people living within these tracts. In this manner, the colonial government narrowed the environmental options which valley residents had to cope with future droughts and disasters. In return, the government substituted famine relief and a few schools and dispensaries. Yet, the government profited through the encouragement of commercial and safari hunting and of tourism. The killing by wildlife rangers of marauding game in villager's fields mushroomed into the wildlife cropping schemes of the 1960s and 1970s. These cropping schemes were designed to show profits for government coffers, but in the end cost more than they returned in cash. The plan called for harvesting large mammals in the game reserves, which were said to be over-stocked, and the shipment and sale of their carcasses into the cities. Although these cropping operations have ceased and the game reserves are now national parks, Zambia continues policies in which urban areas are favoured over rural ones.

Transformation in hunting

Hunting for the Bisa has undergone several transformations in conjunction with changes in their political and economic status. Until the British outlawed 'traditional' weapons, most men were familiar with the use of bows and arrows, and a variety of poisons, spears, and traps. The same weapons were used also in warfare. Responding to British demands for labour, Bisa elders selected a few young men for training as hunters. These men, after an apprenticeship, provided meat and protected crops against raids by wild mammals. Other men spent their time working elsewhere for wages, which when appropriated by the elders went towards meeting local needs for cloth, salt, and weapons.

Under the sanctions of this lineage system, local hunters, using muzzle-loading firearms, could achieve social ascendance through making timely kills and through judicious distribution of meat among kinsmen. Strong norms mitigated against the sale or outside distribution of killed game. The perishability of meat, the isolation of valley communities from consumers elsewhere, and the lack of motorised transport reinforced these local injunctions. Furthermore, an indigenous theory that most large mammals possessed anthropomorphic attributes rendered them dangerous to hunters and their kin should local inhabitants break local norms. Such ideas and circumstances subjugated hunting and its proceeds under the lineage elders who also controlled the use of guns (Fig. 4.2). Operating within these constraints until the mid 1970s, local hunters focused their attention mainly on buffalo, whose numbers slowly increased after its bout with rinderpest, impala (*Aepyceros melampus*), warthog, and a few others as occasioned by opportunity.

Explorers' records and oral sources affirm that buffalo were abundant for most of the latter half of the 19th century. Furthermore, valley people obtained some meat from scavenging buffalo carcases killed by predators, a practice still found today (Fig. 4.3). By the end of the century, colonial administrators observed numerous buffalo skeletons, victims of rinderpest, and noted the shyness of the few surviving buffalo. From this low point, buffalo numbers gradually increased. The threat of rinderpest diminished during the 1950s, the result of government veterinary campaigns to innoculate against the disease in cattle (Sinclair, 1977). As buffalo increased, local hunters shifted their selection increasingly to this species. The large size and predictable behaviour of buffalo offered ample returns for hunters' efforts in terms of local status and prestige (Marks, 1977, 1979).

Two points about local hunting are worth emphasising. First, individuals varied in time spent hunting. Middle-aged men spent more time after game than did either elders or younger men. Despite variable individual efforts, yields were uncharacteristically high for muzzle-loading guns, weapons notoriously ineffective in downing large game. Such high yields indicate substantial numbers of interactions between hunters and prey. During 1966/67, transect counts of wild mammals substantiated this abundance. During this year, hunters in one community killed 101 large mammals. These kills, together with eight elephants killed in the same vicinity by wildlife guards, yielded 91.5 kg (carcase yield) of meat/adult/year for the community. Carcase yield is based upon the average weight for a flayed carcase given the species and sex. In three months during 1973, 50 hunts undertaken by these same hunters, who used breech-loading weapons as well as muzzle-loaders, resulted in 16 mammals killed, yielding 1517 kg (carcase yield). Since these

Fig. 4.2. Butchering the carcase of an adult male buffalo. Initially, the hide is removed and placed on the ground to provide a surface for disarticulating the carcase. Axes and knives are used to disassemble the carcase into portable sizes. The meat is distributed initially among carriers, who further disperse it among their dependents in the village. (Photograph by the author, December 1966.)

kills were made during the dry season, most of the meat was transported elsewhere for sale. Little meat was consumed in the local community.

My second point is that hunters with muzzle-loading guns killed mostly buffalo bulls, a selection which had little effect on herd recruitment. This selection reflects the relative accessibility of bulls, both in bachelor bands and on the periphery of herds. Cows usually form the centre of herds (Mloszewski, 1983). As breech-loading weapons increasingly became used by locals and outsiders, cows, prized for their fatness, were selectively sought and killed. This shift in weaponry, including the use of wire snares, caused noticeable decreases in buffalo and other larger ungulates. Sustained contact with outsiders led to the lapse of the previous norms and to the sale of meat.

The causes and consequences of new influences were evident during a short visit in 1978. Government regulations and the hunting codes were enforced

Fig. 4.3. A Bisa hunter watches from cover as a herd of buffalo moves through an abandoned field. On his shoulder, the hunter carries his muzzle-loading gun and the foreshoulder of a buffalo retrieved from a lion. Meat salvaged from predator kills is consumed and interpreted as an omen of ancestral benevolence. (Photograph by the author, May 1967.)

opportunely on Bisa residents while other offenders, mostly civil servants, were apprehended rarely. Transient and resident outsiders, including government civil servants, provided cash and material rewards for meat and trophies. These inducements transcended local incentives and resulted in the diversion of most meat from the distributional exchange system of the lineage into the cash market. The accessibility of outsiders with transport and cash had changed the context of hunting within the local communities. The commercialisation of meat and game products also affected village social structure and life. A few households with hunters benefited while most received nothing from these transactions.

During 1977, the government assembled a large workforce in the community to build a school, dispensary, and court, and to drill a well. Responding to cash incentives for meat from this group of outsiders on government payroll, some residents set wire snares for buffalo; others became employed as hunting guides. Increased kills from buffalo herds together with increased human traffic and activities around the development project caused a noticeable scarcity of buffalo. Undoubtedly, this level of human disturbance may have caused some buffalo to shift elsewhere. Yet, this large drop in numbers resulted mainly from heavy hunting. In 1978, I encountered signs of only two diminished herds and no bachelor bulls.

These events in one Zambian community demonstrate the main points which necessitate an integrative perspective on hunting. A synthesis of place-based and remotely acting political–economic processes is necessary to interpret observations within small-scale societies. Natural environments and resources do not exist apart from the context of society. In the modern world, beset with inequalities between societies, groups, and individuals, it is no longer acceptable to remain naive about such boundaries and their effects. One is naive to assume a homogeneity within either the social or environmental spheres. Furthermore, with societies everywhere in flux, we can no longer assume constancy in the perception of resources and their uses. Particularly is this true with wild mammalian populations which require land and habitat which competes with an increasing human population for sustenance and for space.

Conclusions

Much of my concern about tropical hunting economies reflects methodological and ideological uncertainties about their structures and transformations. Methodological uncertainty reflects the variety of ways in which subsistence economies have been described. Biologists and social scientists have different approaches to determining what attributes of these small-scale societies are important, what can be safely ignored, and what are the rules of discourse on the subject. These widely differing judgments and evaluations often remain implicit and unexamined. This chapter has sought to provide a conceptual model for subsistence hunting and to make most of its assumptions explicit. As no assemblage of facts speaks for itself, my model makes no claims for neutrality.

Ungulate conservation and management are political–economic issues. In tropical latitudes, these concerns and recognition come from those above (dominant economic classes, often outsiders) and are forced on those below (peasants and other marginal groups). The ideologies of the former are reflected in the activities of consultants, policy-making politicians, senior bureaucrats, and through the distribution of foreign aid. The gathering of information on subsistence economies is also a political–economic act; the information once assembled becomes the object for discussion and debate among those with political and economic power. Such discussions often lead to overt intervention in the foraging activities of marginal societies by institutions of the state.

State institutions are involved already and impinge on the smaller-scale groupings within their boundaries. Any movement toward management of ungulates on a wider scale usually affects the livelihoods of those within smaller-scale societies. Such interventions may involve changes in land tenure, changes in laws and their enforcement, rearrangements in pricing structures and credit arrangements, alterations in cultural structures and subsistence, and degrees of access to resources by variously designated groups. The resolution of these uncertainties remains the problem and the challenge.

References

Ajayi, S. S. (1979). *Utilization of Forest Wildlife in West Africa*. FO: MISC/76/26, Consultant's Report (mimeo). Rome: Food and Agricultural Organization of the United Nations.

Asibey, E. O. A. (1974). Wildlife as a source of protein in Africa south of the Sahara. *Biological Conservation*, 6, 32–9.

Bahuchet, S. & Guillaume, H. (1982). Aka-farmer relations in the northwest Congo Basin. In *Politics and History in Band Societies*, ed. E. Leacock & R. B. Lee, pp. 189–211. Cambridge: Cambridge University Press.

Bennett, C. F. (1968). *Human Influences on the Zoogeography of Panama*. Ibero-Americana No. 51. Berkeley: University of California Press.

Blaikie, P. (1985). *The Political Economy of Soil Erosion in Developing Countries*. New York: Longman.

Chirot, D. (1977). *Social Change in the Twentieth Century*. New York: Harcourt, Brace, Jovanovich.

Cohen, B. J. (1973). *The Question of Imperialism: The Political Economy of Dominance and Dependence*. New York: Basic Books.

de Vos, A. (1977). Game as food: a report on its significance in Africa and Latin America. *Unasylva*, 29, 2–12.

Dwyer, P. D. (1974). The price of protein: five hundred hours of hunting in the New Guinea highlands. *Oceania*, 44, 278–93.

Dwyer, P. D. (1983). Etolo hunting performances and energetics. *Human Ecology*, 11, 145–74.

Ford, J. (1971). *The Role of Trypanosomiases in African Ecology*. London: Oxford University Press.

Fox, R. G. (1969). Professional primitives; hunters and gatherers of nuclear South Asia. *Man in India*, 49, 139–60.

Gamitto, A. C. P. (1960). *King Kazembe, Being the Diary of the Portuguese Expedition to that Potentiate in the Years 1831–1832*. (Translated by I. Cunnison). Lisbon: Estudios de ciencias politicas e socias, No. 42 and 43.

Hames, R. B. (1979). A comparison of the efficiencies of the shotgun and the bow in neotropical forest hunting. *Human Ecology*, 7, 219–52.

Harako, R. (1976). The Mbuti as hunter – A study of ecological anthropology of the Mbuti Pygmies (1). *Kyoto University African Studies*, 10, 37–99.

Harako, R. (1981). The cultural ecology of hunting behavior among Mbuti Pygmies in the Ituri Forest, Zaire. In *Omnivorous Primates*, ed. R. Harding & G. Teleki, pp. 499–555. New York: Columbia University Press.

Hardin, G. J. & Bladen, J. (eds.). (1977). *Managing the Commons*. San Francisco: W.H. Freeman.

Hart, J. A. (1978). From subsistence to market: a case study of the Mbuti Net Hunters. *Human Ecology*, 6, 325–53.

Hart, T. E. & Hart, J. A. (1986). The ecological basis of hunter–gatherer subsistence in African rain forests: The Mbuti of eastern Zaire. *Human Ecology*, 14, 29–55.

Hill, K. (1982). Hunting and human evolution. *Journal of Human Ecology*, 11, 521–44.

Jeffrey, S. (1977). How Liberia uses wildlife. *Oryx*, 14, 168–73.

Lee, R. B. (1979). *The !Kung San: Men, Women and Work in a Foraging Society*. Cambridge: Cambridge University Press.

Leonard, H. J. (ed.) (1985). *Divesting Nature's Capital: The Political Economy of Environmental Abuse in the Third World*. New York: Holmes & Meier.

Marks, S. A. (1976). *Large Mammals and a Brave People: Subsistence Hunters in Zambia*. Seattle: University of Washington Press.

Marks, S. A. (1977). Buffalo movements and accessibility to a community of hunters in Zambia. *East African Wildlife Journal*, 15, 251–61.

Marks, S. A. (1979). Profile and process: subsistence hunters in a Zambian community. *Africa*, 49, 53–67.

Marks, S. A. (1984). *The Imperial Lion: Human Dimensions of Wildlife Management in Central Africa*. Boulder (Colorado): Westview Press.

Marks, S. A. & Robbins, R. D. (1984). *Report of an FAO Consultancy, People in Forestry in Zambia*. Rome: Forestry Division, Food and Agricultural Organization of the United Nations.

Marter, A. & Honeybone, D. (1976). *The Economic Resources of Rural Households and the Distribution of Agricultural Development*. Lusaka: Rural Development Studies Bureau, University of Zambia (mimeo).

Martin, G. H. G. (1983). Bushmeat in Nigeria as a natural resource with environmental implications. *Environmental Conservation*, 10, 125–32.

Mloszewski, M. J. (1983). *The Behavior and Ecology of the African Buffalo*. Cambridge: Cambridge University Press.

Nietschmann, B. (1973). *Between Land and Water: The Subsistence Ecology of the Miskito Indians, Eastern Nicaragua*. New York: Seminar Press.

Palmer, R. & Parsons, N. (eds.). (1977). *The Roots of Rural Poverty in Central and Southern Africa.* Berkeley: University of California Press.

Peterson, J. T. (1981). Game, farming, and interethnic relations in northeastern Luzon, Philippines. *Human Ecology*, 9, 1–22.

Roberts, A. (1976). *A History of Zambia.* New York: Africana Publishing Co.

Saffirio, G. & Scaglion, R. (1982). Hunting efficiency in acculturated and unacculturated Yanomamao villages. *Journal of Anthropological Research*, 38, 315–27.

Sahlins, M. D. (1965). On the sociology of primitive exchange. In *The Relevance of Models for Social Anthropology*, ed. M. Benton, pp. 139–236. London: Tavistock Publications.

Sayer, J. A. (1982). The pattern of the decline of the korrigum *Damaliscus lunatus* in West Africa. *Biological Conservation*, 23, 95–110.

Schrire, C. (1980). An inquiry into the evolutionary status and apparent identity of San hunter–gatherers. *Human Ecology*, 8, 9–32.

Schrire, C. (ed.) (1984). *Past and Present in Hunter Gatherer Studies.* Orlando (Florida): Academic Press.

Silberbauer, G. B. (1981). *Hunter and Habitat in the Central Kalahari Desert.* Cambridge: Cambridge University Press.

Sinclair, A. R. E. (1977). *The African Buffalo: A Study of Resource Limitation of Populations.* Chicago: University of Chicago Press.

Smith, N. J. H. (1976). Utilization of Game along Brazil's Transamazon Highway. *Acta Amazonica*, 6, 455–66.

Tanaka, J. (1976). Subsistence ecology of the central Kalahari San. In *Kalahari Hunter–Gatherers*, ed. R. B. Lee & I. de Vore, pp. 98–119. Cambridge (Massachusetts): Harvard University Press.

Vail, L. (1977). Ecology and history: the example of eastern Zambia. *Journal of Southern African Studies*, 3, 129–55.

5

Northern subsistence hunting economies

DAVID R. KLEIN

Abstract

Subsistence hunting maintains its importance throughout the circumpolar north. In the past, unpredictable changes in movements and numbers of ungulates have forced subsistence hunters to avert starvation by turning to alternate resources, particularly of the sea. Today, northern people live in permanent settlements making periodic forays in search of game which still contributes significantly to northern diets. The introduction of modern firearms and all-terrain vehicles has transformed hunting and placed greater pressure on resources. The development of a cash economy has resulted in some commercialisation through the sale of sport hunting opportunities or animal products. Subsistence hunting is awarded priority by governments, and native people are gaining political influence over management of fish and wildlife resources.

Introduction

Subsistence hunting and gathering was the primary economic basis of virtually all human societies prior to the emergence of agriculture, which led to the organisation of people into non-self-sustaining communities and the concurrent development of market economies. Thus, subsistence hunting and gathering exclusively nurtured the evolution of *Homo sapiens* until at least some 10,000 years BP. Subsistence hunting has remained the primary basis for existence of most human societies in areas unsuitable for agriculture. This pattern is particularly evident in the north (area north of 65°N), where climatic extremes prohibit conventional crop production and associated animal husbandry, and where industrial society has been slow to overcome the geographic and climatic obstacles to transportation and resource exploitation. The only notable exception has been the development of reindeer husbandry across northern Eurasia. Where this occurred, the wild reindeer, that previously had been the primary basis of subsistence economies, were usually eliminated (Klein, 1980).

Caribou in North America and wild reindeer in Eurasia, both *Rangifer tarandus*, have been of primary importance to most northern terrestrial subsistence hunting economies. In some instances, muskoxen (*Ovibos moschatus*) have also supported subsistence hunting economies. Moose (*Alces alces*), mountain sheep (*Ovis* spp.), and deer (*Odocoileus* spp., *Cervus* spp., and *Moschus sibiricus*), among ungulates, also have been important elements in the food base. However, these latter species have generally not been the primary sustaining basis for hunting cultures. Small mammals such as hares (*Lepus* spp.), beavers (*Castor* spp.), muskrats (*Ondatra* spp.) and ground squirrels (*Citellus* spp.); birds, particularly ptarmigan (*Lagopus* spp.) and waterfowl; and fish have served as staple commodities for many northern peoples, particularly during periods of ungulate scarcity. Marine societies have also exploited ungulate populations as supplemental food sources.

Transitory nature of subsistence hunting

Northern subsistence economies based on hunting ungulates have been of a much more temporary nature in both time and space than has been the case of cultures with broader resource dependencies. For example, the caribou-hunting Iñupiat of northern Alaska were not only nomadic so they could stay close to caribou herds as movement patterns shifted over time, but they also were forced periodically to return to dependence on marine resources during periods of caribou scarcity (Burch, 1980). This may have been through direct movements to the coast or through trade, kinship dependencies, or partnerships. Similarly, the 'Caribou Eskimos' or Padlimiut of the Canadian tundra have been characterised by nomadic movements and population fluctuations, including starvation, in response to changing caribou distribution and numbers (Harper, 1964). Archaeological evidence also indicates transitory local presence during prehistoric times of caribou hunters in both northern Alaska (Campbell, 1978; Gerlach, 1982) and in the Canadian North (Harp, 1959; Gordon, 1975). Wide fluctuations in caribou populations in both northern Canada and northern Alaska have been the norm (Kelsall, 1968; Skoog, 1968). Northern subsistence caribou hunters have readily adapted to changes in caribou distribution through their high degree of mobility. However, widespread declines in caribou populations have generally led to corresponding declines in numbers of caribou hunters (Hall, Gerlach & Blackman, 1985).

Subsistence hunting economies based primarily upon hunting of muskoxen have not persisted to the present day, but archaeological evidence records a few such examples. Among the most remarkable is the evidence of two distinct periods during which muskox hunters, apparently arriving from

Ellesmere Island, existed in northernmost Greenland at latitudes as high as 82–83°N (Knuth, 1967). Archaeologists refer to the first period as Independence I Culture. Recent Independence I radiocarbon dates range from 1980 ± 130 BC to 1670 ± 100 BC (Maxwell, 1985), suggesting a duration of about 350 years. Independence II Culture was present about 1000 years later and has been dated at 640 ± 110 BC. Apparently, these people were almost totally dependent upon muskoxen for food; skins were presumably used for clothing and shelter. Fuel was obtained from driftwood, which at the time of Independence II, when North Greenland was locked in sea ice throughout the year, was only available from raised beaches where it had been deposited at least several hundred years earlier. Knuth (1967) also reports on archaeological evidence of more recent human occupation of North Greenland coming from Neoeskimo cultures oriented toward marine mammal hunting.

On Banks Island in the Canadian Arctic, the remains of a muskox hunting camp called 'Umingmak', dating from about 1480 BC, has been found (Müller-Beck, Torke & von Koenigswald, 1971). However, it appears to have been occupied only during summer, and the cultural items present suggest that these people were oriented toward the sea at other times of the year. Archaeological sites with evidence that muskoxen were a major food species have been found in Siberia (Vereschagin, 1959) and Alaska (G. R. Bane, personal communication). The species is believed to have been locally extirpated as late as a few hundred years ago in Siberia and in the late 1800s in Alaska (Campbell, 1978; Uspenski, 1984).

The transitory nature of hunting societies in the past has been a direct product of natural and occasionally anthropogenic fluctuations in the ungulate populations upon which they depend. Without alternative resources to sustain them, except on a short-term basis, or external sources of welfare that later became available through economic, religious and governmental systems, subsistence hunters were faced with the alternatives of hunger, starvation, or abandonment of traditional living areas when their primary ungulate prey populations declined. Movement to other areas was possible only if other ecological niches were available as, for example, when interior Iñupiat caribou hunters of northern Alaska moved to coastal areas when caribou were scarce. The close cultural ties with coastal Iñupiat maintained through language, trade, and intermarriage made this possible (Burch, 1972). Among the 'Caribou Eskimos' of the Canadian North, the Asiagmiut of the Kazan River area exemplify subsistence hunters who traditionally had no other resource sufficient to sustain them when caribou were periodically unavailable (Burch, 1972). Before external sources of welfare were available, hunger and starvation were periodic.

The freedom of movement that has characterised northern subsistence hunting bands in the past no longer exists. Whereas adoption of new means of transport, primarily the snowmobile and outboard motor, have enabled subsistence hunters to range more widely from the home base, the centralisation of subsistence hunting family units into permanent villages with dependence on schools, churches, trading posts, other social services, and permanent housing has brought an end to the nomadism of the past.

Subsistence value of large mammals

Fish and wildlife are essential to the subsistence life styles of many northern peoples. It is obvious that without these local food sources, the subsistence communities could not exist. The value of resources is measured not only by the market value of the products but also by their relative importance to the communities that use them. 'Importance' will include their economic contribution to the communities as well as the more subjective criteria of psychological well-being derived from a sense of economic security, and the cultural traditions or spiritual values that are interwoven with the resources. These subjective values are difficult to quantify and are seldom fully appreciated within the market economy. They cannot be replaced simply by government welfare payments. A commonly held view is that native peoples living by subsistence hunting are doing so out of necessity rather than choice and that they would be better off fully absorbed into the market economy. However, this perspective is not generally shared by the natives of Alaska and northern Canada (Usher, 1976; Berger, 1985).

Although no thoroughly objective basis exists for evaluating subsistence resources, comparisons between subsistence and market economies are frequently made for environmental impact evaluations. In such cases, a cash value is usually assigned to subsistence commodities on the basis of their replacement cost within the local market economy. Using this criterion, Usher (1976) determined that a subsistence household in the western Canadian Arctic might obtain the equivalent of Can$ 6200 worth of meat (1974–75 prices) if 12 caribou, 60 geese and 230 kg of fish were harvested annually. He considered this to be a modest harvest for the region.

In Alaska it is also possible to assign a replacement cost value to meat derived from subsistence activities. Using data from Table 5.1 and a mean replacement cost of US$ 4.40/kg (1982 prices) for all meat and fish, the total annual value of these commodities per household would range from US$ 1314 in Bettles/Evansville to US$ 12,852 at Hughes. The value of moose meat obtained would range from 11–38% of the total subsistence meats in the three village complexes. Replacement costs at 1987 prices would probably exceed US$ 5.50/kg for the least expensive meats and fish available in local stores.

The replacement value of caribou in northwestern Alaska can be estimated from data in Table 5.2. For the village of Anaktuvuk Pass, where caribou are the major subsistence commodity, the number of caribou harvested varies with their availability, which in turn is a product of total herd size and variation in patterns of migration. Patterson's (1974) data from the early 1970s corresponds with a declining caribou herd, while Woolford's (1954) data is from a period of herd increase. Using replacement prices of US$ 2.20/kg and a mean weight of 59 kg of meat per carcase, the value of the

Table 5.1. *Use of subsistence wildlife and fish by Athabascan villages on the upper Koyukuk River, Alaska (based upon edible weight in kg).*[a]

Village	Fish	Birds	Mammals	Total	Moose	(%total)
Bettles/Evansville (2.6)[b]						
Per household	279	6	372	657	250	(38.1)
Per capita	107	3	143	253	96	(37.9)
Alatna/Allakaket (3.9)						
Per household	2864	97	556	3517	400	(11.4)
Per capita	734	25	143	902	103	(11.4)
Hughes (4.3)						
Per household	5326	105	985	6416	868	(13.5)
Per capita	1239	25	229	1493	202	(13.5)

[a]Data for 1982 from Maracotte & Haynes (1985).
[b]Mean number individuals per household in parentheses.

Table 5.2. *Subsistence use of caribou in eight Iñupiat villages within the range of the Western Arctic Caribou Herd in northwestern Alaska in comparison to the inland 'Caribou Iñupiat' village of Anaktuvuk Pass.*

	Caribou/person	Total Harvested	Reference
8 NW Alaska villages[a]	3.20	12,133	Patterson (1974)
Anaktuvuk Pass	26.7-53.3	2000-4000	Woolford (1954)
	10.31	1,000	Patterson (1974)

[a]Includes Point Hope, Kivalina, Kiana, Kotzebue, Noorvik, Shungnak, Noatak, and Selawik.

caribou harvested was US$ 3471–6929 per person and US$ 260,000–520,000 for the entire village during the early 1950s. These heavy per capita harvests were presumably associated with the use of caribou to feed dogs necessary for transportation involved in carrying out subsistence activities. In the early 1970s, with a replacement cost of US$ 3.30/kg, the value of caribou harvested was US$ 2010 per person and US$ 195,000 for the entire village. During 1969–73 the people of Anaktuvuk Pass had the highest annual consumption of subsistence foods in northern Alaska at 755 kg/person (Patterson, 1974), representing approximately 88% of the total diet (Binford, 1978). By 1984 it was estimated that caribou, dall sheep (*Ovis dalli*), moose, and fish made up 70% of their diet (cited by Hall *et al.*, 1985).

The villages of northwestern Alaska have access to caribou of the Western Arctic Herd primarily during migrations; however, fish and marine mammals provide the major subsistence foods to those communities. Their harvest of caribou, while considerably less per capita than at Anaktuvuk Pass, is substantial in view of the large number of villages involved. For the eight northwestern villages in Table 5.2, the meat replacement value of caribou was US$ 624 per person or US$ 2.37 million. Estimated total annual village harvest of caribou from the Western Arctic Caribou Herd during the 1960s and early 1970s was 25,000, yielding a replacement value of nearly US$ 5 million annually.

Availability of game
Caribou

Caribou in North America and wild reindeer in Eurasia experience wide variation in population size. In the past, these population fluctuations occurred largely independently of human harvest. Some authors have postulated a 60–90-year cycle of caribou in North America (Haber & Walters, 1980) and Greenland (Thing, 1984). In Greenland, the fluctuations are believed to be tied to climatic changes associated with variations in sea ice (Vibe, 1967). In North America, the historical record of caribou numbers lacks sufficient precision to distinguish between irregularly spaced fluctuations and a regular cycle. Speculation on causes of past declines of caribou has varied from increased hunting and predation pressure (Bergerud, 1974) to limitations in food availability (Scotter, 1967). Of importance, however, to subsistence hunters has been the relatively long-term fluctuations in their availability. Given the length of time between peaks and lows in numbers, whether they were of a cyclic nature or random events is immaterial in terms of the consequences for survival of subsistence hunters.

Caribou and wild reindeer are relatively easy to hunt (Burch, 1972). When populations are high, their movements are predictable, although annual variations in migration routes are expected. Specialised caribou hunters had preferred hunting locations at traditional water crossings, in mountain passes, or where other terrain features constrain movements (Hall *et al.*, 1985; Sharp, 1977). This preference was more important before the introduction of firearms, which eliminated the need to approach within the effective range of spears or arrows. The widespread use of lead fences in forested areas and lines of stone and sod cairns in tundra to guide caribou during drives or battues was only discontinued when firearms came into general use (McKennan, 1965; Gronnow, Meldgaard & Nielsen, 1983). Archaeological excavations in Greenland (Gronnow *et al.*, 1983) and Alaska (Gerlach, 1982) also confirm use for millenia of specific hunting sites that coincide with present-day movement patterns of caribou. Nevertheless, historical and current caribou subsistence hunting patterns are characterised by both flexibility and mobility to assure continued contact with the caribou consistent with seasonal and annual variations in their movements (Sharp, 1977; Harper, 1964).

Migrations of caribou and wild reindeer revolve around movements between calving grounds and related summer areas and the wintering areas. Calving grounds are relatively discrete and traditional. They provide the one fixed point or area that is the main criterion for distinguishing herds (Skoog, 1968) since caribou cows attempt to use the area in which they calved the previous year. The extent of movement between wintering areas and calving grounds varies with distances generally increasing with herd size. Large migratory herds, often numbering over 100,000 animals, may migrate 800–1000 km between wintering areas and the calving grounds, while small intra-mountain herds may move only tens of kilometres or less in their altitudinal seasonal migrations.

Virtually all calving takes place within a week to ten-day period. Thus, for hunters the most predictable opportunity for encountering caribou and wild reindeer in both space and time has been at the approaches to or on the calving grounds. Ancient encampments are often found in these specific locations. Calving occurs at the interface of the late winter period of easy overland travel, and spring 'breakup' when travel is restricted by melting snow and flooding rivers. However, the focus on this season declined with the advent of firearms and improved methods of transport. The latter, which allow for efficient hunting in other seasons when animals are more dispersed, have made concentrated hunting close to the calving grounds less necessary.

Shortly after calving when caribou are uniformly dispersed over the calving grounds, cow caribou with new calves, and later arriving yearlings and some bulls, aggregate and leave the calving area. Direction of movement may be in response to the emergence of new vegetation, wind direction, or other factors. Although there are consistent long-term patterns in these post-calving movements, there may be considerable variation from year to year. With the emergence of harassing insects in late June or early July, caribou movements become even more directly influenced by weather conditions, with rapid movements in search of relief near the coast or in the mountains during warm, windless periods that favour the insects. Windy, cool periods stimulate the return to more favourable foraging areas. Summer conditions impose restrictions on travel and make the location of caribou and wild reindeer difficult to predict. Hunting becomes largely opportunistic and subsistence hunters have traditionally localised their activities around other resources such as fish and waterfowl.

During the autumn and early winter migration to the wintering areas, animals may become available to hunters from communities along traditional routes. With the onset of cold weather, meat is easily preserved, but the need for winter food and skins for winter clothing, and the desire to obtain fat animals also makes hunting at this time important. Concentrated hunting during the migration period has been a common practice among subsistence hunters.

The caribou-hunting Chipewyan of Canada have traditionally dispersed along the transition zone from summer to winter range at the interface between the tundra and taiga to maximise the likelihood of encountering migrating caribou (Sharp, 1977). Because their hunting effort has been primarily during migration, the Chipewyan have tried to maximise the kill when caribou were encountered to provide for the extended periods at other seasons when caribou are unavailable and to facilitate sharing with other groups. Although successful caribou hunting by the Chipewyan has depended on dispersal of social groups, the subsistence system required a well-defined reciprocity system to distribute meat between those groups that were in the path of the migrating caribou and those that were not.

The Kutchin Athabascans at the village of Old Crow on the Porcupine River in northern Yukon Territory of Canada also hunt mainly during the two annual migrations. Major segments of the Porcupine Caribou Herd winter in the north-central Yukon south of Old Crow. Migration to calving grounds on the coastal plain of northeastern Alaska and the subsequent return migration to the wintering grounds result in large numbers of caribou crossing the Porcupine River near Old Crow. The river provides access for intercepting the migrating caribou. Snowmobiles and dog teams are used during the spring migration when the river is usually still ice covered. Boats with outboard motors are the mode of transport during the autumn harvest. The annual harvest by this village of over 200 people is generally in the range of 500–1000 animals during years when caribou pass close to the village (N. Bartichello, personal communication).

Use of specific wintering areas by caribou varies considerably from year to year. This variability is apparently an evolutionary adaptation to low production rates of lichens, the primary winter forage; to regional variations in snow conditions; and to the influence of forest fires that alter plant succession. Hunting in wintering areas results in wide annual variations in harvest.

Caribou are hunted by the Kutchin Indians of Arctic Village in the southern foothills of the Brooks Range in northeastern Alaska. Historically, these people were semi-nomadic, and they both positioned themselves to intercept migrating caribou of the Porcupine Herd as well as hunting in the wintering grounds used by a portion of the herd that remained in Alaska. Although they are settled in a permanent village today, they use snowmobiles extensively to intercept migrating caribou and to reach caribou wintering relatively distant from the village. Nevertheless, the availability of caribou in winter has always been uncertain, because in some years the entire Porcupine Herd may winter in Canada. The average annual harvest of caribou by this village (110 people in the 1980 census) varies from less than 100 in years when no large components of the herd are wintering in the area or migrating through it, to 1000 or more when caribou are readily available.

In northwestern Alaska, the Western Arctic Caribou Herd, numbering in excess of 200,000 (J. Davis, personal comunication), provides a subsistence base for at least 20 villages. For villages on the periphery of the summer range, such as Barrow and Point Hope, as well as those on the periphery of the winter range, such as Buckland and Huslia, caribou are an incidental part of the total village subsistence economy. They are taken opportunistically. Villages along the migration routes, such as Anaktuvuk Pass, Ambler and Noatak, have a much greater reliance on caribou and generally enjoy a greater assurance of their availability (Table 5.2).

During the 1970s, the Western Arctic Caribou Herd declined precipitously from 250,000 to 75,000 animals. The estimated annual harvest from the herd dropped from *c.* 25,000 prior to 1975 to a low of 2700–3500 in 1976–77 (Davis, Grauvogel & Valkenberg, 1985). Since 95% of the harvest was by subsistence hunters, the effect of the reduced harvest on the local economy was catastrophic. Although the reduced kill resulted primarily from very restrictive harvest quotas and their general acceptance by the local people, many of the villages no longer had access to the caribou as the smaller herd abandoned many of its traditional migration routes and reduced the total range area it used. The hardship caused by the lack of caribou in northwestern Alaska was partially alleviated through increased dependence on fish and other subsistence foods, sharing of meat of marine mammals by coastal villagers, and increased dependence on social welfare.

Muskoxen

Muskoxen are non-migratory throughout most of their distribution with strong fidelity to specific home ranges, although in the High Arctic extensive seasonal movements may occur. In contrast to caribou, their availability to subsistence hunters is more predictable in time and space. In addition, their tendency to form dense groups and stand their ground as a defence against wolves has rendered them vulnerable to hunters accompanied by dogs and equipped with spears, bows and arrows, or rifles. These behavioural and ecological characteristics apparently have been a major factor in their local extirpation and shrinking range throughout the Holocene (Campbell, 1978; Burch, 1980; Klein, 1988). No subsistence hunting societies based primarily on muskoxen have persisted up to the present time, presumably because the species is easily over-exploited. The consequences of over-hunting for subsistence hunters would likely be exodus from the area or possibly starvation if alternate foods were not available. Either alternative would lead to abandonment, as was the case with muskox hunters of the past who occupied North Greenland.

The consequences of over-exploitation for ungulate populations is more variable. If an ungulate population is the single major food source for a society of hunters, one would expect that over-exploitation would result in the decline of the ungulate population and this would lead to starvation or emigration of hunters before the population was extirpated. On the other hand, if the hunters persisted because they had access to alternative food sources, local ungulate populations could be hunted to extinction. This pattern may well have led to the extinction of the muskox in Siberia, Alaska, Yukon Territory, and locally in the Northwest Territories.

Productivity of muskoxen varies markedly throughout their range in relation to range conditions, length of the growing season, winter snow conditions, density of animals, and levels of predation and human harvest. Newly established muskox populations under favourable range conditions and near the southern limits of their distribution have produced annual increments as high as 24% (Jingfors & Klein, 1982). In constrast, populations in the Canadian High Arctic and Greenland have been marginally self-sustaining, with periodic declines and local extinctions in the absence of human harvest. In Northeast Greenland, Inuit hunters from the village of

Scoresbysund annually harvest 300–500 muskoxen from a population of slightly more than 4000 (Thing *et al.*, 1987). This level of harvest approximates the annual herd increment. Wolves have long been absent from the area, but their recent re-establishment may limit the capability of the muskox population to sustain current harvests by subsistence hunters. Although muskoxen are an important subsistence supplement to the Inuit of Scoresbysund, marine resources are their primary subsistence base.

On Banks Island in the Canadian Arctic, the muskox population is experiencing a peak in numbers and productivity. Historically, muskox numbers were generally low there, and local Inuit hunted them only sporadically, preferring the more numerous and traditionally available caribou. With the recent abundance of muskoxen, estimated at $25,700 \pm 2054$ (K. Jingfors, personal communication) in contrast to about 5000 caribou, there is concern that competition for forage may affect the preferred caribou detrimentally. Local people have shown relatively little interest in hunting muskoxen for subsistence use, and limited efforts have been directed toward commercial harvest and export of meat as a way of optimising benefits from the muskoxen and reducing their numbers. The venture has met with little success largely because of obstacles imposed by stringent inspection requirements governing slaughter, difficult logistics, and lack of a developed market for the meat.

In Alaska, herds have been established by transplants both within and outside their historical range (Klein, 1988). These have exhibited high rates of increase, and the State Board of Game has permitted hunting in some areas under a quota system. Although no tradition of subsistence hunting for muskoxen exists among the present-day native people of Alaska, the Yupik Eskimos, living close to muskox populations on Nunivak and Nelson Islands in southwestern Alaska, have expressed divergent viewpoints on how hunting should take place. On Nunivak Island, local people prefer to offer surplus animals for sport and trophy hunting by non-residents, providing an opportunity for income through provision of guide services and accommodations, rental of boats and snowmobiles, and sale of craft items. On Nelson Island, villagers prefer that surplus muskoxen be made available to them for their own subsistence use. An important distinction between the two areas is the presence of a village-owned reindeer herd on Nunivak Island that provides a ready source of favoured meat. On Nelson Island there are no reindeer, nor are there any wild ungulates other than muskoxen.

Management for subsistence use

Management of ungulates for subsistence use by local peoples has been a slowly evolving concept in northern areas. Formerly, the use of fish and wildlife resources by native peoples was largely ignored by governments either by intent or neglect. In many situations, there was no need for management because traditional hunting methods and levels were sustainable. Often the earliest protective regulations were instituted to restrict over-exploitation by explorers, whalers, and traders. This has changed with the adoption of new technologies that have greatly increased the efficiency of hunting and transport, growing human populations, and increasing conflicts between subsistence, commercial, and recreational users.

Among the earliest legal constraints on ungulate harvest by northern peoples in North America was the protection given to Canadian muskoxen in 1917 because of local extirpation through over-hunting. However, not until the 1950s, did wildlife management agencies focus on northern ungulates. Early efforts were superficial and involved primarily reconnaissance surveys. In the early 1950s, the Alaska Game Commission imposed seasons and bag limits for harvest of caribou north of the Arctic Circle. These regulations had little effect on local subsistence harvests, since enforcement was directed towards sport and trophy hunters. By 1959, regulations on caribou hunting north of the Arctic Circle by local residents were completely removed, and harvest quotas and restricted hunting seasons were not imposed again until the drastic decline of the Western Arctic Caribou Herd.

In northern Canada, native people are free to hunt caribou and most other wildlife without restriction because of aboriginal rights granted under the Royal Proclamation of 1763, in treaty agreements, or afforded by non-restrictive government policy. An over-riding exception, however, may be exercised by the federal and territorial governments when a species is considered endangered.

A period of major transition in the legal status of northern aboriginal peoples and their relationship to wildlife for subsistence use was initiated in the late 1960s throughout North America, Greenland, and Scandinavia. This transition was stimulated by growing demands for energy and other northern resources by the industrial world to the south. The discovery of large oil reserves on the northern coast of Alaska in 1968 accelerated action by the United States Congress to settle land claims with native people so that a pipeline could be built to transport oil to southern markets. In 1971, the US Congress enacted the Alaska Native Claims Settlement Act granting land and money to native people in a comprehensive manner designed to stimulate their economic well-being. Companion legislation passed in 1980 established multiple federal land conservation units under the National Park Service, Fish and Wildlife Service, Forest Service, and Bureau of Land Management, with legal protection and preference given to traditional subsistence hunting activities.

As a result of the example set by the Alaska Native Claims Settlement Act and expanding oil and gas exploration, the movement to resolve native claims in northern Canada gained momentum among both the local people and in the federal government. In contrast to Alaska, the Canadian North involves a more extensive area with greater diversity of governmental jurisdiction and a more complex legal history. Thus, the resolution of native claims in Canada has become a complex and continuing process of regionally negotiated settlements based on ethnic and tribal distinctions, previous treaty agreements, and the interplay between federal, provincial, and territorial jurisdictions.

In Greenland, the aboriginal rights movement culminated in 1979 when Denmark granted Home Rule. Although Greenland remains a part of Denmark, Home Rule is somewhat comparable to the status of provincial governments in Canada. All jurisdiction over fish and wildlife management now rests with the people of Greenland, and the protection of subsistence hunting opportunities receives high priority in government.

In northern Scandinavia, Sami (Lapp) cultural organisations have been active in both Norway and Sweden for several years and have gained momentum as large-scale hydroelectric projects have encroached upon lands traditionally used for reindeer herding and subsistence activities. The governments of both Norway and Sweden have recognised that the Sami people are entitled to certain legal rights based upon their unique culture and traditional patterns of land use. Definition of the specific rights to be given legal protection is now the subject of debate in both countries. The tenuous dependence of Sami culture on reindeer husbandry and associated subsistence activities has been emphasised by the contamination of reindeer lichen ranges in the path of atmospheric fallout from the 1986 Chernobyl nuclear power plant disaster.

In North America, the concept of managing wild ungulates for subsistence hunting is relatively new with the possible exception of the use of fire to improve habitats by some Indian groups. The management process is also evolving as new procedures are put into practice, as local people gain political influences, as administrators gain experience, as new laws are passed and tested in the courts, and as public awareness of the subsistence dependencies of northern peoples increases.

In Alaska, responsibility for wildlife rests with the state government on state, federal, and private lands, with the exception of those National Parks and few National Wildlife Refuges where hunting is specifically excluded. Since 1980, subsistence use of fish and wildlife has been given priority when the harvestable surplus is insufficient to fulfil the requirements of all users.

Thus, for management purposes, agencies must obtain data on both the allowable harvest as well as the subsistence requirements of local people. Within the Alaska Department of Fish and Game, responsibility for obtaining data on wildlife populations rests with the Division of Game, while determination of patterns of subsistence use and needs of subsistence users is the responsibility of the Subsistence Division. Final decisions allocating harvests are made by the Board of Game on the basis of recommendations from the Division of Game, the Subsistence Division, local fish and wildlife advisory committees, and the general public. The Board of Game is a seven-member body of residents appointed by the governer.

Major problems in managing subsistence hunting in Alaska have revolved around: (1) the difficulty of deriving an equitable definition of subsistence and in establishing criteria for those who qualify for subsistence priorities; (2) the reluctance to report subsistence harvests by native people long accustomed to unregulated use of fish and wildlife without detailed record-keeping, and their lack of experience with permitting requirements; (3) animosity of urban sport hunters toward those qualifying for subsistence priorities; and (4) inability of the Alaska Department of Fish and Game to obtain the necessary data with the money and personnel available. Many native people in Alaska, frustrated by pressures imposed on their traditional life-styles, support a movement to seek native tribal sovereignty (Berger, 1985). They want total jurisdiction and responsibility for managing and allocating fish and wildlife resources on lands granted to local native communities. Whereas tribal sovereignty may ensure that small native communities control local subsistence resources, it would not secure the broader regional resource base upon which subsistence economies generally depend. There is no simple solution to the problem of protecting traditional subsistence hunting economies with the continuing transition toward a cash economy.

In the Northwest Territories of Canada where native peoples are in the majority, there has been a transition from federal administration of government to increased self-governance. One result has been that local boards consisting of native residents of the Northwest Territories are now participating in policy development and advising on regulations regarding hunting and wildlife management.

Developing effective management systems for northern ungulate populations to assure their continued well-being and productivity for use by subsistence hunters will undoubtedly continue to evolve as northern peoples undergo the transition to increasing self-governance. The process will require a melding of cultural attitudes toward resource management and use from both native and Western cultures.

References

Berger, T. R. (1985). *Village Journey: The Report on the Alaska Native Review Commission. Inuit Circumpolar Conference.* New York: Hill & Wang.

Bergerud, A. T. (1974). Decline in caribou in North America following settlement. *Journal of Wildlife Management*, 38, 757–70.

Binford, L. R. (1978). *Nunamiut Ethnoarchaeology: A Case Study in Archaeological Formation Processes.* New York: Academic Press.

Burch, E. S. Jr (1972). The caribou/wild reindeer as a human resource. *American Antiquity*, 37, 339–68.

Burch, E. S., Jr. (1980). Muskox and man in the central Canadian subarctic. *Arctic*, 30, 135–54.

Campbell, J. M. (1978). Aboriginal human overkill of game populations: examples from interior North Alaska. In *Archaeological Essays in Honor of Irving B. Rouse*, ed. R. C. Dunnell & E. S. Hall Jr, pp. 179–208. The Hague: Mouton Publishers.

Davis, J. L., Grauvogel, C. A. & Valkenberg, P. (1985). Changes in subsistence harvest of Alaska's Western Arctic Caribou Herd, 1940–1984. In *Proceedings 2nd North American Caribou Workshop*, ed. T. C. Meredith & A. M. Martell, pp. 105–18, McGill Subarctic Research Paper No. 40. Montreal: McGill University.

Gerlach, S. C. (1982). A summary of archaeological research at Tukuto Lake, National Petroleum Reserve, Alaska. In *A Review of Cultural Resource Survey and Clearance Activities, 1977–82*, ed. E. S. Hall, pp. 1–25. Anchorage: US Geological Survey.

Gordon, B. H. C. (1975). Of men and herds in Barrenland prehistory. Canada National Museum of Man, Archaeological Survey Paper 28, 74–75.

Gronnow, B., Meldgaard, M. & Nielsen, J. B. (1983). Aasivissiut – The great summer camp: archaeological, ethnographic and zoo-archaeological studies of a caribou-hunting site in West Greenland. *Meddelelser Grønland, Man & Society*, 51, 1–96.

Haber, G. C. & Walters, C. J. (1980). Dynamics of the Alaska–Yukon caribou herds and management implications. In *Proceedings Second International Reindeer/Caribou Symposium, Røros, Norway*. ed. E. Reimers, E. Gaare & S. Skjenneberg, pp. 645–63. Trondheim: Directoratet for vilt og ferskvannesfisk.

Hall, E. S., Jr, Gerlach, S. C. & Blackman, M. B. (1985). In the National Interest: A geographically based study of Anaktuvuk Pass Iñupiat subsistence through time. *North Slope Borough, Barrow, Alaska*, 1, 1–105.

Harp, E. Jr (1959). Ecological continuity on the barren grounds. *Polar Notes*, 1, 48–56.

Harper, F. (1964). Caribou Eskimos of the upper Kazan River, Keewatin. *University of Kansas, Museum of Natural History Miscellaneous Publication*, 36, 1–74.

Jingfors, K. T. & Klein, D. R. (1982). Productivity in recently established muskox populations in Alaska. *Journal of Wildlife Management*, 46, 1092–6.

Kelsall, J. P. (1968). *The Migratory Barren-ground Caribou of Canada.* Canadian Wildlife Service Monograph, No. 3. Ottawa: Queen's Printer.

Klein, D. R. (1980). Conflicts between domestic reindeer and their wild counterparts: a review of Eurasian and North American experience. *Arctic*, 33, 739–56.

Klein, D. R. (1988). The establishment of muskox populations by translocation. In *Translocation of Wild Animals*, ed. L. Nielsen & R. Brown, pp. 298–318. Milwaukee: Wisconsin Humane Society and Caesar Kleberg Wildlife Research Institute.

Knuth, E. (1967). Archaeology of the muskox way. *Contributions du Centre d'Etudes Arctiques et Finno-Scandinaves.* No. 5. Paris.

Marcotte, J. R. & Haynes, T. L. (1985). Contemporary resource use in the upper Koyukuk region, Alaska. *Alaska Department of Fish & Game, Subsistence Division, Technical Paper*, 93, 1–110.

Maxwell, M. S. (1985). *Prehistory of the Eastern Arctic.* New York: Academic Press.

McKennan, R. A. (1965). The Chandalar Kutchin. *Arctic Institute of North America Technical Paper,* 17, 1–156.

Müller-Beck, H. W., Torke, W. & von Koenigswald, W. (1971). Die Grabungen des Jahres 1970 in der Pre-Dorest-Station Umingmak auf Banks Island. *Sonderdruck aus Quartar Bund,* 22, 143–56.

Patterson, A. (1974). *Subsistence Harvest in Five Native Regions.* Anchorage: Joint Federal-State Land Use Planning Commission for Alaska.

Scotter, G. W. (1967). Effects of fire on barren-ground caribou and their forest habitat in northern Canada. *Transactions of the North American Wildlife Natural Resources Conference,* 32, 246–59.

Sharp, H. S. (1977). The Caribou-eater Chipewyan: Bilaterality, strategies of caribou hunting, and the fur trade. *Arctic Anthropology,* 14, 35–40.

Skoog, R. O. (1968). *Ecology of the caribou (*Rangifer tarandus granti*) in Alaska.* Ph.D. Thesis. Berkeley: University of California.

Thing, H. (1984). Feeding ecology of the West Greenland caribou (*Rangifer tarandus groenlandicus*) in the Sisimiut-Kangerlussuaq Region. *Danish Review of Game Biology,* 12, 1–53.

Thing, H., Klein, D. R., Jingfors, K. & Holt, S. (1987). Ecology of muskoxen in Jameson Land, northeast Greenland. *Holarctic Ecology* 10, 95–103.

Usher, P. J. (1976). Evaluating country food in the northern native economy. *Arctic,* 29, 105–20.

Uspenski, S. M. (1984). Muskoxen in the USSR: some results of and perspectives on their introduction. In *Proceedings First International Muskox Symposium,* pp. 12–14, ed. D. R. Klein, R. G. White & S. Keller, Biological Papers of the University of Alaska, Special Report No. 4.

Vereschagin, N. K. (1959). Ovtsebyk na severe Sibiri. *Priroda,* 48, 105–6.

Vibe, C. (1967). Arctic animals in relation to climatic fluctuations. *Meddelelser Grønland,* 170, 1–227.

Woolford, R. (1954). Notes on village economics and wildlife utilization in arctic Alaska. Mimeograph report. Fairbanks: US Fish and Wildlife Service.

SECTION C

Recreational and commercial hunting

ROBERT J. HUDSON & D. H. M. CUMMING

Wild populations have provided meat, special products, and sport since ancient times, providing, now as then, an important incentive for maintaining wild populations and the habitats that support them. Formerly, sanctuaries were essentially hunting reserves in which harvests were limited to a privileged few and trespass or at least inimical land uses were firmly prohibited. Today, such incentives are felt at both the political level in generating public and governmental support and at the individual level by redirecting benefits of conservation to landowners. Hunting, by definition, is an extensive management system so the interests of environmental conservation are better served than by more intensive forms of game husbandry.

However, because of its extensive nature, the logistics of game cropping for meat and special products can be staggering except where it involves reduction of excessive populations in accessible areas such as parks and reserves (Hawley, 1985). Sustained-yield cropping in remote areas often requires heavy mechanisation and incurs high labour costs. Integration of commercial harvest with recreational hunting is an obvious and long-standing solution in Europe. The only problem is the difficulty of maintaining high standards of meat hygiene which are required to open new markets and to meet the increasingly stringent demands of traditional ones.

There also is the problem of administrative complexity, particularly in free-enterprise economies where commercial exploitation is complicated by matters of ownership of land and wildlife. One solution is to remove wildlife from the marketplace, the fundamental basis of the North American system. Although this may be efficient on public lands, it denies a powerful incentive for wildlife conservation and appears best suited to frontier areas where the problem is one of regulating harvests rather than protecting habitats. It is not surprising that revenue from recreational hunting has been directed to landowners where pressures on land are high. Despite protests, this trend has started in America particularly in jurisdictions where a large proportion of the landbase is in private ownership.

This section focuses on formal marketing of meat, by-products, and hunting opportunities from essentially wild populations. The first two chapters contrast European and American traditions of sport hunting. The remaining two chapters describe attempts to provide incentives for nature conservation through commercialised safari hunting in Zimbabwe, and to ensure efficient sustainable use of natural landscapes in the Soviet Union.

These case studies can be considered as representative of formal systems of game administration but many specific examples have not been included. For example, control programmes for wild pigs provide large quantities of meat and form the basis of a modest world industry (Tisdell, 1982). Despite some controversy, kangaroos also provide a harvestible surplus which enters commercial markets (Kirkpatrick & Amos, 1985). But perhaps the most important contributors to world supplies of wildlife products are informal markets.

What is termed poaching or even subsistence hunting in Africa is really commercial hunting in the sense that the products of animals killed are traded rather than consumed in the home or the hunter's community. Elephant poaching, for example, is a highly developed, albeit illegal, commercial hunting enterprise with a continent-wide gross export value for ivory alone of about US$ 40 million annually over the past five years (400 tonnes of ivory at US$100/kg). Similarly, the killing of 8000 black rhino during the last six years will have realised at least US$ 8 million.

References

Hawley, A. W. L. (1985). Commercial meat production from wild cervids. In *Biology of Deer Production*, ed. P. F. Fennessy & K. R. Drew, pp. 327–37. Wellington: Royal Society of New Zealand.

Kirkpatrick, T. H. & Amos, P. J. (1985). The kangaroo industry. In *The Kangaroo Keepers*, ed. H. J. Lavery, pp. 75–102. St Lucia: University of Queensland Press.

Tisdell, C. (1982). *Wild Pigs: Environmental Pest or Economic Resource*. Sydney: Pergamon.

6

Sport hunting in continental Europe

A. B. BUBENÍK

Abstract

Two main hunting systems are used in continental Europe: *revier*, based on the Roman law *res nullius*; and *free-hunt*, based on *Codex Napoleon*. The free-hunt system without strong hunter ethics and governmental supervision does not compare well with the revier system. The efficiency of the revier system lies in co-responsibility and mutual law enforcement by state agencies and revier owners. With the original revier system, two modifications are presented: the Swedish system in which a revier is represented not by area but by a minimum number of moose, and the East German system in which reviers are large units leased by the state to hunter collectives for a defined percentage of harvested game.

History

The roots of sport hunting in Europe go back more than 2500 years to the Ancient Greeks (Anderson, 1985). Their former warrior and later philosopher Xenophon (Marchant, 1925) set the ethical code of hunting. However, the Ancient Romans established the principle of *res nullius*: game is a product of the soil and therefore belongs to nobody. As long as *res nullius* applied, only the Head of State (formerly the Roman Emperor, then the Kings) through the relevant Ministry decided if, when, where, by whom, and to what extent, game should be harvested. Hence, where this principle underpins wildlife policy (most of Europe), harvesting game, or its forcible removal by the landlord, is illegal without government approval (Mantel & Müller, 1935; Balse, 1965; Wolfe, 1970). Historically, royalty usurped hunting rights, or extended them to their favourite noblemen (Lindner, 1937). In feudal Europe, that right was considered *sacro sanctum* for more than a millenium. Its breach, poaching, was considered a serious crime, sometimes equivalent to murder.

Hunting parks, dating to the 12th century, became more numerous, often encompassing thousands of hectares (Lindner, 1937). Foreign game like fallow deer (*Dama dama*) and pheasants have been introduced to those parks since mediaeval times, whereas sika (*Cervus nippon*) and white-tailed deer (*Odocoileus virginianus*) were introduced in the 19–20th centuries (Husák, 1986; Wolf, 1986). On the other hand, bison (*Bison bonasus*) and aurochs (*Bos primigenius*) disappeared from Western and Central Europe in the 15–16th centuries (Heck, 1938), and moose (*Alces alces*) between the late 16th and mid 18th centuries (Hromas, 1986). To produce 'useful' game, gamekeepers trapped, hunted, and poisoned predators; consequently, wolves (*Canis lupus*), lynx (*Lynx lynx*), and bear (*Ursus arctos*) disappeared from Central Europe by the beginning of the 19th century.

In mediaeval times, France contributed many features to an evolving code of hunting ethics (Phoëbus, 1978). To the present day, most European hunting rituals and terms are of French origin. However, during the 14–15th centuries, when Central Europe began to develop its distinctive culture, the momentum shifted to Germany, Poland, Bohemia–Moravia, and Austria where a guild of professional gamekeepers emerged. Because game abundance and harvest depended on their skill, gamekeepers were awarded special status, sometimes higher than the bourgeois. Gamekeepers dressed in distinctive uniforms, and developed a professional jargon and numerous hunting rituals (Frevert, 1936).

The title of *Game Master* required many years of demanding apprenticeship. The Game Master's knowledge of game behaviour was astonishing, even for our time. Among Game Masters were the first authors of game and hunting manuals, treasuries of ecological and behavioural wisdom still not fully exploited by wildlife biologists (Lindner, 1937, 1956). Understandably, the first foresters were recruited from the Game Masters' ranks in the 17th century, as mining, smelting, and, much later, paper production intensified. Since then, hunting has remained the domain of the forestry profession.

With the introduction of firearms in the 16th century, astonishing numbers of game were slaughtered to demonstrate marksmanship and wealth. These excesses culminated with the baroque epoch. The incredible waste of animals and money, and suffering of the serfs (they were obliged to serve as drivers, to keep hunting dogs at their expense, and to let game feed on crops) could not endure. Not surprisingly, the French Revolution terminated royal hunting privileges. In France and countries occupied by French armies where *Codex Napoleon* replaced Roman law, game was proclaimed as property of the people.

The concept of public ownership of game was seductive for farmers and bourgeois. From Portugal to Greece, including Belgium, the suppressed citizenry turned its hatred not only towards landlords but also towards game which they hunted mercilessly to almost total annihilation. Upland game became so scarce that attention shifted to migratory song birds. This practice persists in the Mediterranean area and partly in Belgium. Attempts to protect game by closed seasons, minimum age for hunters, and species, sex and age restrictions on harvests, were unsuccessful in the face of too many hunters and poor hunting ethics. Only the establishment of Game Preserves and National Parks averted total extermination.

In parts of Central Europe not occupied by French armies or reoccupied by the Austro-Hungarian Monarchy (e.g., northern Italy and Slowenia-Croatia), the tension between nobility and farmers exploded a few decades later (mid 19th century). With the abolishment of serfdom, hunting privileges of the nobility were curbed as elsewhere. However, the new laws acknowledged rights of landowners to control access and to hunt on their property. Unfortunately for game, it was hard to establish the 'new order'. Within less than two decades, upland game was endangered.

Evolution of the revier system

The salvation for game came with consolidation of government control in the second half of the 19th century with the introduction of the revier system (RS). The next step, developed after World War I, aimed at improving hunter ethics and knowledge. The RS was redefined and mandatory examination of hunters introduced.

These new ideas originated in Poland, but Germany perfected and legalised the *Reichsjagdgesetz* in the early 1930s (Mantel & Müller, 1935). Regulations concerning ungulates within that law were called classical management (*Klassische Hege*). At the time, it was the best law available to game and hunter. However, with the benefit of hindsight, it was ideologically and sociobiologically biased. The main misconception was that trophies could be improved by systematic culling of apparently inferior trophy-bearers of *inferior Anlage* and that it was possible to shape the trophy in accordance with scoring formulas. Based on present knowledge, it is easily understood why gains of this 'classical' management were not as great as expected. Nevertheless, the notion of the dangerous impact of genetically inferior animals still smoulders in the minds of many hunters and is difficult to replace (Raesfeld, Neuhaus & Schaich, 1985). Another traditional error of *Reichsjagdgesetz* was the discrimination between useful and nuisance game. After World War II, better insight into intraspecific and interspecific relations of game called for revision of the revier system. The present version (third) is only ten years old. Some of its features and achievements are summarised in Table 6.1.

Table 6.1. *Some model hunting systems and their effectiveness.*

	SWITZERLAND		W.GERMANY	E.GERMANY	AUSTRIA	CZECHO-SLOVAKIA	SWEDEN
System (F = Free-hunt R = Revier)	F	R	R	R	R	R	R
Area (1000 km²)	41		248	108	83	128	450
Population (millions)	6.3		61.3	17	7.5	16	8.3
Number of hunters	36,030		265,654	40,000	107,690	150,000*	320,000
% eligible age group	1		1	0.6	2.5	0.4	10
ACCREDITATION							
Minimum age (years)	20		16	18	18	18	18
Discrimination:	none		C	C,R,P	C,R	C,R	none
(Criminal record, Religion, Politics)							
HIC	M		M	M	M	M	M
Valid hunting licence	M		M	M	M	M	M
Hunter's courses		M	M	M	M	M	
Safety courses	M						
Hunters' seminar (hours)			120	100	c. 60	c. 80	c. 50
Examination/shooting test			M	M	M	M	V
HIC valid (years)			1–3	1–3	1[a]	1	1
Shooting test condition of licence							V
Shooting test repetition (years)						3–5	
Fee for hunting licence[b]	29[b]						
Annual fee for HIC[b]	50		75–100	10–100	700–1500	350[a]	
Liability insurance	M		M	M	M	M	V
Hunters Association Membership	V		V	M	M	M	V
Special courses for:							
Trainees and Examinators	M	M	M	M	M	M	M
Gamekeepers	M	M	M	M	M	M	M
with apprenticeship (years)	3	3[a]	3	3[a]	3[a]		
Revier leaseholder with Min. x HIC	1		3		1	3	

HUNTING GUNS						
Purchase free for Shotgun/Rifle	S,R	S,R	X	S,R	X	S,R
Police permit necessary						
Guns kept at Home, with Police	H	H	H,P	H	H	H
Calibre limits (mm)	6.5	5.6 & 6.5+	7.0 +/−	5.6 +	6.5 +/−	6.5 +/−
LAW ENFORCEMENT						
By Police, Gamekeepers, RO	P,G	P,R	P,R	P,R	P,R	P
Number of gamekeepers	1063	600	?			50
Hectares/officer	> 500	250–2000	> 10,000	> 3000		
Poaching is Offence, Crime	O	O,C	C	C	C	
Sentences (Fine, Gaol)[b]	2000	F 1000 G	100	F,J	F,J	
HARVEST PLANNING						
Census from Ground, Air	A,G	G	G	G	G	G
Planning by Government/RO	G	G+RO	G+RO	G+RO	G+RO	RO
Harvest plan according to:						
Number, Sex, Social Class	N,S,C	N,S,C	N,S,C	N,S,C	N,S,C	V
Harvest: Record, Game Check	C	R	C	R	R	Rec. by RO
Trophy Show	V	M	M	M	M	V
Cast Antler Show	V	V	V	V	V	
OPEN SEASONS (days)						
Spring, Summer, Autumn, Winter	A–W	Sp–A–W	Similar to West Germany	Similar to West Germany	S–A–W	
Upland game	45 – 90	90 – 105		30 – 60	60	
Waterfowl	90	60–135/240		30 – 60	30 – 60	
Ungulates	45 – 105	120–150		120–150	30 – 90	
HARVEST						
Gallinaceous birds	6,000	396,400	16,000	294,000	437,000	
Waterfowl	17,000	520,100	35,800	78,000	?	
Other birds	46,500	912,100	?	23,600	?	
Hare, rabbit, woodchuck	18,500	1,297,400	21,100	238,400	208,700	
Roe deer	17,000	678,300	163,000	203,200	87,800	
Sika deer	93	750		100	113	
Red deer	4,500	29,200	21,000	35,800	23,400	500
Fallow deer	420	11,900	12,000	260	2,400	3,000

Table 6.1. (cont.)

System (F = Free-hunt R = Revier)	SWITZERLAND F	SWITZERLAND R	W.GERMANY R	E.GERMANY R	AUSTRIA R	CZECHO-SLOVAKIA R	SWEDEN R
Moose							144,000
Chamois	14,500	1,500	2,800	?	25,750	?	
Mouflon			1,900	1,900	1,450	3,400	
Wild boar	410	270	69,200	109,000	5,750	17,900	100
Small carnivores	18,400	7,600	325,100	139,830	88,000	?	
Bear						40	28
Lynx						100	20
Harvest total	221,000		3,942,150	519,600	994,410	778,313	147,648
ECONOMY (millions local currency)							
Treasury income							
Licences	10.50	0.543					30
Fines	0.140	0.015					
Treasury expenditure							
Law enforcement	10.590						
Game damage	1.130						
Balance	−1.070	+0.558					
Annual hunter expenditures			136	30+	5,000		501
REVIER SYSTEM (date of origin)	1850				1848	1930	1900
Number of hunters	13,315		265,654	40,000	107,670		150,000
Government Reviers (km²)			24,000		16,000	c. 25,000	
Huntable area (km²)			235,740		80,000		320,000
Number of Reviers/ROs		/4,917	106/260	920/			
Ungulate revier size (hectares)			150–2000	4,000	115	>7 moose	
Mandatory lease (years) for:							
Small game			9	9	9	9	6
Ungulates			12	12	12	12	12

Management units (min hectares) for:					
Small game		4–6,000	1,000	1,000	400[a]
Ungulates		10–50,000	4,000	4,000	1–2,000[a]
Number of gamekeepers	1,063	600	210		
Hectares/gamekeeper for:					
Small game revier		1,000	15,000		
Ungulate revier		1–3,000			
Search for wounded game	V	M	M	M	V,M
Number of trained hounds		60,000	20,000	40,000	100,000
Use of dog during hunt	V	M	M	M	V
ECONOMY OF REVIER SYSTEM (local currency)					
Lease/100 hectare for small game / for ungulates		1,500 / 400	2,000		200–5,000 / 500/hunter
Total lease (millions)	1.6	6.2			
Total game damage (millions)					
Venison sale (Legal, Illegal)	L	L	L	L	L
Venison sales (millions)	210				
RANCHING					
Legal, Illegal	L	L	L	L	L
Ranch size (hectares)	3	250–5000	150–16,000		100+
Number of ranches	3	3180	Collectives		
Hunting & Trophy		H,T	H,T	H,T	H,T
Meat, Breeding stock, Non-consumpt.	B,N	M,B,N	M,B,N	M,B,N	M,B,N
PUBLIC ATTITUDE TOWARDS					
Hunting (Positive, Negative)	P	P	90% P	P	75% P
Reviers (Democratic, Socialistic)	D	D	S	D	D

[a] based on personal experience
M Mandatory
V Voluntary
[b] fees in local currency

A *revier* is a registered acreage of greater than 75–150 hectares owned by private individuals, corporations or the state. Reviers can be kept by landowners, or leased to registered hunters. For corporate leases, a minimum acreage per lessee is required to avoid overhunting. However, temporary guests are allowed and generally welcomed. The revier must be leased by the same hunter(s) for at least 9–12 years.

In Sweden, a revier is differently defined as any private ground maintaining a minimum of seven moose. Since ungulates often must be managed in much larger areas than reviers of minimum legal size, several reviers may be aggregated to form a Management Unit (MU) large enough to encompass a manageable population. In Czechoslovakia, revier borders are more or less ecologically defined. In East Germany, reviers are leased to Hunter Collectives for a certain percentage of harvested game which is purveyed to the state enterprise entrusted with selling game (Briedermann, 1981). In the last two decades, Interstate Management Units, such as between the Netherlands and Germany (designed by the author in 1972), or Austria and Yugoslavia (Varićak, 1986), began to operate successfully (Bubeník, 1986a). Other socialistic modifications were developed in Romania and Bulgaria (Bergel, Bencze & Hromas, 1985).

Hunters are officially accredited by a Hunter's Identification Card (HIC) obtained by passing a relatively difficult examination. Depending on the country, it can be taken without or only after attending a preparatory course, or at least a shooting test, and must be renewed every one or three years. Liability insurance usually is mandatory. Applicants must be at least 16, 18, or 20 years old and have a clean criminal record. There is no discrimination towards political orientation in Democratic Countries, but in the Social Democracies, political trustworthiness and atheism may be stipulations. Preparatory courses lasting 60–120 hours deal with all aspects of law, game biology, ecology, hunting strategies, ballistics and practical shooting, use of dogs, and care of meat. In the most advanced systems, the applicant is sent for a one-year apprenticeship to an experienced revier owner (RO). The examination is verbal, practical, and difficult; 40–60% of applicants do not pass on the first attempt. Only two repetitions without restarting the course generally are allowed.

Revier owners can hunt only if they are registered hunters. In some countries, a revier manager (RM) must pass another more difficult state examination whose prerequisites include at least three HIC courses and a special seminar. Revier owners or managers are legally accountable for game mismanagement and their own or a lessee's misconduct. They also are responsible for game inventory, planning, fulfilling harvest quotas, organising collective hunts, and feeding game. The state or landowner has the right to take control of the revier if management is improper or does not comply with the contract.

In view of the high abundance of game and discriminatory culling of sex and social classes, open seasons are generally long: 5–6 months in Central Europe, 1–2 months or two shorter open seasons in Fennoscandia. Retrieving wounded game, or at least searching for it, is generally compulsory. Thus, in most RS countries, a minimum number of experienced hunting dogs are used during or the day after the hunt. Every second or third RS hunter owns a pure-bred hunting dog.

Reviers of more than several thousand hectares must employ a professional gamekeeper(s). For this, a three-year apprenticeship with school attendance and examination is required. Revier owners with several thousand hectares frequently sell licences specifying sex and age, or trophy class. This practice, with a single price for game, is more common in Fennoscandia. In the Socialist States and Western Europe, the fee for trophy game may exceed US$ 10,000.

Primitive weapons such as muzzle loaders or archery equipment are considered unethical and illegal. Using shot on ungulates is forbidden. A minimum rifle calibre, cartridge length, and bullet energy are required. Purchase of weapons is generally restricted to those with a HIC police permit. In Socialist States, all weapons must be registered (Jirkovský, 1987); in East Germany, unused guns are stored at police stations (Schmidt, 1986; Narjes, 1987). To maintain high shooting efficiency, hunters are tested each year or at intervals of several years.

Irrespective of the political system, hunter fraternities are popular and active. Hunters still consider themselves a special social class. Therefore, everyone from common citizens to government leaders willingly takes the strict hunters' examination. In all RS countries, 'diplomatic hunts' are a great annual event. The 'representative reviers' managed for these purposes are considered excellent visit cards of the government. The RS is certainly not a cheap affair, but except for some very expensive leases any responsible hunter can hunt, if only by invitation.

Administration, law enforcement and planning

Generally, hunting in continental Europe is supervised by Departments of Fish and Game (DFG). For free-hunt systems, hunters do not participitate in decisions concerning seasons, licence fees, and supplementary feeding. There are a few state game wardens who conduct censuses or otherwise assess game populations, enforce law, check harvested animals, and feed game. Their areas of jurisdiction include thousands of hectares.

Presently, the revier system is implemented with some local differences in the Netherlands, Belgium, both Germanies, the western part of Switzerland, Denmark, Norway, Sweden, Poland, Czechoslovakia, Austria, Hungary, northern Italy, Romania, Bulgaria, and Yugoslavia. Here, game management and law enforcement are provided almost entirely by ROs and gamekeepers.

The most important administrative feature of the revier system is direct involvement of local ROs; new regulations cannot be implemented without their consent. Responsibility for game inventory and management planning also rests with the ROs. Inventories are conducted mainly from the ground and generally the records must be filed with the DFG each April. Within a month, each RO receives his harvesting plan which may or may not comply with his proposals. Disagreements are resolved by arbitration. In Management Units with central planning, harvest quotas are set per surface unit, generally 100 or 1000 hectares and allocated according to revier size.

Losses due to poaching are minimal; systems in which hunters control themselves are more effective and less expensive than state enforcement. Every registered hunter can be, and professional gamekeeper is, deputised by the police, and has the right to stop and disarm trespassers. Found guilty, the poacher will lose his guns and the vehicle used for poaching, and could be fined or jailed. Registered hunters found poaching (on their own or another revier) lose their HIC for one year to life. In some countries with mandatory membership of the Hunters' Organization, three levels of Courts of Honour exist. Hunters behaving contrary to established ethical codes are sentenced by the local Court of Honour with provision for appeals to the two higher courts. Found guilty, offenders are fined or their HICs are confiscated for the recommended time.

The revier system offers inexpensive game and environmental protection; it produces employment and meat, and is self-sustaining. Consequently, it is considered more democratic than alternative centralised licensing systems, because it does not drain the taxpayer for game protection and supplemental feeding. Because venison can be sold by ROs, the market is well supplied and venison is accessible and favoured by most citizens. The public and government do not consider game management as 'food production first and foremost' as presumed by Geist (1986). The number of registered hunters in all RS countries is relatively stable, averaging between 0.1–10.0% of citizens of eligible age. The reason for low participation is the required dedication rather than its cost.

The trophy cult

Concurrent with these historical changes were changes in attitude to game and hunter ethics. Tribes which invaded Central and Western Europe in many waves before and after Christ settled as farmers. Though becoming more independent of venison, the entrenchment of serfdom prevented them from accessing it. For the nobility, hunting offered physical training and enjoyment. With the Baroque and rise of hunting as an attribute of social status, noblemen competed not only for quantity of game they could offer to their guests, but also for the quality of trophies, particularly red deer antlers. The weight and number of points from antlers were carefully recorded. The best were mounted on beautifully carved heads where they still adorn the walls of European castles (Erbach-Erbach, 1986).

This primitive trophy cult disappeared with the social revolution in the early 19th century because of the scarcity and poor trophy quality of game. However, by the late 19th century, when ungulates again became abundant and virgin hunting grounds became accessible by railway and ship, competition for trophies was renewed. By the late 1920s, with support from the newly established Conseil International de la Chasse (CIC) in Paris, the first international scoring formulas for trophy competitions were developed.

The trophy cult reached its zenith with the 1936 International Hunters Exposition in Berlin (Scherping, 1938). Though inactive during and after World War II, trophy hunting regained momentum in the 1950s (Scherping & Schmincke, 1955). Antlerless females became considered as too desirable to harvest and males were seldom shot before their prime. These hunting preferences created skewed sex ratios and age distributions (Bubeník, 1984). Consequently, hunting organisations in Democratic Countries tried to temper this preoccupation with trophies (Bubeník, 1970a,b; Anon., 1987). It was acknowledged that such a cult discriminated between the common hunter and the rich one. Ironically, the trophy cult is still propagated by Socialist States (Szederjei, 1986) except East Germany. To attract hard currency, they lure rich hunters from the West, offering trophy hunts at exorbitant prices. Depending on scored points, a single trophy costs from several hundred dollars to more than US$ 10,000.

From another viewpoint, preservation of trophies and trophy shows can be a very valuable tool in ungulate management (Anon., 1987; Bubeník, 1986b). Trophies of determined age are reliable parameters of population well-being, a main concern of sociobiological population control (Bubeník, 1986a).

Land use changes

Until the end of the 17th century, vast areas of Europe were covered by virgin forests, harvested only occasionally for lumber but mainly for firewood. The forests were harvested along their edges, or were burned or cleared for new settlements, pastures, and fields. The size of farmland in relation to the forest was small and farming primitive with at least a quarter of the land laying fallow. Frequent wars and epidemics controlled expansion of human settlements so forests were not endangered. Serious damage to forests emerged with the development of firearms, mining, and industrial mills. The demand for lumber rose so quickly that near the end of the 18th century sylvicultural laws and artificial reforestation appeared in Central Europe.

To produce fast-growing trees, coniferous monocultures were preferred and their care improved. However, lacking proper knowledge, the forest communities were brought to phytocenological disorder, with detrimental impacts on soil structure and texture. Coniferous monocultures without understorey and hardwoods disrupted the soil community to such extent that within 150 years it podsolised. On such soils, the forest-tree metabolism must be affected, since such trees are highly susceptible to browsing and bark-stripping, both serious concerns of sylviculture.

Forest damage by ruminants has been documented for two centuries in both cultivated and virgin forests. It was only a question of time and economic pressure before forest damage became intolerable (Burschel, 1979; Plochmann, 1979). Since forest owners and foresters were mostly passionate hunters, they tolerated damage, and tried to minimise it by supplementary feeding. Feeders were stocked with hay behind vertical bars, and trays filled with oats and corn were offered when deep snow covered the ground. Neither those feeders or forage were appropriate for the feeding behaviour of roe deer (*Capreolus capreolus*), red deer (*Cervus elaphus*), mouflon (*Ovis musimon*), or wild boar (*Sus scrofa*). Though they enabled survival, their effectiveness in reducing forest damage was slight.

However, the greatest danger for increasing forest damage accompanied the use of concentrated pellets. Following World War I, F. Vogt of the Czechoslovakian Unilever Company astonished hunters at the 1936 Berlin Trophy Show with the results of his still unique feeding trials (Vogt, 1936, 1947; Vogt & Schmid, 1950). Offering only pressed sesame seed cakes, his penned roebucks and stags exhibited spectacular antler growth. Unfortunately, the impact of pellet feeding on selective browsing and bark-peeling was not recognised immediately (Vogt never mentioned the devastating bark-stripping in his enclosures). During World War II, sesame cakes were not available at all. When economic conditions improved in the 1950s, feeding with concentrated pellets (oilseed meal) was widely adopted in reviers, being advertised by the oilseed industry. Within a few years, browsing and bark-peeling around feeding yards resulted in over US$ 100 million damage annually.

Subsequent research showed that pelleted concentrates stimulated browsing and bark-stripping. Simultaneously, it was found that feeding strategies of wild ruminants tended to be species-specific (Bubeník *et al.*, 1957; Bubeník, 1984; Onderscheka, 1986). Therefore, new feeders with recommendations on their installation were designed (Bubeník, 1984). Studies also revealed that social disorder caused by suboptimal social structure results in non-specific food selection and increased forest damage (Kuen & Bubeník, 1977). A major ten-year field study in the alpine zone of Austria, 'Project Achental', led to the concept of sociobiological game management and a new game-friendly sylvicultural strategy (Schwab, 1979). It has been a successful concept, under which both optimum game fitness (trophies included) and natural forest regeneration are achieved (Rieck, 1986).

Agricultural land south of the 50th parallel provided suitable habitats for small game. With the implementation of the revier system, almost unimaginable densities of 2 hares, 4–5 partridges, and 60 pheasants/hectare were successfully managed (Březina, 1930). Despite such game abundance, crop damage was minimal, except by ungulates in the proximity of forests. Reimbursement of these damages was (and is) regulated by reasonable, well-accepted rules. Unfortunately, with intensive farming practices, densities of many wildlife species dropped to a tenth or less of levels prior and just after World War II. Partridge almost disappeared, hare densities dropped by 50%, and pheasants were confined to hand-reared farms (Bergel, Bencze & Hromas, 1985).

Harvest principles

The culling principles of classical ungulate management, except for moose in Sweden (Cederlund & Markgren, 1987) remained the same until the mid 1970s. Planning and culling was done with respect to the following categories, marked at trophy shows with dots of different colours:

(1) Green dot: 'genetically inferior'.
(2) Red dot: 'anlage good' but juvenile (protected by law), or 'anlage good', but as a trophy-bearer was close to the end of prime age (killing legal).
(3) Blue dot: Anlage dubious (killing dubious).

Doubtfulness was due to questionable age, and/or shape. Frequently, asymmetry was sufficient for a blue or even green dot.

Based on present knowledge, the system is obsolete. Culling only males according to presumable anlage cannot help if the anlage could not be assessed in females. Generally, the genetical variability of trophy shape is greater than the range afforded by the artificial categories of ideal trophies (Bubeník, Raymond & Meile, 1977; Bubeník & Konig, 1985).

Despite the pedantry on which the concept depended, culling results did not reward the effort. Permanent culling of all 'inferior' trophy-bearers, and low interest in shooting juveniles with small trophies, did not allow development of optimum infrastructures, with parity or majority of prime males, a prerequisite of social order and optimum population fitness (Bubeník, 1984, 1986a).

Improvement came with the concept of sociobiological population control (Bubeník, 1984, 1986a), based on species-specific but habitat-dependent social mechanisms. Basically, in solitary species there is a density zenith far below carrying capacity, but almost a zero nadir for social welfare. In contrast, gregarious ungulates have a relatively high, socially limited nadir of density. However, their density zenith is limited only by the carrying capacity of preferred feeding grounds (Bubeník, 1984, 1986a). For example, roe deer should not surpass a density of about 80–100/1000 hectares, whereas red deer density should not drop below 10/1000 hectares.

To maintain optimum social structures, culling must focus primarily on the number of individuals supernumerous to designated social classes (kids, pre-teens [only in bovids], teens as not fully mature animals, primes and post-primes). The species-specific social class ratio improves not only the fitness of game, but also of the habitat. A narrow sex ratio lowers annual recruitment. The mandatory removal of 40–60% of all newborns (depending on species-specific and local recruitment) before weaning encourages better body condition and habitat quality since the number of lactating females, which consume 2–3 times more food than barren females, is reduced. Summer food is not overutilised and more food remains for winter and plant regeneration.

The parity or majority of prime males and females decelerates sexual maturation, stimulates growth, and postpones sexual activity of teens. In this way, they cannot be potential breeding competitors of primes. Thus, breeding and calving periods are abbreviated. Dangerous social interactions are minimised and antlers and horns regain their primary function as organs of threat and individual fitness. Finally, males enter winter in optimal fitness and do not perish by starvation. Under such conditions, the vigour of recruits of prime females is so high that despite long open seasons, hunters have difficulty fulfilling the planned harvest.

Commercialisation

Under the free-hunting system, the direct return from hunting goes to the state as licence fees. Unfortunately, I was unable to obtain records on this matter, except for Switzerland. In view of the low fees, the return probably does not even cover administrative expenses. To my knowledge, no government in Europe without the revier system is able to provide effective law enforcement and appropriate care of game. On the other hand, the industry on which hunters depend, weapons, ammunition, and personal equipment, profits from the high percentage of hunters (up to about 75% of males of eligible age), and poor shooting efficiency.

In RS countries, the situation is quite different. The return to the Treasury from HIC fees, and taxes from revier leases surpasses administration expenses (Table 6.1) and therefore does not rely on government subsidies. However, this represents only a small part of hunters' expenses. The lease for reviers is not negligible: supplementary feeding is costly, and salaries of professional gamekeepers (with a strong union) are at least as high as those of skilled workers. Fees paid for forest damage can be higher than the lease. An accurate saying is that a farmer wills the farm to his oldest son, the revier lease to the second, and fees for crop and forest damage to the youngest, to ensure an equal inheritance.

Few ROs generate a profit from game harvest, except in Fennoscandia, where moose densities are high. When moose populations peaked in 1983, the harvest was 186,000, representing 3.0–3.4 kg moose meat per inhabitant. With the fee for licences and for the meat harvested, it was a great contribution to the Treasury, revier owners, and to the public. However, this population explosion induced considerable forest damage (Lavsund, 1981) from which the forest will not recuperate for 15–20 years.

The situation was different when small game was abundant and a revier of about 2500 hectares could harvest about 1500 hares and several hundred to a thousand Hungarian partridges and pheasants. In Socialist Countries which need hard currency, the return from harvested game is so important that crop and forest damage is frequently neglected. In these countries a distinct percentage of harvested game must be sold to government enterprises, or (e.g., East Germany), delivered *gratis* to the government (Briedermann, 1981; Bergle, Bencze & Hromas, 1985). The return from sale of cast antlers, upper canines of red deer (used for hunter jewellery), hides, and meat covers only a fragment of annual expenses.

Economic opportunities extend to industries. Revier hunters consider themselves a special class whose image is reinforced with custom-made weapons, special clothes for hunting/social events, field glasses, spotting scopes, and ammunition. Most of the annual consumption of cartridges is spent not in reviers, but at shooting stations and competitions.

During the last 20 years, ranching game in fenced areas has begun to develop. Three main types (not necessarily mutually exclusive) exist. *Cultural enterprises* vary from safari parks to small paddocks of a few hectares with resident or other boreal big game species. *Venison enterprises* focus merely on production of palatable meat, mostly of fallow deer. *Breeding and hunting ranches* differ from classical hunting parks by very intensive management. Most of the return comes from selling breeding stock, but if large enough, they offer trophy animals for hunting. The game ranchers of the Democratic States are well organised (Hatlapa & Reuss, 1974) with chapters in each country and a modern representative journal.

Current trends

The World War II and the occupation of Central Europe by foreign armies brought enormous devastation to forests requiring intensive and fast reforestation by plantations. Where growth was accelerated by permanent removal of the ecologically important but economically noxious undercover and ecotone vegetation, the ungulates were forced to browse on planted saplings. Without expensive protection by fencing, reforestation presented economic problems (Plochmann, 1979; Schröder, 1979; Stern, 1979).

Therefore, a reduction of game densities to sylviculturally tolerable densities is still urged by many foresters (Burschel, 1979; Plochmann, 1979; Schröder, 1979). They found some support among ecologists (Festeticz; in DJV, 1986) and media commentators (Stern, 1979) who recommended: (a) replacement of the RS by the North American free-hunt system, in order to reduce game densities, and (b) restraint of supplementary feeding which will lead to self-regulation of game by starvation (Schröder, 1986).

To this fading campaign, support has been added from left-wing political parties. For these, revier hunters represent capitalistic and conservative electors, a great obstacle in socialisation programmes. Their outcry is ironic because in all Social Democracies except Bulgaria, the RS still exists.

Although exaggerated, the only valid arguments used against hunters relate to the trophy cult. Hunters are presented as killers, who see only the trophy and maintain intolerable densities to produce more trophies. The other persistent argument is that most reviers support sociobiologically unsound densities of roe deer and/or chamois. In extreme cases, they should be lowered to a fifth of the present numbers to allow sapling regeneration and restore normal social interrelationships.

The fight between anti-hunters and foresters not favourably disposed to game continues. A great help to hunters is the intellectual level of citizens. They like game and understand that in the absence of large predators, game densities must be controlled by hunting. Secondly, most people know that game management does not drain public funds. Thirdly, they enjoy venison and know that only reasonably high densities will allow them access to that meat. Therefore, it is not surprising that recent polls indicate 65–90% support the revier system.

Results of the Project Achental were invaluable in that debate. In its alpine zone of 10,000 hectares, where for centuries reforestation was possible only behind fences, the forest regenerated in the presence of 30 red deer, 80 roe deer and 80 chamois/1000 hectares. Thus, the aspects of sociobiological game management, together with new sylvicultural strategies (Schwab, 1981) adapted to the presence of ungulates, is a most promising concept.

Another promising development is the present attempt to achieve, through the parliaments in all countries of the European Economic Community (EEC), a unified concept of RS, hunter examination, open seasons and harvest planning, with an EEC hunting card. However, realisation of such an idea will require many legislative changes to achieve a uniform level of hunter knowledge and attitude.

Acknowledgements

Living 58 years in Europe, 40 as a hunter and biologist whose expertise was used many times in my native Czechoslovakia and other countries, I enjoyed writing this chapter. Now living abroad, it was necessary to contact many friends, biologists, and administrators to update present practice and the harvest data. Therefore, I must express my thanks to those who were willing to cooperate: Dr H. Blankenhorn, Federal Department of Fish and Game, Bern, Switzerland; Dr L. Briedermann, Eberswalde, East Germany; Dr G. Frank, President of German Hunters Association (DJV), Munich, West Germany; H. H. Hatlapa, Grossenaspe, West Germany; G. Hennsksen, Directorate for Nature Management, Trondheim, Norway; Professor Ing. K. Ladstater, Secretary General of Austrian Hunters' Association, Vienna, Austria; F. Stalfet, Swedish Sportsmen's Association, Uppsala, Sweden; Dir Dr H. Strandgaard, Viltbiologisk Station, Kalø, Denmark; and P. S. Tunkkari, Department of Zoology, University of Oulu, Finland.

References

Anderson, J. K. (1985). *Hunting in the Ancient World*. Berkeley: University of California Press.

Anon. (1987). Hegeschau statt Knochen-Olympiade. *Die Pirsch*, 30, 45.

Balse, R. (1965). *Die Jägerprüfung*. Jägerprüfung: J. Neumann-Neudamm.

Bergle, J., Bencze, L. & Hromas, J. (1985). Myslivost v europských socialistických státech. Prague: Státní/ zemědělské nakladatelství (in Czech).

Březina, A. (1930). *Československá Myslivost*. Prague: A. Neubert (in Czech).

Briedermann, L. (1981). *Die Jagd in der Deutschen Demokratischen Republik*. Jagdinformationen 1-2, Eberswalde.

Bubeník, A. B. (1970a). Klassische Hege – Klassischer Irrtum? *Die Pirsch*, 22, 90-8.

Bubeník, A. B. (1970b). The conversion used in scoring sporting trophies in the light of behaviour research. *Deer*, 2, 457-9.

Bubeník, A. B. (1984). *Ernährung, Verhalten und Umwelt des Schalenwildes*. München: Bayerischer Landwirtschafts Verlag.

Bubeník, A. B. (1986a). Grundlagen der soziobiologischen Hirschwildbewirtschaftung. In *Rotwild–Cerf rouge–Red deer*, ed. H. Reuss, pp. 97-159. Graz (Austria): CIC.

Bubeník, A. B. (1986b). Über Geweihe–Abwurfzeit und Abwurfzustand als Merkmale des Wohlbefindesns der Hirschartigen. *Der Anblick*, pp. 148-50, 174-6, 271-6, 321-3.

Bubeník, A. B., Lochman, J., Semizorová, I. & Fišer, Z. (1957). Spásání lesních dřevin parohatou zvěří z hlediska jejich fyziologických potřeb. *Lesnictví*, 3, 347-52 (in Czech).

Bubeník, A. B. & König, R. (1985). Morphometry of antlers of the genus *Capreolus* (Gray 1821). In *Biology of Deer Production*, eds. P. F. Fennessy & K. R. Drew, Royal Society of New Zealand Bulletin 22, 273-8.

Bubeník, A. B., Raymond, F. L. & Meile, P. (1977). Morphometry of the horns of chamois (*Rupicapra rupicapra* L.). Preliminary study. In *Proceedings of the IUGB Congress*, Vol. 13, 151-64. Atlanta (Georgia): IUGB.

Burschel, P. (1979). Der Wald als Gesellschaft von Bäumen. In *Rettet den Wald*, ed. H. Stern, pp. 74-93. München: Kindler.

Cederlund, G. & Markgren, G. (1987). The development of the Swedish moose population 1970-1983. In *Proceedings of the Second International Moose Symposium*. Uppsala: Swedish Wildlife Review.

Deutscher Jagschutz-Verband. (1986). Jäger beklagen Informationspolitik des Göttinger Jagdinstitutes. *Wild und Hund*, 89, 8-10.

Erbach-Erbach, F. (1986). Die Erbacher Hirschgalerie: eine Kollektion weltstärkster Trophäen. In *Rotwild–Cerf rouge–Red deer*, ed. H. Reuss, pp. 25-32. Graz (Austria): CIC

Frevert, W. (1936). *Jagdliches Brauchtum*. Berlin: P. Parey.

Geist, V. (1986). Wildlife as a public trust. *Western Sportsman*: 53-5.

Hatlapa, H. H. & Reuss, H. (1974). Wild in Gehege-Haltung, Ernährung, Pflege, Wildnarkose. Hamburg, Berlin: P. Parey.

Heck, L. (1938). Der Ur oder Auerochs. In *Weidwerk der Welt*, ed. U. Scherping, pp. 211-3. Berlin: P. Parey.

Hromas, J. (1986). Puovod a historie rozšíření losa. In *Daněk/Sika/ Jelenec*, ed. F. Husák, R. Wolf & J. Lochman. pp. 286-306. Prague: Státní zemědělské nakladatelství (in Czech).

Husák, F. (1986). Daněk. In *Daněk/Sika/Jelenec*, ed. F. Husák, R. Wolf & J. Lochman, pp. 7-149. Prague: Státní zemědělské nakladatelství (in Czech).

Jirkovsky, P. (1987). Právní úprava drzvení kulových zbraní. *Myslivost*, 65, 102-3 (in Czech).

Kuen, H. & Bubeník, A. B. (1977). Availability of food for red deer (*Cervus elaphus*), roe deer (*Capreolus caporeolus*), and chamois (*Rupicapra rupicapra*) in an Alpine ecosystem. In *Proceedings of the IUGB Congress,* Vol. 13, 393–400. Atlanta (Georgia): IUGB.

Lavsund, S. (1981). Moose as problem in Swedish forestry. *Alces,* 17, 165–79.

Lindner, K. (1937). *Die Jagd der Vorzeit.* Berlin, Leipzig: W. de Gruyter & Co.

Lindner, K. (1956). *Die Lehre von den Zeichen des Hirsches.* Berlin: W. de Gruyter & Co.

Mantel, K. & Müller, P. (1935). *Das Reichsjagdrecht.* 2nd edn. Dresden: Deutscher Jagschutz-Verband.

Marchant, E. C. (1925). *Xenophon: Scripta Minora.* With English translation. Cambridge (Massachusetts): Harvard University Press.

Narjes, F. J. (1987). Mehr Schatten als Licht. *Die Pirsch,* 39, 9–12.

Onderscheka, K. (1985). Ist die Fütterung des Rotwildes in der Kulturlandschaft des alpinen Raumes eine biologische Absurdität oder ein Beitrag zur Erhaltung der Funktion des Ökosystems? In *Rotwild–Cerf rouge–Red deer,* ed. H. Reuss, pp. 386–95. Graz (Austria): CIC.

Phoëbus, G, (1978). *Das Buch den Jagd.* Transl. by G. Biss. Freiburg: R. Mohn.

Plochmann, R. (1979). Wald zwischen Ökologie und Ökonomie. In *Rettet den Wald,* ed. H. Stern, pp. 368–88. München: Kindler.

Raesfeld, F., Neuhaus, A. H. & Schaich, K. (1985). *Das Rehwild.* Berlin: P. Parey.

Rieck, E. (1986). Versuche–Ergebnisse–Massnahmen. *Die Pirsch,* 38, 602–6.

Scherping, U. (ed.) (1938). *Weidwerk der Welt.* Berlin: P. Parey.

Scherping, U. & Schmincke, S. (ed.) (1955). *Jagd und Hege in aller Welt.* Bonn: DJV.

Schmidt, C. (1986). Andere Jagd im anderem Deutschland. *Die Pirsch,* 38, 1685–6.

Schröder, W. (1979). Ändert sich der Wald, ändert sich die Tierwelt. In *Rettet den Wald,* ed. H. Stern, pp 252–79. München: Kindler Verlag.

Schröder, W. (1986). Füttern oder nicht füttern. *Die Pirsch,* 38, 1492–3.

Schwab, P. (1979). Waldgerechte Schalenwildhege im Hochgebirge. *Allgemiene Forstzeitung,* 17/18, 445–55.

Schwab, P. (1981). Das Wild ist nicht an allem schuld. *Die Pirsch,* 33, 16–9.

Stern, H. (1979). Echte und falsche Liebe zum Wildtier. In *Tierliebe und Wildtier,* ed. W. Schröder, pp. 3–18. Tutzinger Studien, Tutzing (Germany): Evangelische Akademie.

Szederjei, A. (1986). Hege und Verbesserungsmasnahmen der Qualität der Rotwildbestände Ungarn. In *Rotwild–Cerf rouge–Red Deer,* ed. H. Reuss, pp. 70–4. Graz (Austria): CIC.

Szederjei, A. & Szederjei, M. (1971). *Geheimnisse des Weltrekordes – Der Hirsch.* Budapest: Terra.

Varićak, V. (1986). Erfahrungen mit der gemeinsamen kärtnerisch slowenischen Rotwildhegegemeinschaften in den Karawanken. In *Rotwild–Cerf rouge–Red deer,* ed. H. Reuss, pp. 35–7. Graz (Austria): CIC.

Vogt, F. (1936). *Neue Wege der Hege.* München: J. Neumann-Neudamm.

Vogt, F. (1947). *Das Rotwild.* Wien: Österreichischer Jagd- und Fischereiverlag.

Vogt, F. & Schmid, F. (1950). *Das Rehwild.* Wien: Österr. Jagd- u Fischereiverlag.

Wolf, R. (1986). Jelen sika. In *Daněk/Sika/Jelenec,* ed. F. Husák, R. Wolf & J. Lochman, pp. 149–227. Prague: Státní Zemědělské Nakladatelství.

Wolfe, M. L. (1970). The history of German game administration. *Forest History,* 14, 6–16.

7

Sport hunting in North America

C. HARVEY PAYNE

Abstract

North America's wildlife management system was founded between 1900 and 1920. In response to overhunting and profound depletion of wildlife in the 19th century, wildlife was removed from the market place and firmly established in public trust. However, in some jurisdictions where demands for sport hunting are high and public lands in short supply, landowners may collect fees from hunters. Ostensibly, this practice induces landowners to enhance wildlife habitats and provide hunting opportunities. Wildlife is managed largely by state, territorial, and provincial governments, although migratory and endangered species fall under the jurisdiction of federal governments. This chapter examines the contrasting history, landscape and settlement patterns which have shaped distinctive wildlife policies in Texas, the Yukon Territory, and Manitoba.

Introduction

Sport hunting in North America is an extension of that practised for centuries in Europe. The major difference is that the state has adopted a more central role in management. Wildlife is essentially a free good and laws regulating its use are designed to perpetuate that status.

Public access to wildlife for hunting is provided on public lands and licence fees are generally low to ensure equal access. Such is the case in theory and in law, but there is some disparity in practice. It is possible to purchase access to private holdings which in some cases are dedicated to the propagation of game for hunting. Technically, it is the right of access that is bartered and the hunter is required to possess a valid hunting licence. Thus, the principle of public ownership is maintained.

In the Canadian Northwest Territories, certain species are allocated to aboriginal people only. However, the law permits them to sell their quotas to other hunters. Fees charged for such hunts are high: Can$ 12,000 will purchase a package which includes a polar bear (*Ursus maritimus*) and a muskox (*Ovibos moschatus*).

Emergence of sport hunting

Wildlife in the context of a free good was never more evident than in the early days of European colonisation (Tober, 1981). North America seemed to abound with wildlife in numbers too bountiful to measure and too numerous to deplete: bison (*Bison bison*), 60 million; barren ground caribou (*Rangifer tarandus groenlandicus*), 5 million; ducks, countless. Yet, in the mid to late 1800s, the passenger pigeon (*Ectopistes migratorius*) and the great auk (*Pinguinus impennis*) were extinct, the bison was virtually extirpated in the wild, and the distributions of most large mammals were greatly contracted.

There were no laws to protect wildlife and appeals to reason and conscience did little to stay the guns. On the contrary, hunters were respected for their bravery and prowess. Anyone could kill any wildlife for any reason at any time. As towns and cities grew, hunting for the market place became widespread. It soon became evident that controls on harvest would have to be implemented.

Many early hunting restrictions in North America focused on the bison, but they were late in coming and enforcement was lax. After all, the Indian wars were dragging on and it had been recognised that the easiest way to end the controversy was to destroy the Indian's means of subsistence, the bison. General Philip Sheridan, of the US Army, proffered (Dary, 1974):

> [the bison hunters] have done in the last two years and will do more in the next year (1876) to settle the vexed Indian question, than the entire regular army has done in the last thirty years. They are destroying the Indians' commissary, and it is a well known fact that an army losing its base of supplies is placed at great disadvantage. Send them powder and lead if you will; ... for the sake of a lasting peace, let them kill, skin and sell until the buffaloes are exterminated.

Sheridan also recommended that a medal be struck for these hunters, depicting a bison on one side and a disillusioned Indian on the other.

Nonetheless, the years of wanton slaughter soon ended. By 1880, it became difficult to find bison where millions existed a half century before. Early in the 20th century, market hunting was successfully curbed and most states and provinces established wildlife agencies, albeit in many areas only the agricultural representative charged with this responsibility.

The great reduction in game populations in the late 19th century came suddenly and unexpectedly to all but the most enlightened conservationists. As local game populations dwindled and the once abundant game of the high country became harder to find, the hunter searched for a scapegoat. He found one – the predator – and a new era began in which the primary conservation effort was extermination of predators, a philosophy welcomed by cattle ranchers and sheepmen.

Wildlife management as a scientific discipline emerged in the 1930s (Leopold, 1933). Wildlife was managed as a crop to provide for growing numbers of recreational hunters. Natural history was studied, biological processes defined, and management strategies developed accordingly. The 1950s and 1960s saw a major increase in wildlife research and management agencies hired biologists. In time, emphasis on biotechnical aspects of management was supplemented with growing recognition of the importance of public involvement. The public responded with enthusiasm and wildlife management became increasingly politicised.

Major shifts in the population of North America, from rural to urban, have occurred. In the United States, the urban population comprised only 40% of the total population in 1900. By 1970, it was over 70%. The shift has placed new demands on wilderness and wildlife and brought changes in philosophy, expectations, and demands by hunters, naturalists, and other organised groups of less clearly defined affinity. Although the long-term goals of these organisations are essentially similar (i.e., they are all conservation orientated), their differences are such that they often thwart the best efforts of managers.

Organisation and administration

The first game laws were passed in the 17th century. Provisions for deer date back to 1677 in Connecticut, 1699 in Virginia, and 1705 in New York (Doughty, 1983). Massachusetts and New Hampshire were first to introduce a warden system in 1850 (Leopold, 1933). The first protection of non-game birds was legislated in New Jersey and Connecticut in 1850 largely in recognition of the agricultural benefits of insectivorous species (Doughty, 1983). The Lacey Act (1900) prohibited interstate trade in game which effectively terminated market hunting.

This section traces the evolution of game laws and current systems of wildlife management in Texas, Yukon, and Manitoba as representative of state, territorial, and provincial jurisdictions (Fig. 7.1). These jurisdictions provide interesting contrasts in geography, history, and land tenure which have shaped management programmes.

Texas

The first Texas legislation for the protection of game was passed in 1860. The Act for the Protection of Game on Galveston Island was local and species specific; hunting of bobwhite quail was closed for two years and closed from March through August each year thereafter (Doughty, 1983). Twenty years elapsed before further attention was given to wildlife conservation and closed seasons for other species were established. The experiences of three generations led to the belief that there was always more game to the west.

By 1877, there were only scattered bands of bison on the plains of west

Fig. 7.1. Except for migratory and endangered species, wildlife resources are administered by territorial or provincial governments in Canada, and state governments in the United States. Because of different histories and patterns of land tenure, interesting contrasts are provided by Texas, Manitoba, and the Yukon Territory.

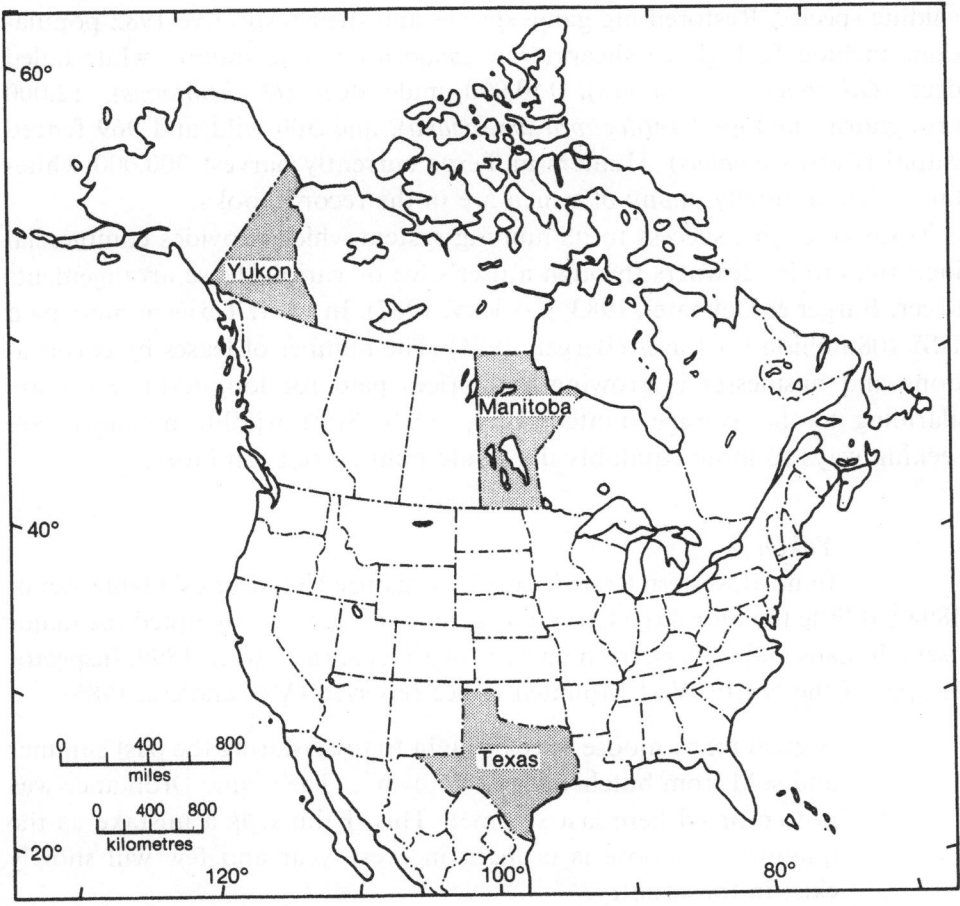

Texas (Dary, 1974). As the bison hunters streamed into northern Texas, there was talk in Austin, the state capital, of protective legislation. But it was only talk and by 1880 bison hunting on the southern plains was history (Dary, 1974: 129).

In 1903, an Act to Preserve and Protect the Wild Game, Wild Birds and Wild Fowl was passed. The law established a five-year closed season for some species and attempted to end commerce in game. But neither the populace nor the judiciary were willing to give the law sympathetic consideration. Eventually, in the 1920s and 1930s, when game populations had been shot out, effective protective measures were implemented.

One of the most important factors shaping wildlife policy in Texas is the dearth of public lands (Teer, 1982). A total of 99% of the rangelands and 97% of the forests are privately owned. Nevertheless, the State of Texas through the Pittman–Robertson programme (established by the Federal Aid in Restoration Act of 1937), transplanted and restored many indigenous wildlife species. Restored big game species and their respective 1982 populations include 50 bighorn sheep (*Ovis canadensis*), 1.98 million white-tailed deer (*Odocoileus virginianus*), 150,000 mule deer (*O. hemionus*), 12,000 pronghorn antelope (*Antilocapra americana*), and 300 wild and 969 fenced wapiti (*Cervus elaphus*). Hunters in Texas currently harvest 300,000 white-tailed deer annually, many of which are in the record books.

Texas owes this success to its hunting system which provides commercial incentives to landowners through a user's fee or various lease arrangements (Teer, Burger & Deknatel, 1983; Kozicky, 1987). In 1971, Texas hunters paid US$ 108 million for leases (Berger, 1974). The number of leases by corporations and businesses is growing and prices paid for long-term leases are alarming to the average hunter (Teer, 1982). State wildlife managers are seeking ways to more equitably distribute hunting opportunities.

Yukon

In northwestern Canada, the Unorganized Territories' Game Act of 1894 did little to control the harvest of game since the law exempted the major users, Indians and settlers, from many provisions of the law. In 1899, Inspector Harper of the North West Mounted Police reported (McCandless, 1985):

> A great many moose were brought to town during the past summer and sold from butcher shops in town ... The Game Ordinance was not enforced here last summer. This, I think, is a mistake as the quantity of moose is brought in every year and few will shortly exist in the country.

The concern was less strongly motivated by conservation ethics than fear that diminishing game supplies could lead to starvation among the Indians and create administrative problems. In an amendment (1900) to the Yukon Act (1898), the Yukon Territory was given jurisdiction over wildlife matters. An Ordinance for the Preservation of Game (1901) restricted annual harvest to six caribou, two moose (*Alces alces*), two sheep and two goats (*Oreamnos americana*) per hunter, with penalties up to Can$ 500 or about 100 days' wages (McCandless, 1985: 34). The penalty had little consequence because hunters and enforcers alike ignored the law.

There was little change for the next 40 years, until the Alaska Highway was completed and the Yukon population began a dramatic increase. There was recognition of the importance of wildlife to residents and the economy of the Yukon. The fur industry maintained its major importance and a complaint regarding the unfairness of the Migratory Birds' Treaty to northerners was registered (this complaint continues to this day). Guiding and outfitting services for big game hunting were licensed in recognition of the contribution of sport hunting to the Yukon economy.

The collapse of the fur industry in the 1940s, resulting from a major shift in fashion, 'brought complete economic dislocation on the scale of a disaster to the Yukon Indian people' (McCandless, 1985). To make matters worse, the wage labour boom resulting from the war effort also ended. Demand for big game hunting was awakening, but in the Yukon the infrastructure was not in place.

Today, big game populations are healthy and once-in-a-lifetime hunting experiences can be enjoyed. Transportation and outfitting costs are high excluding the average non-resident hunter. But the cost of hunting for residents remains relatively low because almost all land is in public owner-ship.

Manitoba

Manitoba was one of the first jurisdictions in western Canada to legislate game laws; the Manitoba Act for the Protection of Game was passed in 1876. The act banned hunting of various species from late spring to early autumn and hares, curiously, were given more protection than other game (Bossenmaier, 1978).

Early legislation was either ignored or poorly enforced. Administration of wildlife was under the Department of Agriculture until 1930, when control of other natural resources was transferred to Manitoba and other provinces by the Government of Canada. The Department of Mines and Natural Resources was formed and a Game and Fisheries Act (1930) was passed which guided management until it was replaced by the Wildlife Act (1963). The Wildlife Act introduced many new aspects. Wildlife Management Areas (WMA), tracts of public land on which wildlife-related activities form the primary use, were established throughout the province. There are now 59 WMAs in Manitoba, encompassing 2.9 million hectares. Most of the WMA land base is located in northern Manitoba where there are fewer land-use conflicts than in the southern part of the province.

'Operation Respect', a programme to encourage good relationships between hunters and landowners, began in 1971 as a joint effort between the government and the Manitoba Wildlife Federation (founded in 1944). The programme also serves to discourage access fees for hunting. The large public land base ensures that there are alternatives to hunting on private lands. This means that paid hunting is unlikely to become widespread, but it also means there are few incentives for landowners to maintain wildlife habitats.

The right of native people to hunt for food on unoccupied public land presents the expected problems of unregulated harvests. This right was enshrined in paragraph 13 of the Manitoba Natural Resources Transfer Agreement (1930) (McNeil, 1983):

> In order to secure to the Indians of the Province the continuance of the supply of game and fish for their support and subsistence, Canada agrees that the laws respecting game in force in the Province from time to time shall apply to the said Indians within the boundaries thereof, provided, however, that the said Indians shall have the right, which the Province hereby assures them, of hunting and fishing game and fish for food at all seasons of the year on all unoccupied Crown lands and on any other lands to which the said Indians may have a right of access.

This legal provision (reflecting rights acknowledged in Treaty) only exists in the Canadian prairie provinces. In comparison, treaties were never signed in the Yukon where aboriginal rights prevail, and in Texas hunting rights are restricted to Indian reservations.

In recognition of the plight of local wildlife populations, Indian people in Manitoba have entered into formal agreements with the province for the joint management of certain species. These efforts have been successful on a local basis, but it will be some time before the concept spreads throughout the province.

In 1980, a new Wildlife Act was passed. Perhaps most significantly, it requires the Minister responsible to provide annual and five-year reports on the status of wildlife resources to the Legislature.

Land tenure in Manitoba is varied. In the south-west, the situation is analogous to Texas with a dearth of public land. The north is almost entirely crown land. In the south-west, a land lease programme has been introduced, whereby landowners agree not to clear land and to limit grazing. The government in turn pays this landowner a lease fee; agreements are for a five-year term. However, access for hunting is neither assured nor prohibited. As on private land, permission to hunt is legally required and the landowner may charge a fee. Fees are modest, if charged at all, and the Texas situation is unlikely to develop in Manitoba due to the abundance of public land in other parts of the province.

Populations and productivity

Ten major big game species are hunted in North America (Table 7.1). Of these, the white-tailed deer is most widely sought, but moose and wapiti rival mountain sheep as the most highly prized trophies.

The white-tailed deer is distributed from the southern shore of Hudson Bay to South America and is found in 45 of the contiguous 48 states and seven of ten Canadian provinces. There are 16 recognised sub-species north of Mexico (Hesselton & Hesselton, 1982). The pre-colonial population of white-tailed

Table 7.1. *Estimated populations and annual harvests of wild ungulates in North America.*

	Populations	Harvests	Carcase yield (tonnes)[a]	Source
White-tailed deer	15,000,000	2,982,000	100,000	Stransky (1984), Hesselton & Hesselton (1982)
Black-tailed deer	1,500,000	500,000	20,000	Connolly (1981)
Wapiti	500,000	85,000	11,000	Thomas & Toweill (1982), Peek (1982)
Moose	900,000	54,000	10,000	Kelsall (1987), Bisset (1987), Prescott-Allen & Prescott-Allen (1987)
Pronghorn	400,000	87,000	2,000	Kitchen & O'Gara (1982), Marchello *et al.* (1985), Prescott-Allen & Prescott-Allen (1987)
Caribou	2,652,000	50,000	2,300	D. Thomas (pers. comm.)
Bison	70,000	10,000	2,200	Hawley (this volume, chapter 19)
Mountain sheep	120,000	1,300	97	Lawson & Johnson (1982)
Mountain goat	70,000	2,200	88	Wigal & Coggins (1982)
Muskoxen	60,000	300	45	Gunn (1982)
Peccary		27,500	440	Bissonette (1982), Sowls (1984)

[a]Calculated using representative carcase weights

deer was estimated at 40 million by Seton (1929). Current populations are about 15 million and well over two million are harvested annually (Hesselton & Hesselton, 1982).

Mule deer extend westward from the main distribution of white-tailed deer, overlapping in the Midwest. They are present south of the Mexican–United States border and north as far as Alaska and the Yukon. Seven subspecies are recognised (Mackie, Hamlin & Pac, 1982). Seton (1929) estimated that there may have been as many as 13 million mule deer (including black-tailed deer) in North America during pre-colonial times.

Wapiti, with five extant and two extinct subspecies in North America, are distributed throughout the mid-western states and provinces in montane and aspen parkland regions (Peek, 1982). Introductions have been made as far east as Florida (Bryant & Maser, 1982). At the time of settlement, populations were as high as ten million, but they declined to a profound low about 1922 by which time two subspecies were already extinct and three more were threatened (Bryant & Maser, 1982). Populations recovered to about 500,000 by the late 1970s providing an estimated annual harvest of 85,000.

The moose is the largest member of the deer family with adult males weighing over 500 kg (Coady, 1982). They are holarctic in distribution. In North America, the moose is largely restricted to areas north of the 48th parallel, but small pockets may occur as far south as Utah. Four subspecies are distinguished in North America (Coady, 1982).

Four subspecies of caribou are recognised in North America. Peary's caribou (*R. t. pearyi*) number no more than 10,000–15,000; barren ground caribou (*R. t. groenlandicus*) are estimated at over 1 million; Grant's caribou (*R. t. granti*) have been estimated at 250,000 in recent years (down from 600,000 in 1968) (Miller, 1982). Woodland caribou (*R. t. caribou*) probably number less than 200,000. Caribou hunting in northern Canada is largely limited to native people who harvest approximately 50,000 annually.

Pronghorn antelope are distributed throughout the Midwest, north from Texas to Alberta and Saskatchewan and occasionally the south-west corner of Manitoba. Pronghorns reached their lowest numbers of approximately 13,000 in the 1920s but recovered through management efforts to over 400,000 by 1976 (Kitchen & O'Gara, 1982).

Bison once dominated the continent from Pennsylvania to Oregon and from Mexico to well into Canada's Northwest Territories. Perhaps once numbering in excess of 60 million, bison presently number about 70,000–100,000, most of which are in captive herds. Two subspecies, the plains bison (*B. b. bison*) and the wood bison (*B. b. athabascae*) are recognised in North America (Reynolds, Glaholt & Hawley, 1982). The largest wild herd (approximately 4000) is in Wood Buffalo National Park, in northern Canada. Unfortunately, this is a diseased hybrid population, the result of the unfortunate introduction of plains bison in the 1920s.

Mountain sheep are distributed throughout the western mountain ranges from Alaska to Mexico: bighorns (*O. canadensis*) occur in the southern ranges, whereas thinhorns (*O. dalli*) are restricted to areas north of the Peace River, British Columbia. Mountain sheep may have once numbered 2 million but are now reduced to one-tenth that number. The major cause of the decline has been disease and competition with domestic livestock (Lawson & Johnson, 1982).

About 90% of all mountain goats occur in the rugged coastal mountains of British Columbia and southern Alaska, but they are also found in the Yukon, Alberta, and in pockets south to Idaho. Population estimates are vague; the present estimate for British Coumbia is 20,000–60,000, and for Alaska, 15,000–25,000. No more than 2500 are harvested annually (Wigal & Coggins, 1982).

Two subspecies of muskox are recognised in North America, one on the arctic mainland and the other on the islands of the arctic archipeligo. They are found on the north shore of Alaska and throughout the Canadian arctic, with the exception of Baffin Island (Gunn, 1982). They have been introduced to northern Quebec. Populations suffered drastic declines following European exploration, fur trading, and whaling in the arctic. Pristine populations have not been estimated but present numbers are 50,000–60,000 and increasing. In the past two decades, the muskox has become recognised as a trophy game animal.

Collared peccaries are found from Argentina north to New Mexico, Arizona, and Texas. Two subspecies are recognised in the United States. In 1978/79, over 76,000 hunters harvested 27,488 peccaries, mostly in Texas (Bissonette, 1982).

Economic value

In 1980, 17 million hunters in the United States spent US$ 5.6 billion (i.e., 1000 million) on their sport (44% on big game hunting) (USBSFW, 1982). Of this, licence fees accounted for US$ 300 million. In addition, American hunters have made voluntary contributions of over US$ 6 billion for wildlife conservation programmes since 1923. In Canada, 1.8 million hunters spent the equivalent of US$ 1 billion in 1981 (Filion *et al.*, 1983). In comparison, 3.6 million Canadians spent US$ 1.6 billion on primary non-consumptive trips (Filion *et al.*, 1983). About 150,000 tonnes (carcase weight) of wild ungulates are harvested annually (Table 7.1). The meat replacement value (beef is approximately US$ 3/kg) can be calculated as US$ 450 million.

Future of sport hunting

Recreational hunting has become very much a tradition in North America. Although a challenge to the morality of hunting has been mounting, it still maintains its place as 'American as Apple Pie' in much of society. Hunters have played a major role in the development of wildlife policies in North America (Reiger, 1975). In some areas, they were the first to decry market hunting, albeit some of their own 'sporting' practices were no more consistent with conservation than were the practices they opposed. In a national survey in the United States, Keller (1980, in Ahrns, 1982) found that 67.7% of respondents had fed birds at some time, whereas only 25.5% had hunted at some time in their lifetime and only 14.5% had hunted in the past two years. In Texas, 8.27% of the population hunted in 1978/79. A Canadian survey revealed that 9.8% of Canadians hunted in 1981. Generally, the rural provinces such as Manitoba reported higher rates of participation in hunting (11.2%) than did the more urbanised provinces. Although participation is high by international standards, interest in hunting seems to be declining in both relative and absolute terms. Sport hunting carries a connotation, wrong in my view, that the killing of animals is sport. Krutch (1957) concluded that 'killing for sport was the perfect type of that pure evil for which metaphysicians have sometimes sought'.

In recent years, big game hunters have been forced to defend hunting in the face of growing opposition (Baker, 1985). Recent surveys of public opinions and attitudes have startled many hunters, a minority of whom are of the egocentric view that without their interest, dollars and management by culling, wildlife populations would virtually disappear.

Animal rights lobbyists threaten the continuance of hunting as a respectable North American pastime. Their philosophy may not be logical, pragmatic or scientific, but their salesmanship is only rivalled by evangelists. They reach the public by creating 'large splashes in small ponds' which, in turn, receive unworthy public exposure in the popular daily tabloids. In the public mind, they are the crusaders; the seal hunter and trapper, the villain. Perception is reality in the mind of the public and the 'antis', as they are often known, are masters at moulding perception. The animal rights groups are highly organised employing the highest technology and best strategists available.

The anti-hunting movement spends over US$ 50 million each year to promote its views. This has been countered by the Wildlife Legislative Fund of America supported by hunters. This fact is far more alarming than the revelation that naturalists annually spend more money on their interest than do hunters. The attitude of North Americans is being moulded by animal rights activists; consumers are on the defensive. In the United States, a national survey revealed that, although 82% approved of subsistence hunting for meat, 80% disapproved of trophy hunting for heads, horns, and other valued artifacts.

References

Ahrns, J. (1982). Nonconsumptive Wildlife Issues. In *Texas Wildlife Resources and Land Use*, pp. 83–91. Austin: The Wildlife Society.

Bailey, J. A., Elder, W. & McKinney, T. D. (1974). *Readings in Wildlife Conservation*. Washington: The Wildlife Society.

Baker, R. (1985). *The American Hunting Myth*. New York: Vantage Press.

Berger, M. E. (1974). *Texas Hunters: Characteristics, Opinions and Facility Preferences*. Ph.D. Thesis. College Station: Texas A & M University.

Bisset, A. R. (1987). The economic importance of moose (*Alces alces*) in North America. In *Proceedings of 2nd International Moose Symposium*. Uppsala: Swedish Wildlife Research.

Bissonette, J. A. (1982). Collared peccary. In *Wild Mammals of North America: Biology, Management and Economics*, ed. J. A. Chapman & G. A. Feldhammer, pp. 841–50. Baltimore: Johns Hopkins University Press.

Bossenmaier, E. F. (1978). *100-Plus Years of Wildlife Protection and Development in Manitoba*. Information Series 78–3. Winnipeg: Manitoba Department of Renewable Resources and Transportation Services.

Bryant, L. D. & Maser, C. (1982). Classification and distribution. In *Elk of North America: Ecology and Management*, ed. J. W. Thomas & D. E. Toweill, pp. 1–59. Harrisburg: Stackpole Books.

Coady, J. W. (1982). Moose. In *Wild Mammals of North America: Biology, Management and Economics*, ed. J. A. Chapman & G. A. Feldhammer, pp. 902–22. Baltimore: Johns Hopkins University Press.

Connolly, G. E. (1981). Limiting factors and population regulation. In *Mule and Black-tailed Deer of North America*, ed. O. C. Walmo, pp. 245–85. Lincoln: University of Nebraska Press.

Dary, D. A. (1974). *The Buffalo Book: The Full Saga of the American Animal*. Chicago: Swallow Press.

Doughty, R. W. (1983). *Wildlife and Man in Texas: Environmental Change and Conservation*. College Station: Texas A & M University Press.

Filion, F. L., James, S. L., Ducharme, J.-L., Pepper, W., Reid, R., Boxall, P. & Teillet, D. (1983). *The Importance of Wildlife to Canadians: Highlights of the 1981 National Survey*. Ottawa: Canadian Wildlife Service, Environment Canada.

Gunn, A. (1982). Muskox. In *Wild Mammals of North America: Biology, Management and Economics*, ed. J. A. Chapman & G. A. Feldhammer, pp. 1021–35. Baltimore: Johns Hopkins University Press.

Hesselton, W. T. & Hesselton, R. A. M. (1982). White-tailed Deer. In *Wild Mammals of North America: Biology, Management and Economics*, ed. J. A. Chapman & G. A. Feldhammer, pp. 878–901. Baltimore: Johns Hopkins University Press.

Kelsall, J. P. (1987). The distribution and status of moose (*Alces alces*) in North America. In *Proceedings of 2nd International Moose Symposium*. Uppsala: Swedish Wildlife Research, Supplement (in press).

Kitchen, D. W. & O'Gara, B. W. (1982). Pronghorn. In *Wild Mammals of North America:*

Biology, Management and Economics, ed. J. A. Chapman & G. A. Feldhammer, pp. 960–71. Baltimore: Johns Hopkins University Press.

Kozicky, L. (1987). *Hunting Preserves for Sport or Profit.* Kingsville: Texas A & I University.

Krutch, J. W. (1957). A damnable pleasure. In *Readings in Wildlife Conservation*, ed. J. A. Bailey, W. Elder & T. D. McKinney, pp. 77–81. Washington: The Wildlife Society.

Lawson, B. & Johnson, R. (1982). Mountain sheep. In *Wild Mammals of North America: Biology, Management and Economics*, ed. J. A. Chapman & G. A. Feldhammer, pp. 1036–55. Baltimore: Johns Hopkins University Press.

Leopold, A. (1933). *Game Management.* New York: Charles Schribner's Sons.

Mackie, R. J., Hamlin, K. L. & Pac, D. F. (1982). Mule deer. In *Wild Mammals of North America: Biology, Management and Economics*, ed. J. A. Chapman & G. A. Feldhammer, pp. 862–77. Baltimore: Johns Hopkins University Press.

Marchello, M. J., Berg, P. T., Slanger, W. D. & Harrold, R. L. (1985). Cutability and nutrient content of antelope. *Journal of Food Quality*, 8, 209–18.

McCandless, R. G. (1985). *Yukon Wildlife: A Social History.* Edmonton: University of Alberta Press.

McNeil, K. (1983). *Indian Hunting, Trapping and Fishing Rights in the Prairie Provinces of Canada.* Saskatoon: University of Saskatchewan Native Law Centre.

Miller, F. L. (1982). Caribou. In *Wild Mammals of North America: Biology, Management and Economics*, ed. J. A. Chapman & G. A. Feldhammer, pp. 923–59. Baltimore: Johns Hopkins University Press.

Peek, J. M. (1982). Elk. In *Wild Mammals of North America: Biology, Management and Economics*, ed. J. A. Chapman & G. A. Feldhammer, pp. 851–61. Baltimore: Johns Hopkins University Press.

Prescott-Allen, C. & Prescott-Allen, R. (1987). *The First Resource: Wild Species in the North American Economy.* New Haven: Yale University Press.

Reiger, J. F. (1975). *American Sportsmen and the Origins of Conservation.* New York: Winchester Press.

Reynolds, H. W., Glaholt, R. D. & Hawley, A. W. L. (1982). Bison. In *Wild Mammals of North America: Biology, Management and Economics*, ed. J. A. Chapman & G. A. Feldhammer, pp. 972–1007. Baltimore: Johns Hopkins University Press.

Seton, E. T. (1929). *Lives of Game Animals.* Garden City (NY): Doubleday.

Sowls, L. K. (1984). *The Peccaries.* Tucson: University of Arizona Press.

Stransky, J. J. (1984). Hunting the whitetail. In *White-tailed Deer: Ecology and Management*, ed. L. K. Halls, pp. 739–80. Harrisburg: Stackpole Books.

Teer, J. G. (1982). Texas wildlife: now and for the future. In *Texas Wildlife Resources and Land Use*, pp. 9–20. Austin: The Wildlife Society.

Teer, J. G., Burger, G. V. & Deknatel, C. Y. (1983). State-supported habitat management and commercial hunting on private lands in the United States. *Transactions of the North American Wildlife & Natural Resources Conference*, 48, 445–56.

Thomas, J. W. & Toweill, D. E. (1982). *Elk of North America: Ecology and Management.* Harrisburg: Stackpole Books.

Tober, J. A. (1981). *Who Owns the Wildlife: The Political Economy of Conservation in Nineteenth Century America.* Westport (Connecticut): Greenwood Press.

United States Bureau of Sport Fisheries and Wildlife (1982). *1980 National Survey of Fishing, Hunting and Wildlife-associated Recreation (Initial Findings).* Washington: United States Bureau of Sport Fisheries and Wildlife.

Wigal, R. A. & Coggins, V. L. (1982). Mountain goat. In *Wild Mammals of North America: Biology, Management and Economics*, ed. J. A. Chapman & G. A. Feldhammer, pp. 1008–20. Baltimore: Johns Hopkins University Press.

8

Commercial and safari hunting in Zimbabwe

D. H. M. CUMMING

Abstract

Safari hunting is economically and ecologically attractive. Like game ranching for meat production, safari hunting is a relatively recent development in Zimbabwe dating to the early 1960s. Safari hunting is conducted under a variety of arrangements on a variety of land bases. Landowners are responsible for wildlife management on their lands. Public land available for hunting is leased to Safari Operators or Hunters' Associations, or individual hunts are sold. Three types of safari are distinguished: big game safaris, plains game safaris, and ranch hunts. A 21-day big game safari earns about US$ 23,000 in foreign exchange. Individual hunts auction for Z$ 2600–Z$ 24,000. Trophy and lease fees earn Z$ 1.00/hectare in Safari Areas and Z$ 0.40/hectare from Communal Lands. Gross foreign currency earnings of the commercial safari industry were Z$ 7 million in 1985.

Introduction

Commercial hunting has a long history in Africa. Elephant have been hunted for centuries and their ivory exported to Europe and Asia (Alpers, 1975; Parker, 1983). During the 19th century, hunting supported a thriving export trade in ivory and hides (Pringle, 1982). Tales of hunting and adventure by Harris (1847), Cumming (1850), Selous (1895), and others did much to stimulate sport hunting which was usually linked to commercial hunting in Africa. A few month's 'sport hunting' generally returned a fair profit from ivory and hides sold at the end of an expedition to the 'far interior' (Tabler, 1960).

By the turn of the century, vast herds of game in eastern and southern Africa had been all but eliminated by hunting and the great rinderpest epidemic of 1896 (Plowright, 1982). Legislation to protect large mammals and establish game reserves and national parks followed (Cumming, 1983). As herds recovered, resident sportsmen were able to buy licences and travel into wild areas on hunting expeditions. Foreign sportsmen either hunted with friends or hired professional hunters. The glamour of the modern African hunting expedition, or safari, probably began with Theodore Roosevelt's 'African Game Trails' and was further stimulated by Ruark and Hemingway. This trend culminated in the classic East African safari and the celebrated corps of professional 'white hunters' so well described by Dyer (1979).

Commercial hunting is economically and ecologically attractive because it can provide good financial returns with minimal investment and environmental management. Safari hunting involves a low off-take of trophy animals from wild populations with greater financial returns per animal taken than is possible when cropping for meat. Safari hunting also can yield relatively high returns from wildlands and wildlife populations which are not otherwise commercially exploitable. An analysis by Clarke *et al.* (1986) suggests that commercial hunting, in conjunction with harvesting for meat and live sale of animals, can yield as much if not more than extensive cattle ranching. More importantly, wildlife can produce higher returns at lower levels of dry matter consumption. Child & Child (1986) reported a gross return of Z$ 0.18/kg of game carried on a ranch in the midlands of Zimbabwe as opposed to Z$ 0.06/kg from cattle ranching in the same area. Net revenue earned from wildlife was estimated at Z$ 6.35/hectare as opposed to Z$ 3.78/hectare from cattle.

Safari hunting has become a controlled, structured, and profitable form of land use in many African countries (Fig. 8.1). The commercial safari hunting industry in Zimbabwe uses land under state, private, and communal ownership and so provides a useful case study.

Structure and scale of the safari industry

Commercial safari hunting is a relatively recent development in Zimbabwe. During the 1950s, sport hunting was confined to private land although the Hunters' Association was able to lease small areas of what was then classified as Crown Land. Attempts were made in 1957 by the fledgling Wild Life Conservation Department to establish a 'hunting safari scheme' but suitable land was not available. Proposals to establish a buffer zone on the boundary of Wankie (now Hwange) National Park in which hunting could occur were not accepted by the government of the day (Fraser, 1959).

The Wild Life Conservation Act (1960), however, provided for the establishment of Controlled Hunting Areas (CHAs). The first CHAs were established in the Zambesi Valley and opened to sportsmen in 1961. The demand for hunting camps greatly exceeded the supply and hunts were drawn in a lottery. In 1962, a second hunting area was opened temporarily to sportsmen before it became a tsetse control hunting area. In 1963, two further CHAs were declared, Wankie CHA on the northern boundary of Hwange National Park, and the Tuli Circle CHA in the south of the country. The demand for hunting continued to exceed the supply despite some increase in hunting opportunities (Table 8.1).

Game ranching also began in Zimbabwe in the late 1950s but focused for many years on meat production which was unable to compete with the

Fig. 8.1. African countries with commercial safari hunting enterprises (1, Cameroon; 2, Central African Republic; 3, Ethiopia; 4, Tanzania; 5, Zambia; 6, Zimbabwe; 7, Botswana; 8, Namibia; 9, South Africa).

safari hunting in the form of the 'mini-safari' (as distinct from the classic East African big game safari) in conjunction with meat production – an innovation which proved to be increasingly profitable (Johnstone, 1975).

By the late 1960s, the Government began to take an interest in wildlife production systems (Fraser, 1970). A depressed cattle ranching area in north-western Zimbabwe was purchased and assigned to the Department of National Parks and Wild Life Management (NPWLM) for safari hunting. Although many of the farms were using wildlife, most were too small for viable game ranching or safari hunting enterprises. The land was divided into seven concessions which were leased in 1973. The Safari Operators were obliged to reside on their concessions and contribute to their management by maintaining hunting tracks, fire guards, and game water supplies. The area became the Matetsi Safari Area under the Parks and Wild Life Act (1975). This experiment greatly stimulated safari hunting and encouraged the lease of further Safari Area (SA) concessions to commercial Safari Operators. The major components of the commercial safari industry are wildland (and owners of wildland), huntable populations of large mammals and game birds, safari operators (outfitters) and hunters prepared to pay to hunt (Fig. 8.2).

Table 8.1. *Early development of commercial sport hunting in Zimbabwe.*[a]

Year	Hunting Areas	Total revenue[b]	Hunters		Revenue (1986 Z$)[c]
			Number	% non-res.	
1961	Zambesi Valley (ZV)	5,432	98	—	50,699
1962	ZV, Nagupande	10,502	148	—	98,019
1963	ZV, Wankie CHA	24,382	221	—	227,563
1964	ZV, Wankie CHA	30,682	227	—	280,140
1965	ZV, Wankie CHA	31,420	213	—	281,974
1966	ZV, Wankie + Tuli	34,788	254	—	306,953
1967	ZV, Wankie + Tuli	31,763	250	—	275,061
1968	ZV, Wankie + Tuli	25,556	172	—	216,402
1969	ZV, Wankie + Tuli	42,398	135	36.3	174,923
1970	ZV, Wankie + Tuli	53,740	207	36.2	217,865
1971	ZV, Wankie + Tuli	65,628	265	51.7	258,088
1972	ZV, Wankie + Tuli	88,151	371	42.6	332,347
1973	ZV, Wankie + Tuli	76,117	285	47.0	276,550

[a]Data are from Annual Reports of the Department of Wild Life Conservation and the Department of National Parks and Wild Life Management (1963–73).
[b]Revenue for 1961–68 is £ Sterling, 1969–73 is Zimbabwe dollars.
[c]1Z$ = 0.60 US dollars.

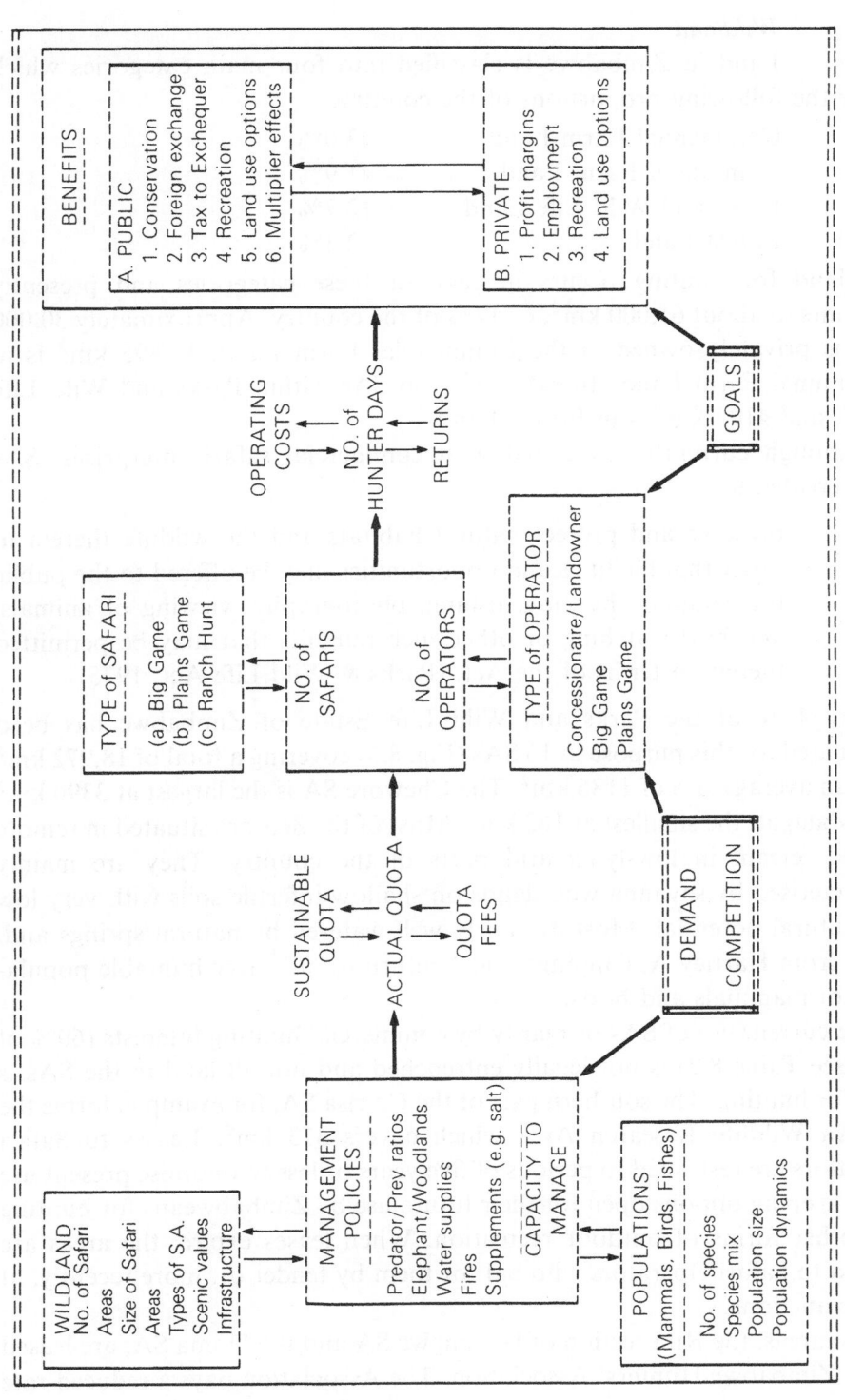

Fig. 8.2. Major components of the safari industry in Zimbabwe.

Wildland

Land in Zimbabwe is classified into four main categories which cover the following proportions of the country:

Commercial Farm Land	43.0%
Communal Farm Land	41.9%
Parks and Wild Life Land	12.7%
Forest Land	2.4%

Wildland for hunting occurs in each of these categories and presently amounts to about 65,000 km^2, or 17% of the country. Approximately 30,000 km^2 is privately owned in the Commercial Farm Land, 13,895 km^2 is in Communal Farm Land, 16,945 km^2 is in SAs within Parks and Wild Life Land, and 4105 km^2 is in Forest Land.

Although currently associated with commercial safari enterprises, SAs were created to:

> preserve and protect natural habitats and the wildlife therein in order that facilities and opportunities may be offered to the public for camping, hunting, fishing, photography, viewing of animals, and bird watching or other such pursuits that may be permitted therein in terms of the Act. (Parks & Wild Life Act, 1975)

Nearly 40% of the Parks and Wild Life Estate of Zimbabwe has been designated for this purpose in 16 SAs (Fig. 8.3) covering a total of 18,972 km^2 with an average size of 1186 km^2. The Chewore SA is the largest at 3390 km^2 and Malapati the smallest at 162 km^2. Most of the SAs are situated in remote rugged terrain in low-lying arid parts of the country. They are mainly characterised by savanna woodlands on shallow infertile soils with very low agricultural potential. Most are fairly well watered by natural springs and, apart from Hartley A, Chipinge and Umfurudzi, all carry huntable populations of mammals and birds.

The current use of SAs primarily by commercial hunting interests (60% of SAs, see Table 8.2) is not legally entrenched and not all land in the SAs is used for hunting. The southern part of the Chirisa SA, for example, forms the Sengwa Wildlife Research Area which covers 373 km^2. Leases to Safari Operators are restricted to periods of five years or less to optimise present use while keeping options open for their future use by Zimbabweans for hunting and other forms of outdoor recreation. When leases expire, the areas are offered to Safari Operators who bid for them by tender or, more recently, at public auctions.

Two areas, the Rifa section of Hurungwe SA and the Doma SA, are leased to the Zimbabwe Hunters' Association. The Association pays a reduced rate for animals on quota and allocates hunting to their members in these lease areas. Hunting in the remainder of the Hurungwe SA (Nyakasanga section and the Makuti section), Sapi SA, and Tuli SA is sold by NPWLM. The available quota is packaged into 10 or 14 day hunts, each of which is allocated to a particular camp site. During the 1970s, hunts were sold at a fixed price on a draw system since demand always exceeded supply. This changed to a tender system and, more recently, to public auction with 33% of hunts reserved for Zimbabwe residents.

Forest areas of indigenous woodland in western Zimbabwe adjoin Hwange National Park or the Matetsi SA and carry huntable populations of large mammals. Safaris on these areas are run by the Forestry Commission

Fig. 8.3. The major elements of the National Parks & Wild Life Estate of Zimbabwe and the distribution of Communal Land concessions, Forest Area hunting and areas of Commercial Farm Land where wildlife utilisation is a major land use.

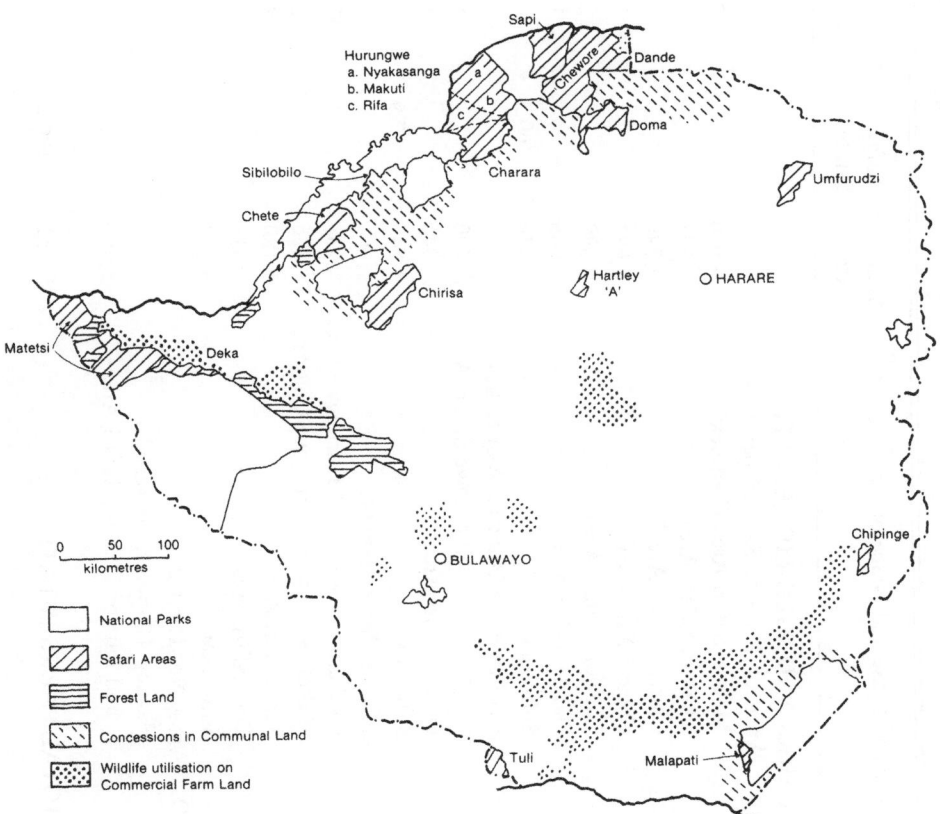

Table 8.2. *Safari areas and hunting concessions (leases) under different categories of use or leasehold and returns of revenue.*

Type of Lease and Name of Area	Area km²	% Safari Estate	No. of Leases	Species on Quota	Annual Lease Fees	Trophy Fees	Total Revenue	Revenue per ha
PARKS AND WILD LIFE ESTATE								
A. Areas on Lease to Safari Operators								
Matetsi Safari Area Complex[a]	4,040		7	17	96,000	398,845	494,845	1.22
Charara Safari Area	933			15		38,320	38,320	0.41
Chirisa Safari Area[b]	1,340		1	15	70,000	60,540	130,540	0.97
Chete Safari Area	1,081		1	13	61,000	38,030	99,030	0.92
Chewore Safari Area	3,390		2	16	90,000	152,450	242,450	0.72
Dande Safari Area[c]	1,046		1	19	46,000	68,385	114,385	1.09
Sub-Total	11,830	60.3			363,000	756,570	1,119,570	0.95
B. Areas Leased to Hunters' Association								
Rifa Section of Hurungwe Safari Area	600		1	14		53,330	53,330	0.89
Doma Safari Area	945		1	14		34,700	34,700	0.37
Sub-Total	1,545	7.9	2			88,030	88,030	0.57
C. Public Tender for Hunts[d]								
Nyakasanga Section of Hurungwe SA	1,138		42	11		258,000	258,000	2.27
Makuti Section of Hurungwe/Charara	836		12	8		59,900	59,900	0.72
Sapi Safari Area	1,180		18	13		148,900	148,900	1.26
Tuli Safari Area	416				0	0	0	0.00
Sub-Total	3,570	18.2				466,800	466,800	1.31
D. Areas Not Hunted								
Hartley, Chipinge, Umfurudzi, and Part of Charara Safari Areas	2,687	13.7				0	0	0.00
TOTAL PARKS ESTATE	19,632	100.0			363,000	1,311,400	1,674,400	0.85

COMMUNAL LAND HUNTING CONCESSIONS

A. Areas on Long Lease							
Binga (No. 3)	1,246	1	15		42,650	42,650	0.34
Binga/Manjolo Communal Land	1,400	1	13		29,310	29,310	0.21
Omay/Gokwe (No.1)	1,095	1	15		82,580	82,580	0.75
Omay (No. 2)	1,015	1	11		62,140	62,140	0.61
Sibilobilo/Dela/Mapangolo	1,089		15		76,085	76,085	0.70
Sub-Total	5,845	42.1		0	292,765	292,765	0.50
B. Areas on Annual or Short-Term Leases							
Gokwe Communal Land	1,200	1	5		28,350	28,350	0.24
Mukwichi Communal Land	850	1	15	17,000	37,535	54,535	0.64
Dande/Mazarabani Communal Land	5,000	1	14	16,000	71,650	87,650	0.18
Chiredzi (including Malapati SA)	1,000	1	12	50,000	41,520	91,520	0.92
Sub-Total	8,050	57.9		83,000	179,055	262,055	0.33
TOTAL COMMUNAL LANDS	13,895	100.0		83,000	471,820	554,820	0.40
GRAND TOTAL	33,527			446,000	1,783,220	2,229,220	0.66

[a]Area of 4,040 km² includes Deka SA and Forest Lands of Kazuma and Panda-Masuie.
[b]The Sengwa Wildlife Research Area of 373 km² is not hunted. Total area of Chirisa is 1,713 km².
[c]The concession area includes 523 km² of Communal Land.
[d]Values under trophy fees are the returns from public auctions of hunts in these areas.

although some animals are made available to other commercial operators. The forest areas within the Matetsi complex have formed an additional hunting area ('pool area') for the Matetsi concessionaires.

Hunting concessions in the Communal Lands are administered by NPWLM although the land itself is administered by the Ministry of Local Government and the District Councils. Lease and trophy fees are collected by government but returned to District Councils from central revenue in the form of grants for approved development projects.

Until recently, Communal Land concessions were remote, largely undeveloped, uninhabited, and often adjacent to Parks and Wild Life Land. Expanding human populations and immigration following the eradication of tsetse fly in the Zambesi Valley led to rapid expansion of subsistence farming and the influx of domestic livestock. These developments now threaten the viability and continuity of commercial wildlife utilisation in these concessions. The number of concessions in the Communal Lands has declined from 17 in 1981 to 11 in 1986. Appropriate land-use planning and programmes such as the Communal Areas Management Programme for Indigenous Resources (Martin & Taylor, 1984; Martin, 1986) seek to involve communal farmers in wildlife management and to develop institutional mechanisms which allow revenue from wildlife to go directly to the community.

Under the Parks and Wild Life Act (1975), landowners or occupiers are responsible for the management and utilisation of wildlife resources on their property. This gives commercial farmers the latitude and incentives to manage their wildlife resource on a sustainable and commercially productive basis. Many extensive cattle ranches have either incorporated wildlife into their management systems or switched from cattle ranching to wildlife systems which presently depend on commercial hunting and safari operations for their profitability (Child, unpublished report). Some 30,000 km^2 of ranch land in the commercial farming sector of Zimbabwe is used for game ranching and safaris.

Types of safari

Safari Operators, Professional Hunters, and Guides must register under the Development of Tourism Act. Safari Operators are required to meet prescribed standards in equipment and service to clients. These standards cover such matters as vehicles, field accommodation, and cuisine. Learner Professional Hunters serve a two-year apprenticeship and both they and Professional Hunters undergo practical and written examinations conducted by NPWLM and the Safari Operators' and Professional Hunters' Association. Three types of safari are distinguished by the species they

offer and the size and category of land on which hunting takes place. These
are:

Big Game Safaris (28 registered companies/operators)
Offered only by operators who lease state land or who hunt an
area of at least 130 km^2 on which elephant (*Loxodonta africana*),
lion (*Panthera leo*), and buffalo (*Syncerus caffer*) must be
present.

Plains Game Safaris (14 registered operators)
Offered by operators hunting on a minimum of 60 km^2 on which
plains game are available with *no domestic stock present* and on
which dangerous game may sometimes be present.

Ranch Hunts (13 registered operators)
Offered by operators hunting less than 60 km^2 where only plains
game is present or more than 60 km^2 where domestic stock is
also present.

The quota set for a concession is packaged by the operator into 21, 15, or 10-
day hunts. Ranch hunts may be 5–10 days depending on the species offered.
A typical 21-day big game safari might comprise the following:

Dangerous game: 1 elephant or 1 lion, plus 1 leopard (*Panthera
pardus*), and 2 buffalo

Plains game: 1 sable (*Hippotragus niger*) or 1 waterbuck (*Kobus
ellipsiprimnus.*), 3 impala (*Aepyceros melampus*), 1 warthog
(*Phacochoerus aethiopicus*), 1 bushbuck (*Tragelaphus scriptus*), 1
zebra (*Equus burchelli*), 1 duiker (*Cephalophus* spp.), 1 klipspringer (*Oreotragus oreotragus*), 1 eland (*Taurotragus oryx*) or 1
wildebeest (*Connochaetes taurinus*)

Other game: 1 hippopotamus (*Hippopotamus amphibius*) and/or 1
crocodile, 1 hyaena (*Crocuta crocuta*) and perhaps one or two
other small carnivores.

A typical 10-day hunt might comprise a buffalo or a leopard and a bag of
plains and other game similar to that on a 21-day hunt.

Daily rates charged by outfitters vary considerably but as a general guide
the Safari Operators' Association indicates the following price ranges for
different types of hunt:

Big Game Safaris: US$ 650–700 per day plus trophy fees.
Plains Game Safaris: US$ 250–400 per day plus trophy fees.
Ranch Hunts: less than US$ 250 per day.

Table 8.3. *Government trophy fees (Z$) for animals hunted on Parks and Wild Life Land and Communal Land and mean of trophy fees charged by Safari Operators (US$) in Zimbabwe.*

Species	Government trophy fees				Safari Operators trophy fee 1986[a]
	1981	1982	1984	1986	
Elephant					
Male	2,000	2,000	2,000	3,500	5,806
Female	700	700	700	1,200	—
Young male	—	—	—	2,000	—
Buffalo					
Male	350	350	500	500	1,353
Female	150	150	300	350	—
Lion					
Male	1,000	1,000	1,000	1,500	3,311
Female	600	600	750	1,000	—
Leopard	600	600	600	700	2,452
Zebra	250	250	250	400	710
Wildebeeste	130	130	130	200	638
Sable	500	500	600	800	1,545
Waterbuck	300	250	250	300	876
Eland	350	350	350	400	915
Nyala	—	—	600	600	—
Kudu					
Male	200	200	200	250	760
Female	75	75	75	100	—
Impala					
Male	30	30	50	50	160
Female	20	20	30	30	—
Reedbuck	100	100	100	150	383
Bushbuck	100	100	100	150	395
Grysbok	20	20	20	20	121
Duiker	5	5	15	15	105
Steenbok	15	15	15	15	92
Klipspringer	75	75	75	100	307
Warthog	25	25	40	50	136
Bushpig	25	25	40	40	104
Giraffe	400	400	400	400	1,372
Hippopotamus	500	500	500	600	1,428
Crocodile	300	300	300	300	1,142
Hyaena	60	60	150	150	171
Baboon	10	10	10	10	40

[a]V. Booth (personal communication).   (1Z$ = 0.6 US$)

The licence or trophy fee is paid by the safari client. The trophy fees set by NPWLM (Table 8.3) are quoted in US$ by the operator resulting in average trophy fees which are higher than the NPWLM rates. Historically, this discrepancy developed from the changing rates of exchange between the Z$ and the US$ and trophy fees have generally been advertised in US$. Combining daily rates and trophy fees the charge for a 21-day Big Game Safari is about US$ 23,000 which must be paid in foreign currency.

The holders of concessions pay a lease fee or a right to hunt fee which (except in the case of Matetsi) has been determined by the tenders submitted by operators in bidding for the concessions. In addition, the concession holder must pay the trophy fees for the animals he has on quota. These fees and payments for the Safari Areas and Communal Land concessions for 1986 are summarised in Tables 8.2–8.5.

Hunts sold by National Parks

Hunts in the Nyakasanga and Makuti sections of the Hurungwe SA are sold directly to the hunter under tender or public auction (the Tuli SA, which has not been hunted for the last two seasons, is usually included in this scheme).

In 1986, NPWLM auctioned 36 fourteen-day hunts and 42 ten-day hunts which were conducted from ten camps in the Nyakasanga, Makuti, and Sapi areas. All hunts included one buffalo male, one buffalo female and, depending on the camp at which the hunt was based, varying species and numbers of antelope, warthog, bushpig (*Potamochoerus porcus*), baboon, and game birds. In addition to the basic bag, some 20 hunts included additional species such as elephant, lion, leopard, hippopotamus, and nyala (*Tragelaphus angasi*). Auction floor prices for ten-day hunts varied from Z$ 1900 for a hunt with a basic bag to Z$ 12,500 for a hunt with a basic bag plus an elephant. The basic bag for both hunts was:

Buffalo, male	1	Grysbok	1
Buffalo, female	1	Baboon	1
Impala, male	6	Guinea fowl	4
Impala, female	5	Francolin	4
Duiker	1	Doves/pigeons	4

The reserve price for such a hunt was Z$ 1600 while that for the basic bag plus an elephant was Z$ 5100.

Auction floor prices for the 14-day hunts varied from Z$ 2600 to Z$ 24,000. The bag varied from camp to camp but included species such as nyala, leopard, hippopotamus, lion and elephant, in addition to the basic bag. Floor prices in excess of Z$ 16,000 were for camps which included hippo and leopard plus a bull elephant. The highest bid of the auction was Z$ 24,000 for a 14-day hunt which comprised 1 elephant bull, 1 trophy lion, 1 hippopotamus, 1 leopard and a basic bag of buffalo, impala, duiker, grysbok, baboon, and game birds.

The distribution of dangerous game (elephant, lion, and buffalo) and of plains game (antelope) in Zimbabwe is such that few areas other than Matetsi can offer a fully balanced hunt comprising both dangerous and plains game. This imbalance (see quotas listed in Tables 8.4 and 8.5) has encouraged links between the concessions in Safari Areas or Communal Lands which carry dangerous game and private ranches which carry plains game.

Management of safari areas and concessions

Safari Areas are managed in much the same way as National Parks in Zimbabwe, i.e., with minimal interference to ecosystems. The only direct population management is culling elephants to protect woodlands (Cumming, 1981). Culling of wildebeest and impala also occurred in the Tuli SA during the 1982–84 drought. Fire management is aimed at protection from man-made fires by preparing fire breaks. Cool, early season burns are used to encourage re-growth of woody vegetation in the areas of *Brachystegia* woodland under pressure from elephants.

Infrastructural development in SAs is largely limited to dry season dirt roads and tracks. Permanent buildings are a feature of the Matetsi concessions because this area was a former farming area and many of the buildings were in place when the land was acquired. Safari Operators also are obliged, under the Matetsi lease agreements, to reside on their concessions throughout the year. Occupation of other SAs is seasonal and involves temporary and usually tented accommodation. Hunting is permitted throughout the year but occurs mainly during the cool dry season from April through August and the hot dry season of September through October or mid November.

Safari Areas in the Parks and Wild Life Estate are under the direct control of parks staff who patrol these areas and oversee the operation of professional hunters. Such controls are not present in the Communal Land concessions or on private land.

Wildlife resources in Communal Land concessions are not managed apart from the control of problem animals and occasional anti-poaching operations carried out by National Parks staff and by the Safari Operators themselves. Problem animals are most often crop-raiding elephants and lions killing livestock. Increasing settlement and human/wildlife conflicts are resulting in a decline in wildlife resources in these concessions.

Table 8.4. *Hunting quotas, trophy fees and revenue for Safari Areas during 1986.*

Safari Area	Elephant			Buffalo		Lion		Leop.[a]	Sa.	Wa.	El.	Ze.	Ny.	Kudu		Impala		Bb.	Rb.	Grb.	Du.	St.	Kl.	Wh.	Bp.	Gir.	Hi.	Cr.	Hy.	Ba.	Total Values
	M	F	YB	M	F	M	F							M	F	M	F														
Matetsi complex / Deka	21	—	—	112	40	28	—	28	88	87	14	88	—	100	50	113	113	6	—	70	7	—	—	111	70	14	—	—	7	unlimited	398.845
Chete	4	—	—	10	5	2	2	6	1	—	—	4	—	10	—	20	—	4	—	7	10	—	7	7	6	—	—	—	—	—	38.030
Chirisa	4	4	3	18	5	3	—	8	1	5	1	10	—	12	—	40	—	7	5	—	6	—	6	8	—	—	—	—	—	30	60.540
Charara	2	—	—	21	7	4	—	4	—	2	—	—	—	7	—	28	28	14	—	6	6	—	14	6	7	—	—	—	7	14	38.820
Hurungwe – Rifa	3	—	—	30	20	1	3	4	4	4	—	—	—	2	8	50	50	4	—	5	46	—	—	15	2	—	6	—	6	100	53.330
– Nyakasanga	8	—	—	50	50	2	6	5	3	3	—	—	—	—	—	320	260	—	—	46	46	—	—	5	—	—	—	—	10	184	112.900
– Makuti	2	—	—	20	20	—	—	6	—	—	—	—	—	2	—	15	—	—	—	30	30	—	—	10	6	—	—	—	—	—	30.490
Sapi	3	—	—	36	30	2	6	5	5	—	—	—	4	4	18	140	130	—	—	66	33	—	—	15	—	—	4	10	10	120	75.265
Chewore	9	—	22	60	20	4	4	8	8	4	—	—	2	12	15	100	100	20	—	10	10	—	12	20	15	—	8	—	10	—	152.450
Dande	3	3	6	40	4	3	—	4	4	2	2	1	—	8	—	15	—	15	—	15	15	—	2	10	4	—	3	1	4	30	68.385
Doma	1	—	—	10	2	3	2	4	—	—	5	2	—	20	20	10	—	5	—	10	10	—	—	20	10	—	—	—	—	20	34.700
Umfurudzi																															
Chipinge																															
Tuli																															
Totals	60	7	31	407	203	52	23	82	97	107	22	105	6	177	111	836	581	75	5	265	173	0	41	227	120	14	21	1	54	498	
Animal value (Z$)	3,500	1,200	2,000	500	350	1,500	1,000	700	800	300	400	400	600	250	100	50	30	150	150	20	15	15	100	50	40	400	600	300	150	10	
Values (Z$ × 1,000)	210	8.4	62	204	71.1	78	23	57.4	77.6	32.1	8.8	42	3.6	44.3	11.1	41.8	17.4	11.3	0.75	5.3	2.6	0	4.1	11.4	4.8	5.6	12.6	0.3	8.1	4.98	1063.755
Values (US$ × 1,000)	126	5.04	37.2	122	42.6	46.8	13.8	34.4	46.6	19.3	5.28	25.2	2.16	26.6	6.66	25.1	10.5	6.75	0.45	3.18	1.56	0	2.46	6.81	2.88	3.36	7.56	0.18	4.86	2.99	638.253

[a]Species abbreviations: Sa = Sable, Wa = Waterbuck, El = Eland, Ze = Zebra, Ny = Nyala, Bb = Bushbuck, Rb = Reedbuck, Grb = Grysbok, Du = Duiker, St = Steenbok, Kl = Klipspringer, Wh = Warthog, Bp = Bushpig, Gir = Giraffe, Hi = Hippopotamus, Cr = Crocodile, Hy = Hyaena, Ba = Baboon, YB = Young Bull.

Table 8.5. *Hunting quotas, trophy fees, and revenue for Communal Land concessions during 1986.*

Concession Area	Elephant M	Elephant F	Elephant YB	Buffalo M	Buffalo F	Lion M	Lion F	Leop.[a]	Sa.	Wa.	El.	Ze.	Ny.	Kudu M	Kudu F	Impala M	Impala F	Bb.	Rb.	Grb.	Du.	St.	Kl.	Wh.	Bp.	Hi.	Cr.	Hy.	Ba.	Total Values
Binga No. 3	3	3	5	5	5	2	—	3	2	2	2	4	—	5	—	25	10	8	—	10	10	—	—	10	—	—	3	—	—	42,650
Binga/ Manjolo	2	3	2	10	—	2	—	2	1	—	1	2	—	4	—	2	—	4	3	3	—	—	—	4	—	—	3	—	—	29,310
Omay No. 1	3	4	5	10	2	2	—	5	2	3	2	2	—	10	4	20	10	6	4	6	10	—	6	10	—	—	—	—	—	48,670
Omay No. 2 Deia/	4	4	3	18	5	3	—	8	3	5	1	10	—	12	—	40	—	7	5	6	6	—	6	8	—	—	—	—	30	62,140
Mapongolo	3	1	—	6	2	1	—	3	2	7	1	6	—	1	—	25	5	6	1	3	3	—	3	5	—	—	—	—	—	28,855
Sibilobilo	2	2	5	10	3	2	—	6	4	9	1	6	—	4	1	35	10	10	1	4	4	—	4	8	—	—	—	—	14	47,230
Gokwe Unit 1	3	4	4	5	—	1	—	2	1	—	1	2	—	4	2	10	4	4	2	4	4	—	1	5	—	—	—	—	—	33,910
Gokwe Communal Land	6	—	—	3	—	—	3	3	3	—	—	—	—	—	—	6	—	3	—	—	—	—	—	—	—	—	—	—	—	28,350
Mukwichi Dande/	2	2	5	10	—	2	—	2	3	2	2	2	—	5	—	—	—	10	—	2	3	—	3	10	10	—	—	—	10	37,535
Mazarabani	6	—	6	20	10	4	—	8	2	6	—	4	—	12	—	30	10	12	—	10	30	—	—	10	—	—	—	4	20	71,650
Chiredzi	4	6	—	10	—	1	—	3	—	1	2	10	4	10	—	16	10	1	—	1	10	—	1	—	—	—	—	1	5	41,520
Totals	38	29	35	107	22	20	3	45	20	35	13	48	4	67	7	209	59	71	16	43	80	0	24	70	10	1	6	5	79	471,820
Animal value (Z$)	3,500	1,200	2,000	500	350	1,500	1,000	700	800	300	400	400	600	250	100	50	30	150	150	20	15	15	100	50	40	600	300	150	10	
Values (Z$ × 1,000)	133	34.8	70	53.5	7.7	30	3	31.5	16	10.5	5.2	19.2	2.4	16.8	0.7	10.5	1.77	10.7	2.4	0.86	1.2	0	2.4	3.5	0.4	0.6	1.8	0.75	0.79	471.82
Values (US$ × 1,000)	79.8	20.9	42	32.1	4.62	18	1.8	18.9	9.6	6.3	3.12	11.5	1.44	10.1	0.42	6.27	1.06	6.39	1.44	0.52	0.72	0	1.44	2.1	0.24	0.36	1.08	0.45	0.47	283.09

[a]Species abbreviations as for Table 8.4.

Hunting quotas

Annual aerial censuses of elephant using standard techniques are conducted during the dry season by NPWLM. These censuses also cover some of the other more conspicuous large herbivores such as buffalo, sable, waterbuck, eland, and zebra. The only SA in which the full range of species is censused each year using both road strip counts and aerial counts is the Matetsi SA. This is because Matetsi has a more complete system of access tracks and is better staffed.

Quotas for trophy animals in each of the SA and Communal Land concessions are set by NPWLM on the basis of aerial censuses, ground reconnaissance by their staff and reports from professional hunters. Quotas for elephant are set at 0.5% of the population in a given area whereas quotas for larger ungulates, such as buffalo, sable, waterbuck, eland, and zebra are set at approximately 2%. Quotas for large cats are set at 8% of the estimated population. These offtakes have resulted in a sustained yield of high quality trophies. Where the levels have been exceeded, through incorrect estimates of population size, trophy quality has declined (Booth & Jones, unpublished report). For the most part, detailed information on population size, structure, and dynamics is lacking. Quotas are set conservatively and refined adjustments from year to year are only possible in the Matetsi SA. Quotas are usually set by the end of October for the following year. Safari Operators then market their hunts during the off-season period of December through March, but they would prefer quotas to be set a year in advance to provide more time for marketing.

Economic returns

The direct revenue earned by government from trophy and lease fees for SAs is about Z$ 1.00/hectare (Table 8.2, Z$ 1,674,400 from 16,945 km² of SA hunted = Z$ 0.99/hectare). Direct revenue from the Communal Lands is about Z$ 0.40/hectare. The discrepancy between the SA and Communal Land revenue is due to higher wildlife densities in the SAs and because many of the Communal Land concessions include large areas with substantial human settlement and little wildlife. The direct revenue from January 1981 to December 1986 from trophy and lease fees in the Communal Lands amounted to Z$ 3,440,283. This figure includes the revenue earned from Chirisa and Dande SAs. Both areas, prior to the Parks and Wild Life Act (1975), were game reserves in the Communal Land and revenue earned in these areas is returned to the local authorities.

The economics of the transition from meat production to safari hunting in the late 1960s is illustrated by Johnstone's (1975) data from Rosslyn Game Ranch which is now part of the Matetsi SA. Gross income between 1967 and

1973 on Rosslyn, in 1986 Z$, varied between Z$ 2.06 and Z$ 4.90/hectare while profits varied from Z$ 0.60 to Z$ 3.15/hectare (Table 8.6). Earnings from safari hunting rose from less than 2% of income in 1967/68 to 75% of income in 1972/73; a change which coincided with increasing profitability.

A recent financial analysis of three Matetsi concessions (including the area of the former Rosslyn Game Ranch) by Taylor (unpublished report) shows similar levels of gross income but lower profits; Z$ 1.39/hectare in 1983

Table 8.6. *Gross income, operating costs, and profits on Rosslyn Ranch in north-western Zimbabwe[a].*

Year	Income					Operating Costs		Profit	
	Safaris	Meat	Skins	Total	Total /ha	Total	Total /ha	Total	Total /ha
1967/68	847	33,608	13,989	48,444	2.10	25,247	1.10	23,197	1.01
1968/69	1,271	28,401	17,740	47,412	2.06	33,484	1.45	13,928	0.60
1969/70	21,049	21,596	13,046	56,051	2.43	35,286	1.53	20,765	0.90
1970/71	62,528	15,443	606	78,577	3.41	44,505	1.93	34,072	1.48
1971/72	57,081	12,472	3,902	73,455	3.19	42,229	1.83	31,226	1.35
1972/73	84,879	17,912	10,286	113,077	4.90	40,481	1.76	72,596	3.15

[a]Area of Ranch = 23,060 ha and is now included within the Matetsi Safari Area. Data in 1986, in 1986 Z$, adapted from Johnstone (1975).

Table 8.7. *Average income, expenditure and profit of three Matetsi Safari Area concessions during 1983[a].*

	Z$/hectare	%
Gross income	7.60	
Expenditure:		
Operating costs	2.66	42.8
Labour	1.03	16.6
Marketing	0.26	4.1
Administration	0.46	7.4
Trophy fees	1.80	29.1
Total Costs	6.21	100.0
Net Profit	1.39	

[a]Taylor (unpublished report). Values in 1986 Z$.

compared with Z$ 3.15 in 1972/73 (Tables 8.6, 8.7). The lower profitability of the safari concession is partly a result of Government trophy and other concession fees which amounted to Z$ 1.80/hectare. No trophy or licence fees are charged by government for animals killed on private land.

An earlier analysis by Taylor (1982) showed that hunting provided 93% of the income on the Matetsi concessions while operating costs (40%), labour (21.1%), and concession and trophy fees (28.8%) accounted for the bulk of expenditure. Vehicle and aircraft expenses amounted to 59% of operating expenses. An essentially similar pattern of costs was reflected in Taylor's (1984) analysis (Table 8.7).

Buffalo Range Ranch, in south-east Zimbabwe, has maintained an 8000 hectare game section and a 10,000 hectare cattle section since 1960. The large herbivore biomasses and effects of game and cattle on vegetation in these two sections were examined by Taylor & Walker (1978), who demonstrated certain ecological advantages of wildlife. A recent examination of revenues derived from the cattle and wildlife sections has indicated that wildlife was about three times more profitable than cattle over the period 1978–84 (Child, unpublished report). This profitability is, however, partially confounded by earnings from a Communal Land safari concession which are also reflected in the financial returns of the wildlife section.

Returns from a proposed wildlife and dryland cropping system, as opposed to a cattle ranching and dryland cropping system in a Communal Land area of the Zambesi Valley, were examined by Brunt, Clarke & de Greling (1986). They found that gross revenues from wildlife and cropping were likely to be Z$ 16.30/hectare as opposed to Z$ 12.80 for cattle. The returns to the cattle system included the value of draught power, milk and manure, while returns to the wildlife system included deduction for the cost of fertilizer and ploughing in the absence of cattle. The wildlife system included the full spectrum of large mammals which occur in the Zambesi Valley and included safari hunting, cropping for meat, and live capture and sale of game.

The gross foreign currency earnings of the commercial safari industry in Zimbabwe were Z$ 7 million during 1985. This figure compares with beef exports of Z$ 31,322,000. Government expenditure in support of the safari industry is less than Z$ 1 million, whereas the beef industry is supported by veterinary services which received approximately Z$ 35 million in vote appropriations from government plus a Cold Storage Commission which reported a loss of Z$ 45 million for 1984–85. These inputs do not include an EEC-supported aid programme for foot-and-mouth disease control to support the export of beef to Europe.

The overall value of the domestic stock industry is, of course, very much higher than the simple export figures suggest. The gross value of domestic stock slaughterings in Zimbabwe for 1985 was Z$ 148,908,000 (Central Statistical Office, 1986) and with added values for draught power, manure, and milk production the value of the industry is more than Z$ 1000 million. On the other hand, the full value of the wildlife industry in terms of tourism (generally valued at Z$ 100–200 million), commercial and safari hunting, and meat production has not been critically evaluated for Zimbabwe. A high proportion of meat consumed in rural areas in Africa is obtained from wild mammals. Although some true subsistence hunting and trapping occurs, most of it enters informal rural markets.

Constraints to growth and land-use implications

The two most important constraints to growth of the safari industry in Zimbabwe are the availability of wildland and huntable populations of large mammals. The extent of these, and other, constraints varies with land category.

Significant expansion of the Parks and Wild Life Estate is unlikely. Opening National Parks to hunting or changing land presently designated as National Parks to SAs is also unlikely. Substantial increases in the quotas of existing SAs can be achieved only with more intensive management which may conflict with other conservation and aesthetic goals for these areas. The Matetsi experiment clearly shows, however, that land marginal for cattle ranching can be economically viable under safari hunting (Fraser, 1970). Some State Land (commercial farm land purchased for resettlement and other undesignated state land) is presently being managed for wildlife production and this area could be increased.

In the Communal Lands, areas where elephant, lion, and buffalo can exist without serious conflict with subsistence farmers are declining rapidly. Linked to this are institutional arrangements which presently prevent the landowner/occupier from reaping the full benefits of wildlife on his land. The Campfire project (Martin, 1986) seeks to develop ways in which peasant farmers can benefit more directly from wildlife.

Present constraints to the expansion of hunting on Commercial Farm Land are mainly size of property and lack of a sufficiently wide spectrum of species. Many areas will never be able to carry elephant and lion because of the conflicts with other land uses; however, farming areas adjacent to National Parks can make use of dispersing animals. Present demand for various antelope for restocking farms exceeds the supply (mainly from Parks and Wild Life Land) several fold. Other major constraints are the lack of research and marketing services which have been a key feature in the development of agriculture in Zimbabwe. The recent formation of the Wild Life Producers' Association as a commodity association under the Commercial Farmers Union in Zimbabwe is a step towards alleviating some of these handicaps.

Safari hunting, like all tourism, is susceptible to political instability or isolated incidents which pose a potential threat to the personal safety of visitors. Such incidents (e.g., armed banditry) can lead to large-scale cancellations by prospective clients. This raises the legitimate concern that land-use practices which are subject to the vagaries of political dissidents or leisure fashions are too vulnerable to be taken seriously in rural development strategies. (Recent trends in the market values of minerals and key Third World agricultural products such as coffee indicate that even the 'safest' products are vulnerable to changing markets.) Linked to this potential weakness are the pressures from anti-hunting lobbies in Europe and North America. I am not aware of any evidence to suggest that these pressures have adversely affected the safari industry in Zimbabwe, but they have done so in Kenya (Parker, 1983). Most countries allow the import of trophies, but some (such as Australia) demand management and monitoring programmes which most developing nations cannot execute.

In some two-thirds of Zimbabwe, commercial hunting could substantially improve the viability of animal production systems. This land falls into Agro-ecological zones III, IV, and V of Vincent & Thomas' (1960) classification. Zone III covers 18% of the country, has an annual rainfall of between 650–750 mm and is suitable for drought-resistant crops and livestock. Zone IV covers 37% of the country, has a rainfall of between 450–650 mm, is subject to periodic seasonal droughts and is generally unsuitable for dryland cropping, but suited to livestock production. Zone V covers 27% of the country and generally lies below 900 m with an erratic rainfall below 650 mm. It is suited to extensive livestock production or game ranching. Although much of the land in Zones IV and V in the commercial and communal farming sectors is unlikely to support big game safaris, a wide range of alternative safari hunting is feasible. Possibilities include a range of mini-safaris based on antelope, game bird, and waterfowl hunting. Safari Operators and commercial farmers have not yet explored the potential for joint wildlife ventures with peasant farming communities.

Livestock in Africa serve many functions and their husbandry is deeply embedded in most cultures (Doran, Low & Kemp, 1979; Roberts, 1980; Sandford, 1983). The displacement of livestock by wildlife production systems, however profitable, is unlikely. However, the potential for running both livestock and wildlife is considerable. In those areas where livestock is precluded by the presence of tsetse fly, wildlife production, and particularly safari hunting, is an obvious alternative.

Within southern central Africa there are still large areas of land occupied by tsetse and which are unavailable to domestic stock. In the Southern African Development Coordination Countries (Angola, Botswana, Malawi, Mozambique, Tanzania, Zambia, and Zimbabwe), the area occupied by tsetse is some 1,850,700 km² or 38% of the region. In both Zambia and Botswana, large areas have been declared wildlife management areas; over 100,000 km² or 20% of Botswana (Botswana Government Paper, 1986) and 163,853 km² or 22.3% of Zambia (Pullan, 1983). Much of this area is being used for commercial and safari hunting. There is, however, considerable potential for growth and for greater economic returns, both to local communities and governments, from commercial and safari hunting.

Acknowledgements

I am grateful to Dr Russell Taylor and Meg Cumming for their critical comments on earlier drafts of this chapter, to Claud Masaraure for drawing Fig. 8.3 and to Vernon Booth for data on average trophy fees.

References

Alpers, E. A. (1975). *Ivory and Slaves in East Central Africa*. London: Heinemann.
Botswana Government Paper. (1986). *Wildlife Conservation Policy*. Government Paper No. 1 of 1986, Republic of Botswana. Gaborone: Government Printer.
Brunt, M. A., Clarke, V. J. & de Greling, C. (1986). Tsetse area development and land use planning in the Zambesi Valley. Harare: FAO Consultant Report.
Child, B. A. & Child, G. (1986). Wildlife, economic systems and sustainable human welfare in semi-arid rangelands in southern Africa. *FAO/Finland Workshop on Watershed Management in Arid and Semi Arid Zones of the Southern African Development Coordination Conference (SADCC) Countries*. Maseru (Lesotho).
Clarke, V. J., Cumming, D. H. M., Martin, R. B. & Peddie, D. A. (1986). The comparative economics of African wildlife and extensive cattle production, pp. 87–96. *Proceedings of FAO African Forestry and Wildlife Meeting*. Bamako (Mali).
Cumming, D. H. M. (1981). The management of elephants and other large herbivores in Zimbabwe. In *Problems in Management of Locally Abundant Wild Mammals*, ed. P. A. Jewell, S. Holt & D. Hart, pp. 91–118. New York: Academic Press.
Cumming, D. H. M. (1983). The decision making framework with regard to culling large mammals in Zimbabwe. In *Management of Large Mammals in African Conservation Areas*, ed. R. N. Owen-Smith, pp. 173–86. Pretoria: Haum.
Cumming, R. G. (1850). *Five Years of a Hunter's Life in the Far Interior of South Africa*. London: J. Murray.
Doran, M. H., Low, A. R. C. & Kemp R. L. (1979). Cattle as a store of wealth in Swaziland: implications for livestock development and overgrazing in eastern and southern Africa. *American Journal of Agricultural Economics*, 61, 41–7.
Dyer, A. (1979). *The East African Hunters: The History of the East African Professional Hunters' Association*. Clinton (New Jersey): Amwell Press.

Fraser, A. D. (1959). *Annual Report of the Department of Wild Life Conservation.* Salisbury: Department of Wild Life Conservation.

Fraser, A. D. (1970). *The Matetsi Committee Report, 1969.* Typescript, pp. 1–165. Causeway, Salisbury: Ministry of Lands.

Harris, C. W. (1847). *Wild Sports of Southern Africa.* London: Macmillan.

Johnstone, P. A. (1974). Wildlife Husbandry on a Rhodesian Game Ranch. In *The Behavior of Ungulates and its Relation to Management,* ed. V. Geist & F. Walther, pp. 888–92. Morges: IUCN Publications New Series No. 24.

Johnstone, P. A. (1975). Evaluation of a Rhodesian game ranch. *Journal of the Southern African Wildlife Management Association,* 5, 43–51.

Martin, R. B. (1986). *Communal Areas Management Programme for Indigenous Resources (Campfire).* Branch of Terrestrial Ecology, Campfire working document No.1/86. pp. 1–110. National Parks & Wild Life Management. Harare, Zimbabwe.

Martin, R. B. & Taylor, R. D. (1984). Wildlife conservation in a regional land-use context: the Sebungwe region of Zimbabwe. In *Management of Large Mammals in African Conservation Areas,* ed. R. N. Owen-Smith, pp. 249–70. Pretoria: Haum.

Parker, I. S. C. (1983) *Ivory Crisis.* London: Chatto & Windus.

Plowright, W. (1982). The effects of rinderpest and rinderpest control on wildlife in Africa. *Symposium of the Zoological Society of London,* 50, 1–28.

Pringle, J. A. (1982). *The Conservationists and the Killers.* Cape Town: Books of Africa.

Pullan, R. A. (1983). The use of wildlife in the development of Zambia. In *Natural Resources in Tropical Countries,* ed. Poi Jin Bee, pp. 267–325. Singapore: University of Singapore Press.

Roberts, R. S. (1980). African cattle in pre-colonial Zimbabwe. *NADA Annual of the Division of District Administration (Zimbabwe),* 12(2), 84–93.

Sandford, S. (1983). *Management of Pastoral Development in the Third World.* London: Overseas Development Institute, and Chichester: John Wiley & Sons.

Selous, F. C. (1895). *A Hunter's Wanderings in Africa.* 4th edn. London: Richard Bentley & Son.

Tabler, C. E. (1960). *Zambesi and Matabeleland in the Seventies. The narrative of Frederick Hugh Barber 1875 and 1877–78 and the Journal of Richard Frewen 1877–1878.* London: Chatto & Windus.

Taylor, R. D. & Walker, B. H. (1978). Comparison of vegetation use and herbivore biomass on a Rhodesian game and cattle ranch. *Journal of Applied Ecology,* 15, 565–81.

Taylor, T. A. (1984). The financial implications of quota allocations. Paper presented to a Safari Industry Workshop, Harare.

Vincent, R. & Thomas, R. G. (1961). *An Agricultural Survey of Southern Rhodesia.* Salisbury: Government Printer.

9

Commercial hunting in the Soviet Union

V. E. SOKOLOV & N. L. LEBEDEVA

Abstract

Commercial hunting in the Soviet Union is of substantial economic value. Over 550,000 ungulates are harvested annually providing to the state 250,000 hides and over 10,000 tonnes of meat for a collective value of 25 million roubles (Rb) (approximately US$ 40 million). The most important commercial species, representing 20% of total game production, are moose (*Alces alces*), saiga (*Saiga tatarica*), wild boar (*Sus scrofa*), red deer (*Cervus elaphus*), roe deer (*Capreolus capreolus*), musk deer (*Moschus moschiferus*), and wild reindeer (*Rangifer tarandus*). Game management is vested in the Agro-Industrial Complex with hunting areas allotted to national, cooperative, and social organisations for commercial and/or sporting purposes. The vast hunting grounds of the Soviet Union, their mosaic pattern, and diverse fauna create favourable conditions for successful development of integrated game management.

Introduction

Wildlife production continues to play a significant role in the national economy of the Soviet Union. Hunting and trapping still involve a considerable number of people and, after some decline in the 1930s and 1940s, the importance of hunting and trapping is again increasing. In the Far North, most males of ethnic minorities are involved in hunting. These people (Evenks, Dolgans, Kets, Khants, Mansis, Nenetses, Nganasans, and others) account for 50–80% of the harvest. Profits from hunting represent 52–58% of household income (Karelov & Semkin, 1981; Zabrodin *et al.*, 1983).

Prospects for commercial hunting are good. There are large areas of relatively low productivity – tundra, taiga, deserts, and mountains – where only range livestock husbandry and game management can be practised. Fortunately, the process of destruction of natural environments was checked early. Although logging, oil production, and other activities still seriously threaten wildlife, the Soviet Union has avoided the profound destruction of natural resources experienced by some densely populated countries of Europe, Asia, and Africa.

Certain aspects of the country's social development are relevant to commercial hunting. Foremost is the retention of hunting traditions by a number of ethnic minorities as well as by the native Russian population. Another factor is the concentration of people in large settlements which minimises human impacts. A third feature is the active conservation policy that led to the recovery of such commercially important species as moose, wild reindeer, saiga, and wild boar.

This chapter describes the organisation of commercial ungulate hunting in the Soviet Union and its economic importance with a view to evaluating future prospects and directions. It is based on the personal experience of V. E. Sokolov as a consultant to game management organisations and also on a survey of major Soviet studies.

Administrative organisation

In the Soviet Union, game management is an important branch of the national economy (Table 9.1). Its main objective is development of commercial hunting through rational use of national game resources, establishment of hunting grounds, recovery and enhancement of game resources, and regulation of populations. Funds allocated for the census, conservation, and propagation of game animals increased from Rb 9.3 million in 1978 to Rb 26.3 million in 1982 (Visyaschev, 1979, 1983).

Game resources are national property. Wild mammals and birds inhabiting hunting grounds, irrespective of the latter's affiliation, belong to the State Game Fund, whose use for hunting is conditional on the observance of laws stipulating the objects, dates, and methods of hunting.

The State Land Fund contains 2231 million hectares including 1717 million hectares of hunting grounds which include open lands, forest land, and aquatic areas serving as habitat for mammals and birds and usable for game management. About half of the country is forested.

General control of game management is vested in the Agro-Industrial Complex, the national body in charge of agriculture, agricultural processing, and game management. The *Tzentrosoyuz* (Consumers' Cooperative Society) is the main purveyor of game products. In the Soviet Republics, game management agencies either function independently as part of republican governments (as in the Russian Soviet Federation of Socialist Republics, RSFSR), or are affiliated with national ministries and committees of forestry, ministries of agriculture, or committees of nature conservation. In the RSFSR, game management is vested in the Main Board for Game Management and Reserves affiliated with the RSFSR Council of Ministers and its local bodies. In most Soviet Republics, game management is administered by republican boards of the National Game Inspection Agency (Dezhkin, 1976).

State and cooperative commercial game management units, reindeer-breeding state farms and collective farms, state game units (reserve and non-reserve), grounds of hunters' societies, including those of the military society and *Dynamo* sports society, are the basis of the socialist system of game management in the Soviet Union (Table 9.2). Hunting grounds are allocated for a period of no less than ten years with the option of renewal. Although many game management units are dedicated to sport hunting, 75% of hunting grounds are commercial units, the bulk of them in Siberia (Melnikov, 1982).

Table 9.1. *Zonation of the Soviet Union with special reference to game management units.*[a]

	Average area (million hectares)	Costs of game management (Rb/1000 hectare)	Hunting grounds per hunter (ha/hunter)	Mean output (1971-75)	
				Fur (Rb/1000 ha)	Meat (kg/ha)
Tundras and forest-tundras	7.8	1.5	163.7	10.8	0.2
North taiga		1.8	155.4	8.6	0.8
Middle taiga	4.1	3.8	101.7	18.5	1.9
South taiga	2.8	4.3	118.9	27.5	2.1
South montane forests	1.2	10.6	32.7	37.0	2.6
Forest-steppe and steppes (Siberia and Kazakhstan)	0.6	22.7	59.4	10.4	1.0
Mixed forests and forest-steppe (European USSR)	0.2	4.8	63.0	18.1	2.5
Deserts and semi-deserts (Kazakhstan and Central Asia)	1.0	28.3	16.6	150.0	—

[a]Stakhrovsky (1978).

Commercial hunting is conducted largely by state and cooperative game management units (Table 9.2). In 1982, the number of state management units reached 100, and the value of yearly production exceeded Rb 90 million. Output from state commercial management units increased from Rb 112 million during 1966–70 to Rb 450 million during 1981–85.

Sport hunting units which provide opportunities for public hunting are distinguished somewhat conventionally from commercial hunting units, since they also purvey wildlife products in most Soviet Republics. In the RSFSR alone, hunting societies purvey annually about 4000 tonnes of meat from wild ungulates (Sitsko, 1981).

By the 1980s, over 3 million hunters were registered. Of this number, over 300,000 people purveyed game products, including 20,000 staff hunters and 80,000–100,000 seasonal hunters. Game management units are now staffed by 3200 trained specialists (Poletzky, 1982).

Licences are of two types: (1) sport – purchased by the hunter who is entitled to the entire product, and (2) commercial – animals killed are purveyed to the state and the hunter is paid for the hide and meat. To achieve a selective harvest, licences are priced according to the age and sex of the animal. For example, a licence to kill a bull moose during the rut is as high as Rb 150, which makes it possible to limit the kill of best sires (Visyaschev, 1983).

Game management is a planned economic activity. Plans developed by

Table 9.2. *Number of game management units in 1980.*

Commercial Units	
State-owned	
Commercial game management units	103
Reindeer-breeding farms	187
Cooperatives	
Fur-bearer units	128
Reindeer-breeding farms	25
Sport Hunting Units	
State-owned	
State game management and reserved game management units	119
Sylvicultural game management units	334
Hunters' societies	
Republic hunter societies	6629
Military hunter societies	265
Dynamo sports society	67

game management bodies and by voluntary societies of hunters and fishermen in the republics, territories, and regions outline measures for improving conservation and recovery of the fauna, habitat, and other measures aimed at intensifying game management. Game reserves (*zakazniki*) play an important role in conserving game resources and augmenting productivity of hunting grounds.

Efficiency is greatly increased by hunting in teams. Depending on the local conditions, an organised team of 10–20 hunters is allotted hunting grounds for periods of no less than five years, with renewal options. A team is assigned a harvest plan setting the quotas both in terms of revenue and products. All-terrain vehicles reduce the hunting period, enhance product quality, reduce labour requirements, and minimise loss of meat (Lukashenko & Maksimuk, 1981; Poletzky, 1982). In the Kalmytzky commercial unit, 15 teams (5–7 people in each) removed about 80,000 saigas in one month. The best teams obtained over 90 tonnes of meat. Every team has a truck, motorcycle, and repeating rifles. High labour productivity enabled the Kalmytzky unit, where over 80% of the total output is accounted for by saigas, to obtain over Rb 400 thousand profit (Feldman, 1979).

Populations and harvests

The Soviet Union is home to 21 indigenous wild ungulate species (including 60 subspecies) and two introduced species. Licences are issued to harvest such valuable species as moose, maral (*Cervus elaphus sibiricus*), Manchurian wapiti (*C. e. xanthopygus*), European red deer (*C. e. elaphus*), Caucasian red deer (*C. e. maral*), acclimatised sika (*Cervus nippon*), reindeer, roe deer, mouflon (*Ovis musimon*), west-Caucasian tur (*Capra ibex caucasica*), Siberian ibex (*Capra ibex siberica*), saiga, wild boar, and musk deer. But only eight are commercially important: moose, saiga, wild boar, red deer, roe deer, musk deer, reindeer and, to a lesser extent, Siberian ibex (Fig. 9.1). These ungulates yield over 20% of gross game production which is estimated at Rb 200 million.

The annual harvest in the Soviet Union is 550,000–600,000 wild ungulates, yielding over 10,000 tonnes of meat and 250,000 hides, for a total value of Rb 25 million (Kolosov, 1975; Bannikov *et al.*, 1982). In 1981, 74,700 moose, 94,400 wild reindeer, 8400 deer of other species, 96,900 saigas, 54,600 wild boars, and 24,100 roe deer were harvested providing 14,600 tonnes of meat. From this harvest, 75% of moose, 61% of deer (excluding reindeer), 39% of wild boars, 31% of roe deer, and almost all reindeer and saigas were purveyed to the state (Poletzky, 1982).

Moose

Due to long-term bans on hunting, subsequent harvest limitations, and widespread rejuvenation of forests, moose populations in the 1950s reached commercial levels (Lavov, 1978). By 1983, the population in the RSFSR reached 750,000 and exceeded 50,000 in other republics.

Moose popluations can increase by 45% annually, but the actual increment varies with protection regime, population structure, predator populations, and habitat conditions. In reserved areas, increments vary from 25–33% and in unreserved areas 15–20% (Bannikov, 1978; Bannikov *et al.*, 1982). Harvest quotas are set at 25–30% of the total population if the population density is close to the optimum (3–4 head/1000 hectares) (Bannikov *et al.*, 1982). Total harvests have increased steadily from 30,500 animals in 1970 to 74,700 in 1981. With an average carcase weight of about 150 kg (Filonov, 1983), this represents recent yields of about 11,000 tonnes.

In most parts of the RSFSR, moose hunting opens on 1 October and ends on 15 January, though biologists maintain that the season should be opened earlier and the duration reduced (Yazan, 1975). Special licences are issued by game agencies.

The most widely used hunting methods are battue hunting, driving, stalking (Kheruvimov, 1969), baying with dogs, and calling during the rut.

Fig. 9.1. Distributions of the main ungulates harvested in the Soviet Union.

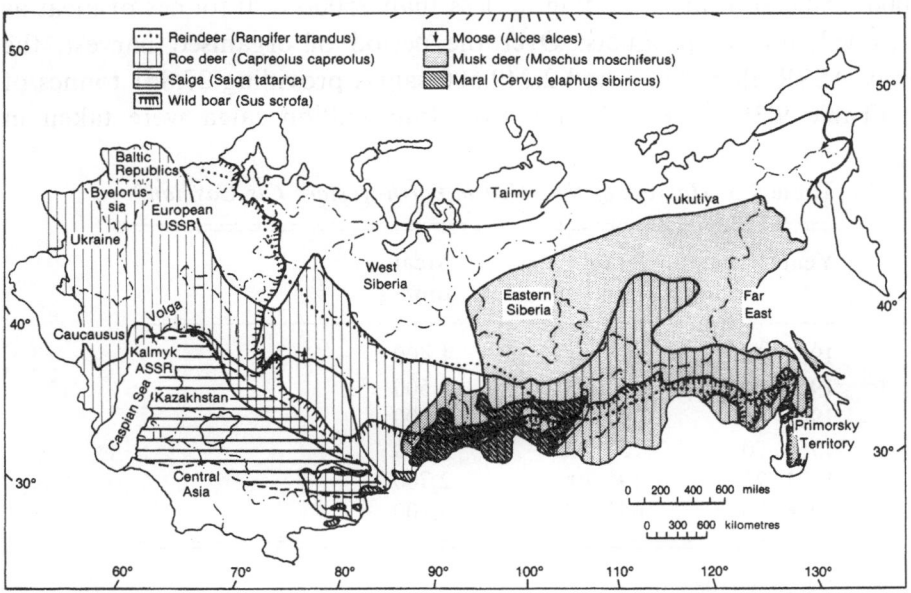

Newer methods of harvesting moose rely on helicopters (in taiga regions) and corrals set up on migration routes (Yazan, 1975; Filonov, 1982; Sitsko, 1983).

Saiga

By the 1920s, the saiga was on the verge of extinction. Owing to subsequent conservation measures, the saiga population and range expanded rapidly. In the early 1950s, the hunting ban was lifted and planned exploitation was initiated, beginning on the east bank of the Volga in 1951 and Kazakhstan in 1954.

The saiga population is characterised by sharp fluctuations under the impact of natural and anthropogenic factors. At its peak in 1957–60, the population reached two million at an average density of 8 head/1000 hectares. The saiga forms four distinct populations: one in the Kalmyk ASSR and the rest in Kazakhstan (Fadeev & Sludsky, 1982). In 1980, the overall stock of the Western Precaspian was about 400,000 head and that of Kazakhstan was 600,000–800,000 head.

The annual herd increment is 40–50% of the spring population in favourable years, but during droughts and epizootics it is no higher than 10–20%. During population increases, 20–25% of the stock can be harvested without detriment, while during a decline, hunting should be either totally banned or conducted on a limited basis (Zhirnov, 1982b). With a total population of 1–1.3 million, conflicts with agriculture are only moderate. Such a population can be managed with an annual removal of 200,000–250,000 head; providing no less than 2500–3120 tonnes of meat as well as valuable by-products. Over the period of organised harvest, the Kalmyk ASSR alone harvested 2,249,500 saigas providing 27,851 tonnes of meat (Table 9.3). In Kazakhstan over four million saiga were taken in

Table 9.3. *Harvest of the saiga in north-western Cis-Caspian.*[a]

Years	Total numbers	Meat tonnes
1951–55	103,000	1,400
1956–60	612,000	8,500
1961–65	347,000	4,800
1966–70	52,000	700
1971–75	195,700	2,700
1976–80	653,100	9,100

[a]Zhirnov (1982a)

1956–81, yielding over 70,000 tonnes of meat (Table 9.4) (Fadeev & Sludsky, 1982).

Saigas are harvested from 1 October–1 December. Longer hunting seasons increase the percentage of barren females (Fadeev & Sludsky, 1982). The main harvest methods are night hunting from vehicles, or by driving to a shooting line. Occasionally herds are driven into transportable corrals (Zhirnov, 1982a). When the harvest is well organised, primary processing is done in special abattoirs; carcases are skinned, dressed, and hung for several hours in a tent or spaceous room for cooling. Before being sent to the consumer, the carcases are subjected to close sanitary inspection (Zhirnov, 1982a).

Wild reindeer

Wild reindeer have been successfully re-established in a number of regions. In some areas, they are sufficiently abundant to allow harvesting (Razmakhnin, 1986). The two largest populations are in the Taimyr Peninsula (575,000) and between the Yana and Indigerska Rivers in Yakutiya (120,000) (C. Gates, personal communication). The total population of wild reindeer in 1980/81 was 924,000. The annual increment is 15%.

Originally, reindeer hunting met the needs of northern peoples and provided meat for state and collective fur farms. Commercial harvests began in 1961 when the population was estimated at 200,000. During the first decade (1960–70), 96,100 reindeer were taken. By 1971, the population of wild deer exceeded 600,000 head. In 1981/82, 98,581 were removed, providing 5103 tonnes of meat (Table 9.5).

In 1971, the Taimyrsky commercial game management unit was estab-

Table 9.4. *Saiga numbers and harvest output in Kazakhstan.*[a]

Years	Spring numbers	Animals harvested	Meat (tonnes)	Hides	Revenue (Rb)	Profit (Rb)
1956–60	690,000	257,700	4,608	102,000	4,443,000	320,000
1961–65	610,000	723,000	13,050	628,000	12,915,000	910,000
1966–70	570,000	411,000	7,438	359,000	8,046,000	1,078,000
1971–75	1,052,000	1,595,000	27,355	1,459,000	30,382,000	10,702,000
1976–80	588,000	957,000	15,551	882,500	22,607,000	9,507,000
1981	820,000	193,100	3,038	185,600	3,808,000	2,145,000
Total	—	4,136,800	71,040	3,616,100	82,201,000	24,662,600

[a]Fadeev & Sludsky (1982).

lished to harvest the largest population. Over the first five years, this unit harvested over 78,000 reindeer and increased its output from Rb 353,000 to Rb 3,365,000. During 1980–86, the harvest increased 2.5-fold and large amounts of high quality meat were provided to the native population of the North.

The optimum stock of the Taimyr population is estimated at 475,000 with a harvest quota of 70,000 (Dragan & Karelov, 1983). In Taimyrsky, the mean annual harvest rates are 31%, profitability 20%, and product value Rb 336,000/1000 hectares (Tarasov, 1979).

Hunting of wild reindeer has always been associated with the largest concentrations of populations during spring or autumn migrations. Two major types of hunting are employed: mechanised hunting based on new technologies, and traditional hunting based on age-long experience. Mechanised methods of hunting using river vessels at water crossings, aircraft, and snowmobiles are more effective. The traditional methods of hunting employ dog or deer teams, stalking with a tame deer (*uchug*), killing animals at water crossings and migration routes, etc. (Nagretsky, 1976).

Large-scale commercial hunting of wild reindeer is patterned after traditional methods of hunting – spearing the deer at water crossings during mass autumn migrations. More females than males are taken, since adult males and females with calves leave summer grounds on different dates (Novikov, 1983). The optimum sex structure is one male to 2.5–3 females. In the Taimyr, the ratio is 1:1.9 which results in a considerable reduction in the population's growth potential (Syroechkovsky, 1982). Hunters also stalk, drive, or hide at sites used predictably during migration. Larger-scale harvest necessitates artificial concentrations of reindeer herds with selective removal of different age and sex classes. Permanent or portable enclosures, or nets (Novikov, 1983) are effective.

Table 9.5. *Harvests of wild reindeer in the Soviet Union.*[a]

Years	Planned harvest	Actual harvest	Meat (tonnes)
1971–75	228,012	140,054	4,609
1976–80	439,320	352,162	13,783
1980–81	101,872	75,675	2,892
1981–82	117,050	98,581	5,103

[a]Novikov (1983).

Wild boar

Over the last 30 years, wild boar have expanded their numbers and range. By the late 1960s, wild boar had widely populated the northern and eastern European Soviet Union, having long crossed the northern limits of its historical range. One of the reasons is the establishment of a wide network of game management units with rigorous protection, intensive management, and transplants. In parts of its range (Byelorussian SSR, Lithuanian SSR, etc., populations increased 18–30-fold (Bannikov *et al.*, 1982). The total population in the Soviet Union is roughly 350,000–400,000 with densities varing from 0.5–60 head/1000 hectares.

Under favourable conditions, populations may double over a single breeding season. Populations with densities exceeding 10 head/1000 hectares are tolerant of periodic harsh winters; if some portion of the population perishes, numbers fully recover in one or two years (Fadeev, 1973, 1978). The recommended harvest is 25–40% of the stock of each age and sex class. Annual national harvests are in the order of 50,000.

Throughout the RSFSR, the wild boar is hunted from 1 October–15 January. Hunting laws provide for harvests outside the above period where there are conflicts with agriculture and forestry. Methods of harvesting are diverse. Hunters may stalk (especially in winter), or lie in wait at natural or supplemental feeding sites, watering sites, croplands, or baits (fish or muskrat carcasses). Battue hunting with dogs has become popular. Shotguns (12–16 gauge) are normally used, the cartridge being loaded with slugs or buck-shot.

Roe deer

Due to overhunting, roe deer declined to dangerously low levels by the 1920s–30s and their range was reduced to scattered patches. In response to conservation measures, the isolated parts of the range merged. Over the last two decades, the roe deer population in the RSFSR has stabilised at the relatively low level of 300,000–500,000 animals. The population of European roe deer grows, if only at a slow rate, while numbers of Siberian roe are declining. Nationwide, roe deer number 630,000–730,000 (Sokolov & Danilkin, 1981; Danilkin, 1982). Densities range from 0.1–170 head/1000 hectares, with an optimum of 20–100 head/1000 hectares of suitable habitat (Danilkin, 1985).

Roe deer are highly productive in good habitats. The biological increment in the Soviet Union is about 20% which permits an annual kill of 100,000–150,000. The harvest is much greater than officially registered. The harvest does not exceed 0.8–6.8% of the population, although according to existing quotas, 10–15% of the stock may be removed. Currently in Estonia, 22% of the stock may be removed. Over the last 20 years, the annual harvest of the roe deer has ranged from 13,000–52,000, (about 5% of the total stock). The weight of a roe carcase averages 20 kg (Lavov, 1978).

Red deer

Most of the country's red deer (130,000–150,000) are located in Siberia and the Far East; the European Soviet Union and the Caucasus are populated by 50,000–70,000 for a national stock of about 200,000. The European subspecies is presently not hunted, since most are concentrated in reserves. A wider and more regular distribution of the eastern subspecies has allowed more extensive exploitation (Bannikov & Pivovarova, 1983). With rigid conservation measures including population control and supplemental feeding during harsh winters, populations increase rapidly and attain high densities. In reserves, densities approach 25–30/1000 hectares which is five to ten times that outside reserves.

The increment on hunting grounds with different protection and biotechnical regimes is 4–18% (Bannikov, 1979). In recent years, the harvest of maral and Manchurian wapiti has been 3000–5000 head.

Stags are hunted for velvet antlers from 25 May–15 July, and all age and sex classes, except calves, are hunted for meat and hides from 1 October–15 January.

Musk deer

The estimated current stock of musk deer is 100,000. Population densities range widely from 0.5–80 head/1000 hectares, averaging 6/1000 hectares. Protective management sharply increases densities and the annual increment may approach 64% (Bannikov *et al.*, 1978, 1982). In some populations, 6–8/1000 hectares can be harvested. The harvest in recent years has been about 5000 animals, representing 5% of the stock. In most areas, musk deer are hunted from 1 September–1 March. They are taken with dogs, by stalking, or by driving.

Economic aspects

In recent years, meat production of ungulates in the state commercial game management units of the RSFSR has stabilised at 3400 tonnes. Up to 72% of this production is purveyed by the Taimyrsky and Kalmytzky units. In 1983, 50,000 reindeer were harvested on Taimyr and 40,000 saigas in Kalmykiya, providing 2460 tonnes of meat. The remaining 98 state commercial game management units purveyed only 940 tonnes. Reindeer account for up to 58% of total production, saiga 14%, moose 20%, and other species (roe deer, wild boar, Manchurian wapiti, etc.) 8% (Tarasov & Feldman, 1984).

In 1982, the commercial value of wildlife products was Rb 32.68/1000 hectares, (Estonia Rb 351, Latvia Rb 338, Lithuania Rb 263, Kazakhstan Rb 131, Beylorussia Rb 65, the Ukraine Rb 53, Russia Rb 27). Production from individual units has exceeded Rb 700; e.g. the Nyaumyatis unit of the State Committee for the Conservation of Nature of the Lithuanian SSR (Rb 1046), the Kedaiteiskoye unit of the Republican Society of Hunters and Fishermen of the Lithuanian SSR (Rb 736), the Zagorskoye State Sylvicultural–Game Management Unit of the Agro-Industrial Complex (Rb 2600) (Visyaschev, 1983).

The saiga is currently harvested by the Astrakhansky and Kalmytzky commercial game management units in the Kalmyk ASSR. The saiga management units are highly profitable; those in Kazakhstan receive Rb 1.5–2.5 million in net profit annually. During 1957–67, commodity output of the saiga harvest in the north-western Cis-Caspian averaged Rb 270/1000 hectares and when the harvest was most intense (1957–59) this index was Rb 350–530/1000 hectares (Zhirnov, 1982b). The total value of the entire saiga production exceeds Rb 100 million (Zhirnov, 1982b).

Product values

In the Tomsk Region, the retail price of moose meat is about Rb 1.20/kg. Yazan (1975) estimated the value of a single moose at Rb 270–275, taking into account the retail price of the meat, hide, and internal organs. Dezhkin (1975) estimated a corresponding value of Rb 350 and, adding the recreation value, Rb 500–550. The fixed prices of wild boar and saiga meat are Rb 1.05–2.50/kg and Rb 0.50/kg, respectively. In the Amur Region, roe deer venison is sold at Rb 1.50/kg.

The most valuable product from marals is velvet antlers. The drug obtained from the antlers is a tonic stimulating cardiac activity, augmenting nitrogen and carbohydrate metabolism, and lessening fatigue. Saiga horns are valued for manufacturing various drugs. From their outer cover, the drug *saitarin* has been prepared for use as a tranquilliser, anti-convulsive, and anaesthetic agent (Zhirnov, 1982a).

The secretion from the abdominal gland of the musk deer is regarded as the best natural musk and is used in medicine and perfumes as a fixative. The dry weight price of the musk gland is Rb 35 for 31 g (Ustinov, 1978). An adult musk deer yields up to 10 kg high-quality meat and 20 dm^2 perfect suede.

Prospects for increased production

Harvest can be considerably increased through rational utilisation and intensified game management. Current off-takes of wild boar are about 14% despite a recommended annual harvest of 25–40%. In the Baltic Republics alone, the sustained harvest of moose and wild boar is at least 50% of the stock. The existing population of wild boar in the RSFSR could support an annual harvest of 36,000–40,000, providing another 700 tonnes of high-quality meat (Visyaschev, 1979).

According to Sokolov & Danilkin (1981), the stock of roe deer in the Soviet Union could attain several million head. At a mean population density of 10 head/1000 hectares, the stock of roe deer could be increased to 10 million, almost 15 times its current abundance. Harvest rates can be up to 30% of the total stock based on all age and sex classes (Bannikov *et al.*, 1982).

Given a mean density of 10 animals/1000 hectares for roe, maral, or Manchurian wapiti, and a minimum admissible removal quota of 10% of autumn stock, these species can provide 100–120 kg meat/1000 hectares of hunting grounds in southern Siberia. Planning the population sex structure would raise this index to 150–180 kg, rendering it 10–12 times higher than current off-take. This is well demonstrated by the experience of management units of the RSFSR Glavokhota (state game management agency) in a number of regions. The maximum yield of meat in Estonia and Latvia reaches 400 kg/1000 hectares (Melnikov, 1982) while in the Primoriya Territory it is 8.8–10.4 kg/1000 hectares. With supplemental feeding, maral densities of 3–10 head/1000 hectares may be increased to 8–20 head/1000 hectares (Sviridov, 1977).

Increased harvest of reindeer could be attained with more structured management, modern technology, harvesting forest populations, later harvest dates for tundra populations, and artificial concentrations of reindeer herds with selection for various age and sex classes. The above programme can be accomplished using permanent and portable drift fences, and by driving deer into enclosures or nets (Novikov, 1983). Maximum sustainable exploitation of the Taimyr–Evenk herd can provide about 2500 tonnes of meat, 60,000 hides, and 240,000 kamus (leg skins) (Zabrodin, 1979). In addition, commercial reindeer breeding is of much importance to nature conservation, since it promotes conservation of the unique tundra ecosystems, preventing their unnecessary transformation and destruction (Dezhkin & Men'kova, 1981).

Exploitation of different age classes of ungulates also is expedient. When harvesting calves is prohibited and their numbers are high, winter mortality can be high (Filonov, 1983). When young are harvested, meat yield is reduced during the first two years, and is thereafter compensated by increased production of offspring (Melnikov, 1982). If harvest of calves is permitted, the output of meat of wild ungulates in the Soviet Union may be increased by 3000 tonnes (Poletzky, 1982). With further improvement of conservation, predator control, and more rational exploitation of ungulates, their harvest and meat yield can be increased several times.

A mean yield of meat of 100 kg/1000 hectares of hunting grounds is a realistic goal in the European Soviet Union, southern zone of Siberia, and the Far East. The highest values are reached in Estonia, where in recent years over 400 kg meat/1000 hectares have been produced. Most commercial game management units have considerable potential in exploiting available resources.

The vast hunting grounds of the Soviet Union and diverse fauna provide favourable conditions for successful development of game production based on intensive management of natural resources. Renewable resources are of great importance to the national economy now, and will play an even greater role in the future.

References

Bannikov, A. G. (1978). Wild ungulates of the USSR, their use and conservation. In *Voprosy ratzionalnogo ispolzovaniya i okhrany dikikh kopytnykh i drugikh zhivotnykh*, Vol. 97, pp. 3–6. Sbornik nauchnikh trudov Moskovskoi Veterinarnoi Akademii imeni K.I. Skryabina, Moscow (in Russian).

Bannikov, A. G. (1979). The modern state of the Bokhara deer. In *Okhrana prirody Turkmenii*. Vol. 5, Ashkhabad (in Russian).

Bannikov, A. G. & Pivovarova, E. P. (1983). The red deer in the USSR. In *Biologicheskie osnovy ispolzovaniya i okhrany dikikh zhivotnykh*, pp. 34–39. Sbornik nauchnikh trudov Moskovskoi Veterinarnoi Akademii imeni K. I. Skryabina, Moscow (in Russian).

Bannikov, A. G., Pivovarova, E. P. & Fandeev, A. A. (1982). The principles of rational use of some ungulate species in the USSR. In *Structure, tovarno-tekhnologicheskiye svoistva i ratzionalnoe ispolzovanie zhivotnogo syrya i produktov zhivotnovodstva*, Vol. 125, pp. 88–99. Sbornik nauchnikh trudov Moskovskoi Veterinarnoi Akademii imeni K. I. Skryabina, Moscow (in Russian).

Bannikov, A. G., Ustinov, S. K. & Lobanov, P. N. (1978). Musk deer in the USSR. In *Voprosy ratzionalnogo ispolzovaniya i okhrany dikikh kopytnykh i drugikh zhivotnykh*, Vol. 97, pp. 6–36. Sbornik nauchnikh trudov Moskovskoi Veterinarnoi Akademii imeni K. I. Skryabina, Moscow (in Russian).

Danilkin, A. A. (1982). Roe-deer resources in the USSR and their rational utilization. In *Promyslovaya teriologiya (voprosy teriologii)*, pp. 108–115. Moscow: Nauka (in Russian).

Danilkin, A. A. (1985). Roe deer in the USSR: measures for conservation and population recovery. *Okhota i okhotnichiye khozyaistvo*, 3, 16–18 (in Russian).

Dezhkin, V. V. (1975). Eco-economic foundations of game management. In *Okhotovedeneviye*, pp. 3–106. Moscow (in Russian).

Dezhkin, V. V. (1976). A word about hunting. In *Seriya Chelovek i priroda*, Vol. 9, pp. 8–82. Nauchnye osnovy okhotnichiyego khozyaistva. Moscow: Znanie Publishers (in Russian).

Dezhkin, V. V. & Men'kova, N. V. (1981). Comparative aspects of exploitation of the resources of wild and farm ungulates. In *Ekonomika, organizatziya i ispolzovanie resursov okhotnichiego khozyaistva RSFSR*, pp. 7–40. Sbornik nauchnikh trudov tsentral'hoi nauchno–issledovatel'skoi laboratorii Glavokhoty RSFSR. Moscow (in Russian).

Dragan, A. V. & Karelov, A. M. (1983). A resource estimate of hunting grounds of the Enisei north. In *Okhrana, ratzionalnoe ispolzovanie biologicheskikh resursov Krainego Severa*, pp. 228–35, Moscow (in Russian).

Fadeev, E. V. (1973). Number dynamics of the wild boar (*Sus scrofa*) in European Russia. *Zoologichesky Zhurnal*, 52, 1214–9 (in Russian).

Fadeev, E. V. (1978). Wild boar. In *Krupnye khischniki i Kopytnye zveri*, pp. 256–93. Moscow (in Russian).

Fadeev, V. A. & Sludsky, A. A. (1982). *Saiga in Khazakhstan: Ecology and Economic Importance*. Institute of Zoology, Kazakh SSR Academii nauk, Alma-Ata: Nauka (in Russian).

Feldman, M. (1979). Effectiveness of harvest should be raised. *Okhota i okhotnichie khozaistvo*, 2, 4–6 (in Russian).

Filonov, K. P. (1983). *Moose*. Moscow: Lesnaya Promyshlennost (in Russian).

Karelov, A. M. & Semkin, S. G. (1981). Refinement of the exploitation of game resources in the Extreme North region. In *Ekonomika, organizatziya i ispolzovanie resursov okhotnichiego khozyaistva RSFSR*, pp. 79–82. Sbornik nauchnikh trudov tzentral'noi nauchno–issledovatel'skoi laboratorii Glavokhoty RSFSR, Moscow (in Russian).

Kolosov, A. M. (1975). *Conservation and Enrichment of the USSR Fauna*. Moscow: Lesnaya Promyshlennost. (In Russian).

Lavov, M. A. (1978). Roe-deer. In *Krupnye khischniki i Kopytnye (Large carnivores and ungulates) (Les i ego obitateli. Forest and its dwellers)*, pp. 191–220. Moscow: Lesnaya Promyshlennost (in Russian).

Lukashenko, M. A. & Maksimuk, A. V. (1981). Preliminary results of the work of integrated teams in game management units and their role in the use of game resources. In *Ekonomika, organizatziya i ispolzovanie okhotnichiego khozyaistva RSFSR*. pp. 7–40. Sbornik nauchnikh trudov tsentral'noi nauchno–issledovatel'skoi laboratorii Slavokhoty RSFSR, Moscow (in Russian).

Melnikov, A. (1982). *Eco-economic Foundations of the Organization of Game Management of Siberia*. Doctoral thesis. Moscow (in Russian).

Melnikov, V. K. (ed.). (1981). *Foundations of Game Management*. Irkutsk: Irkutsk Agricultural Institute (in Russian).

Nagretsky, L. M. (1976). Concerning the problem of the methods of harvest of reindeer. In *Dikii severnyi olen' (biul. nauchno-tekhn Informatzii)*, pp. 55–8. Norilsk (in Russian).

Novikov, B. (1983). Problems of deer harvest. *Okhota i okhotnichiye khozyaistvo*, 2, 3–5 (in Russian).

Poletzky, V. (1982). A fuller use should be made of game resources. *Okhota i okhotnichiye khozyaistvo*, 10, 1–3 (in Russian).

Razmakhnin, V. Y. (1986). Wild reindeer in the USSR, their protection and utilization. *Rangifer, Special Issue*, 1, 347–9.

Sitsko, A. (1981). Advances and problems of game management in Russia. *Okhota i okhotnichiye khozyaistvo*, 1, 1–2 (in Russian).

Sitsko, A. (1983). Raising the standard of ungulate harvest. *Okhota i khotnichiye khozyaistvo*, 3, 1–3 (in Russian).

Sokolov, V. E. & Danilkin, A. A. (1981). *Siberian roe deer*. Moscow: Nauka (in Russian).

Stakhrovsky, E. V. (1978). Zonation of game management in the USSR. In *Sbornik nauch. tekh. informatzii (okhota–pushnina–dich)*. Vol. 61, 3–13. Vsesoyuznyi nauchno-issledovatel'skyi Institut Okhotnichyego Khozyiastva i Zverovodstva imeni Proffessora Zhitkova. Kirov: Volgo-Vyatskoe (in Russian).

Sviridov, N. S. (1977). Conservation and reproduction of the maral in the Siberia. In *Organizatziya i tekhnologiya proizvodstva v okhotnichyikh khozyaistvakh vostochnoi Sibiri*, pp. 3–9. Irkutsk: Irkutsk Agricultural Institute (in Russian).

Syroechkovsky, E. E. (1982). Wild reindeer. In *Promyslovaya teriologia*, pp. 51–71. Moscow: Nauka (in Russian).

Tarasov, S. M. (1979). Northern commercial game management units of the RSFSR Glavokhota and their prospects. In *Okhotnichiye promyslovoe khozyaistvo Severa*, pp. 19–30. Moscow: Kolos.

Tarasov, S. & Feldman, M. (1984). State commercial game management units of the RSFSR: objectives and problems to be solved. *Okhota i okhotnichiye khozyaistvo*, 12, 6–7 (in Russian).

Ustinov, S. K. (1978). Musk deer. In *Krupnye khischniki i kopytnye zveri*, pp. 230–55. Moscow: Lesnaya Promyshlennost (in Russian).

Visyaschev, G. (1979). Results of the year 1978. *Okhota i okhotnichiye khozyaistvo*, 12, 4–5 (in Russian).

Yazan, Yu. P. (1975). The problem of moose harvest. *Okhotnichiye khozyaistvo*, 10, 1–2 (in Russian).

Zabrodin, V. A. (1979). Management and conservation of wild reindeer of the Tundra populations. In *Problemy okhrany i khozyaistvennogo ispolzovaniya resursov dikikh zhivotnykh Eniseiskogo Severa*. Vol. 26. Novosibirsk: Trudy Krainego Severa SO Vaskhnil (in Russian).

Zabrodin, V. A., Karelov, A. M., Dragan, A. V., Semkin, S. T. & Popov, G. R. (1983). Rational technology of harvest and recovery of game stocks of the extreme north. In *Okhrana i ratzionalnoe ispolzovanie biologicheskikh resursov Krainego Severa*, pp. 221–8. Moscow: Kolos (in Russian).

Zhirnov, L. V. (1982a). *Revival*. Moscow: Lesnaya Promyshlennost (in Russian).

Zhirnov, L. V. (1982b). Modelling of the dynamics of a saiga population. In *Izvestiya Timiryazevskoi selskokhozyaistvennoi Akademii*, Vol. 5, pp. 157–61. Moscow: Kolos (in Russian).

SECTION D

Herding

L. M. BASKIN

In the vast expanses of tundra, dry steppes, deserts, and mountains, range livestock husbandry is the only realistic means of developing agricultural potential. Although pastoralism is a widespread production mode for domestic animals, it is less common for semi-domestic and wild species. Except for experimental attempts to integrate oryx (*Oryx beisa*) into traditional pastoralist systems in East Africa (King *et al.*, 1977), herding is largely limited to reindeer (*Rangifer tarandus*) as described in the following chapters. The purpose of this introduction is to establish the basic ethological principles of herd control.

The most important behavioural characteristics governing biocenotic relationships (interactions of animal populations with plant cover) in extensive herding systems relate to *ethological* and *spatial* population structure. The former comprises such characteristics as group composition, hierarchy, leaders, distances between animals, and the phenomenon of territoriality. The latter reflects the use of home range and seasonal pastures, each including several home ranges and migration routes.

Ethological structure

Herd control is based largely on the defensive behaviour of animals, but other forms of activity are used. Responses in which one urge predominates contrast with those operating when several urges are combined. In the former case, the animal responds in a much more automatic 'need–stimulus–action' sequence (Craig, 1981). The range of possible reactions is predictably narrowed by properly streamlining behaviour. For example, a quietly grazing herd will not follow an artificial leader – they must be disturbed so they gather into a compact mass.

In order of decreasing magnitude, motivations form the following hierarchy: defensive–sexual–maternal–feeding–comfort. A given motivation is easier to replace by a stronger motivation than a weaker one, although exceptions occur. For example, sexual excitement and associated aggressiveness often eliminate shyness of people. Feeding dominance in reindeer during spring attenuates responses to calls and threatening gestures. At that time, only setting a dog makes it possible to group the herd. Changes in the activity and attenuation of reactivity also can be accomplished by driving animals to moderate exhaustion.

Leaders

The tendency of herds to follow the first animal to leave the herd (leader) makes it possible to control movement by leading an artificial leader in front of the herd. Goats, asses, a dog, or the herdsman can serve as artificial leaders in sheep flocks; and riding deer serve a similar function in herds of domestic reindeer.

Many herd members are potential leaders and it is possible to determine which ones (Fig. D.1). The propensity of certain individuals (leaders) to leave the herd in a dangerous situation is related to their sensitivity to the stimulus

Fig. D.1. Selection of leader in a reindeer herd: a, reindeer are grazing quietly; b, orientation reactions to sudden danger; c, gathering into a compact group (knot); d, movement of deer on the spot; e, the leader is found and the herd is following it. Solid arrow, direction of danger; dashed arrow, direction of movement of deer; circle, calf.

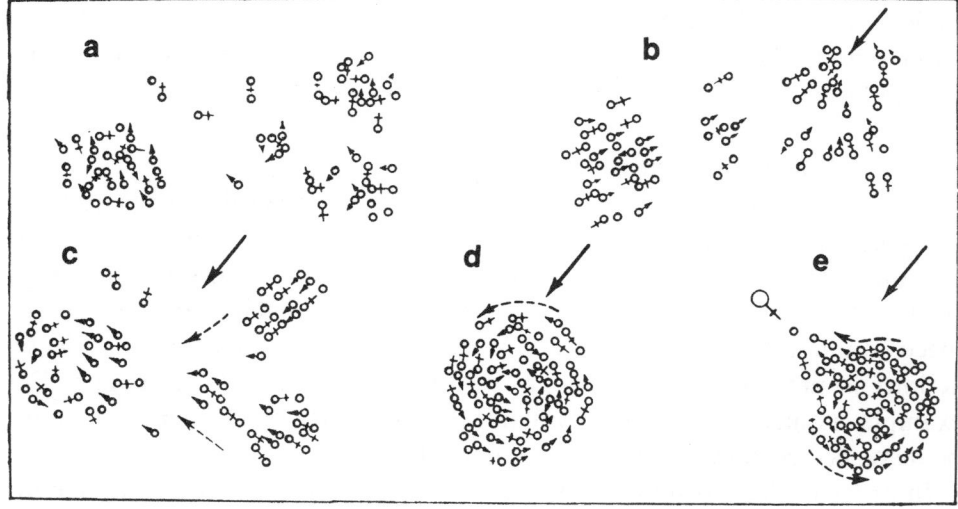

which may arise from general shyness and/or specific experience. Their propensity to leave the herd also is determined by herd size. In a large herd, animals feel safer and are better protected from blood-sucking insects.

Although any animal may leave the herd, its subsequent behaviour is labile. Young and, to some degree, adults are governed by natural reactions formed via imprinting, operant learning, and imitation. A natural reaction to danger is the tendency to escape upwind or upslope, which enhances perception of olfactory cues and visibility, respectively. Also typical is movement along visible landmarks: ravines, river beds, or even foot-tracks and the urge to return to the previous feeding site. Of course, these responses are species-specific.

It is almost impossible to manage a herd without a leader. But in some situations, herds simply may not have potential leaders. Such is the case in northern reindeer husbandry when animals are driven from herd to herd, to butchering points, etc. Some herds, such as those composed of calves or young alone, move only in a definite direction when frightened by man (upwind, upslope, to a definite place). If a single female or several riding males are added to the herd, or an adult male is led in front of the herd, the herd becomes manageable.

Experienced herdsmen understand the necessity and have the ability to identify leaders enabling them to control the herd even in unfavourable situations. Under insect attack on hot summer days, reindeer gather in a compact mass, which continually mills in a circle. At that time they are difficult to handle, and generally can only be moved upwind. However, by identifying and directing potential leaders, the herdsmen can often succeed in moving the herd downhill. The complexity of the situation is that both male and female potential leaders are normally within the herd, while the young rotate in a circle in the outer rows. Herdsmen attempt to move potential leaders to the forage, whereupon they are driven in the needed direction (Fig. D.2).

Rigid and flexible management

Rigid herd management is applied in capturing wild or domestic ungulates. Saigas can be captured by the *bound circle* method, in which a vehicle circles the herd to gather animals in a compact mass, permitting a man to approach within 12–20 m. Horses, sheep, and yaks are captured by gathering the herd into a compact mass, in which the defensive distance is small. Reindeer are more shy and they are captured using their urge to associate with a larger herd.

Rigid management is often used when animals must be counted. Use is made of their urge to run from where they are harassed to join their companions standing quietly aside. In reindeer, several riding deer are leashed aside from the herd. When the herd is harassed, the animals make a short cut to the tethered deer, passing through a gap where they can be easily counted.

Flexible herd management is applied in a quiet situation when different forms of activity are combined. Without ceasing grazing, animals strive to keep a safe distance from the herdsmen, responding to changes in the direction of the herd movement, etc. The relationship of different forms of activity changes continually. With satiation, food reactions of the animals are inhibited, and their defensive behaviour varies with the action of the herdsman.

Using different techniques, the herdsman may affect the relationship of activity elements, enhancing some and inhibiting others. A good herdsman achieves an equilibrium of defensive and feeding reactions. With an inexperienced or inattentive herdsman, the herd either scatters freely over the pasture using areas different from those planned, or badly frightened, discontinues grazing to gather into a compact mass (Baskin, 1974).

Flexible management is very important in rotating herds which are grazing on the move. In reindeer herds, the rear and front parts of the herd differ in the number of potential leaders. Hence, the front part of the herd, more

Fig. D.2. Movement of reindeer herd by herdsmen from the rotation site during the activity of blood-sucking insects: 1, reindeer; 2, potential leader; 3, herdsman; 4, direction of driving by herdsman; 5, direction of herd rotation; 6, wind direction.

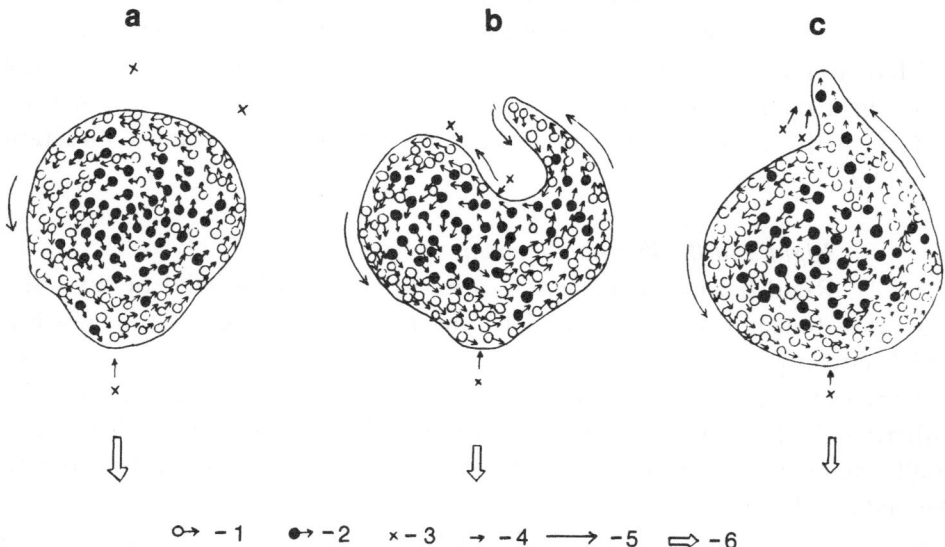

excitable and less definite in its actions, tends to rotate quite readily. The rear portion is quieter and more docile. To rotate the herd, one herdsman puts pressure on the rear portion of the herd from the side, while the other holds back the front deer causing the herd to reverse its direction. At the end of the manoeuvre, the front-line deer (normally young and leaders without calves) pass through the herd to the front rows again.

Spatial distributions
Ethological distances

Several ethological distances are relevant to herd control (McBride, 1971; Syme & Syme, 1979). *Individual distance* is two or three lengths. At this distance, signal postures are demonstrated and vocal signals are exchanged between a male and female or a dominant and a subdominant. *Herding distance* is one at which herd members coordinate their behaviour. Also at this distance, they communicate and receive signals with regard to the ambient environment and follow leaders. *Species-specific distance* is the

Table D.1. *Distance (metres) between animals during grazing and resting.*

Species	Grazing		Resting	
	Individuals	Herds	Individuals	Other species
Asiatic wild ass (*Equus hemionus*)	7.1 ± 0.9^a	45 ± 17	—	—
Domestic horses (*Equus caballus*)	5.1 ± 0.4	50 ± 11	9.5	500
Arabian camel (*Camelus dromedarius*)	7.7 ± 2.0	130 ± 45	—	3000
Reindeer (*Rangifer tarandus*)	$3.0 \pm —$	25	—	500
Yak (*Bos grunniens*)	7.6 ± 1.3	—	—	—
Saiga (*Saiga tatarica*)	5.0 ± 1.4	—	2.7 ± 0.5	600
Argalis (*Ovis vignei*)	2.6 ± 0.3	—	2.5 ± 0.3	1200
Domestic sheep (*Ovis aries*)	1.7 ± 0.3	25	1.2 ± 0.3	200

[a]Means ± standard errors.

distance at which conspecifics are distinguished and mating partners are sought. There is considerable interspecific variation in these measures (Table D.1).

Herding patterns of domestic sheep, argalis, and horses are influenced by individual distances (Fig. D.3). The scattering of animals is discontinued when the space between the neighbours on the right and on the left reaches the individual distance. However, familial and personal relationships somewhat complicate the pattern and one can often see that the arc is formed not by individual animals but by groups. Among horses, a compact mass is split up in an arc pattern into individual harems (Fig. D.4).

At the herding distance, animals best react to the manoeuvres of the man – the movements of his hands on his lasso. The species distance determines the possibility of herds (*kazakhs*) of horses and camels approaching one another without male aggression, particularly during the rutting period (Baskin, 1982).

Predicting movements

Siting corrals and searching for lost animals requires the ability to forecast behaviour. Long-term (weeks and months) forecasts are possible if the seasonal pastures of the population, major routes, and probable dates of migration are known in relation to prevailing weather conditions.

Fig. D.3. Movement of the herd in an arc pattern: a–e, sequence stages; 1, shape of the herd; 2, direction of the movement; 3, wind direction.

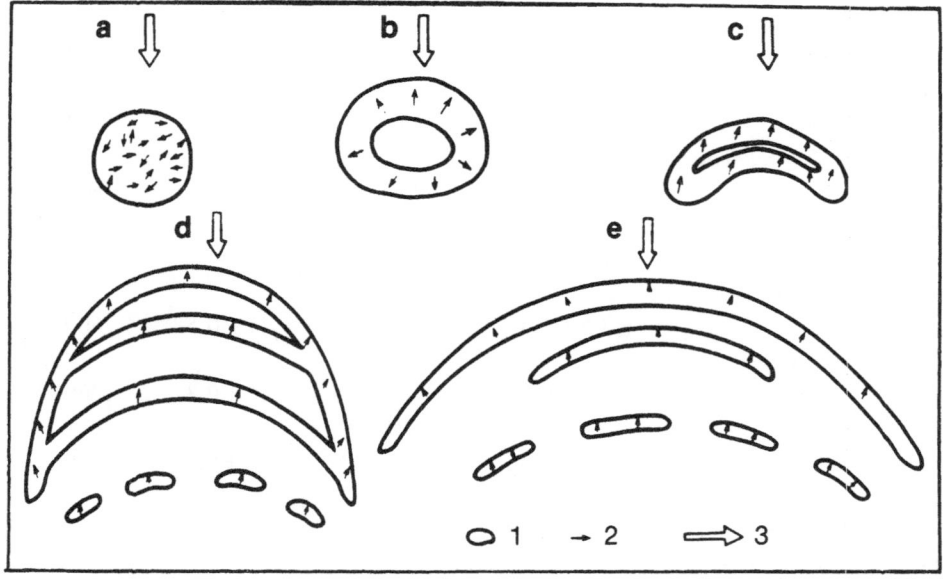

The diet of all domestic and wild ungulates of temperate and high latitudes is characterised by sharp seasonal differences in nutritional value (Kilgour & Dalton, 1984). In autumn and snow periods, they normally consume feeds which are rich in carbohydrates but poor in proteins and minerals. The growth of animals and restoration of skeletal and muscular tissues are only possible by consuming green plants. Hence, the season of feeding on green plants, particularly new growth, should be as long as possible.

In temperate and high latitudes where seasonal changes vary with latitude, altitude, or distance to the coast, the duration of green plant consumption can be extended through migration. Domestic reindeer herds start their migration northward to the coast or mountains roughly 20 days after the

Fig. D.4. Dispersal of a taboon during grazing in summer. 1, numbers of stallions; 2, outlines of definite groups; 3, the wind direction.

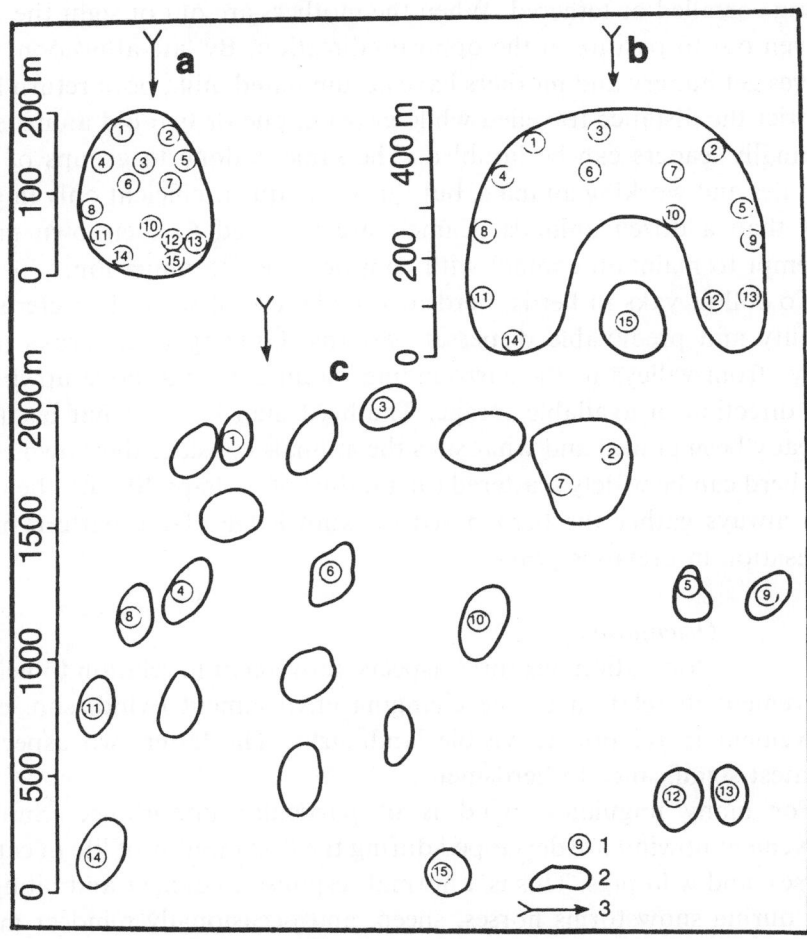

snow has melted. On summer ranges, the vegetation develops 15–20 days later than on winter ranges. In autumn, the herds return to use ranges where wilting of vegetation is delayed. The capacity for remembering habitat features and responding with migration is best developed in horses, camels, and yaks, so that when pasture is allotted correctly the herds hardly need to be tended. Domestic sheep are less likely to return on their own to the sheep-cote where they spend the night or to watering points.

Patterns of range use can be reinforced by herdsmen. In camel husbandry, a number of techniques are used to encourage animals to return on their own. It is not difficult during the watering season when natural waterbodies dry up and the camels come to the well on their own at one or two-day intervals. In spring, when the green plants contain much moisture and there are occasional ponds, camels drink only rarely. To make them return each day for milking, the females are driven out to pasture after the sucklings are fed, the calves being corralled or tethered. When the mothers are out of sight the calves are driven out to pasture in the opposite direction. By late afternoon, when the calves get hungry and mothers have accumulated milk, both return home. To restrict the distance travelled while grazing, one or two old animals who are normally leaders can be hobbled. The same is done to groups of pregnant females and working animals, but this technique is efficient only in groups of less than a dozen animals. Camels are attached to their own group and attempt to maintain contact with considerable determination.

To collect yaks in herds, herdsmen make use of their characteristic range fidelity and predictable dispersal patterns. Grazing freely, yaks invariably move from valleys to the surrounding mountains and move upwind and in the direction of available forage. The herdsman knows what pastures have already been grazed and what sites the animals consider their own. Although the herd can be widely scattered (in a radius of perhaps 40 km), the herdsman can always gather the herd based on knowledge of the pattern of pasture utilisation in previous years.

Orientation

Orientation has three aspects: movement in relation to neighbours, movement in relation to the changing environment (wind, sun, etc.), and movement in relation to visible landmarks. The latter two aspects are of greatest significance to herdsmen.

For many ungulates, wind is of particular importance. Sniffing and movement upwind are developed during the first months of life in cattle, deer, horses, and wild pigs. This is a normal response to danger and biting insects, but during snowstorms horses, sheep, and occasionally reindeer may move downwind. At lower latitudes animals may turn their backs to the sun when it becomes hot. Therefore, herdsmen often graze sheep and cows early in the morning in a southerly direction, so later, when the sun rises high, the herd will willingly turn home.

Relief, and above all slope, is of great importance. The possibility of improving observation by rising to an elevation is used by many animals – mountain sheep, goats, reindeer, camels, and many antelopes. Reactions of ungulates to visible landmarks have been long known and widely used by both herdsmen and hunters. Other things being equal, they move along river

Fig. D.5. Location of corrals based on terrain features. 1, direction of movement of deer; 2, general direction of migration; 3, 'rose of winds'; 4, rivers; 5, lakes; 6, walls of corral; 7, horizontal contours; 8, butchering.

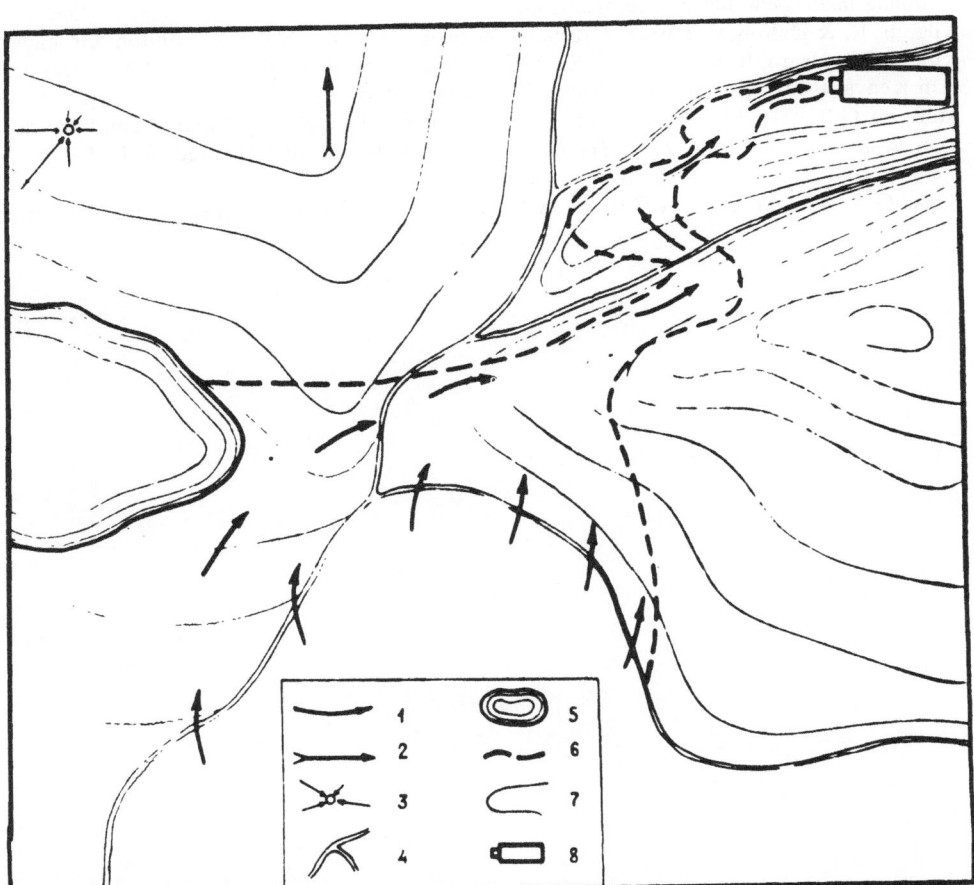

beds, forest edges, along paths or following tracks. Animals can be directed, by fences or, in open country, piles of sod, stones, or scarecrows.

Reactions of animals to visible landmarks also determine the siting of corrals (Fig. D.5). Corrals for handling reindeer should be oriented with their mouth downslope or downstream since a herd is always easier to drive upwind and uphill.

References

Baskin, L. M. (1974). Management of ungulate herds in relation to domestication. In *The Behaviour of Ungulates and its Relation to Management*, ed. V. Geist & F. Walter, pp. 530–41. IUCN publications; new series, 24. Morges: IUCN.

Baskin, L. M. (1982). Behaviour studies as a basis for horse breeding zootechnology in kazakhs. In *Production Pastorale et Societe*. Bulletin de l'equipe ecologie et anthropologie des societes pastorales, pp. 29–44. Paris: Maison des sciences de l'homme.

Baskin, L. M. (1984). A method of mammal behaviour description. *Acta Zoologica Fennica*, 171, 229–31.

Craig, J. V. (1981). *Domestic Animal Behavior: Causes and Implications for Animal Care and Management*. New Jersey: Prentice-Hall.

Kilgour, R. & Dalton, C. (1984). *Livestock Behaviour. A Practical Guide.* London: Granada.

King, J. M., Heath, B. R. & Hill, R. E. (1977). Game domestication for animal production in Kenya: theory and practice. *Journal of Agricultural Science (Cambridge)*, 90, 445–57.

McBride, G. (1971). Theories of animal spacing: the role of flight, fight and social distance. In *Behaviour and Environment. The Use of Space by Animals and Man*, ed. A. H. Esser, pp. 53–68. New York: Plenum Press.

Syme, G. L. & Syme, L. A. (1979). *Social Structure in Farm Animals*. Amsterdam: Elsevier.

10

Reindeer husbandry in the Soviet Union

L. M. BASKIN

Abstract

Over the last 50 years, the Soviet reindeer (*Rangifer tarandus*) population has been stable at approximately 2.3 million, yielding 41,900 tonnes liveweight annually. Reindeer are grazed on cooperative production units in herds of 1200–3500 head. The best performance is shown in Kamchatka with survival rates of 97%, and calving rates of over 87/100 cows. Gradual transformation of ranges has occurred due to depletion of lichen cover and replacement of moss/lichen tundras with shrub/grass tundras. In response, diets and foraging ecology of reindeer have changed. The main systems of management include close herding, free camp herding, loose herding, seasonal herding, and fenced systems. To intensify reindeer breeding, the proportion of females should be increased to over 60% and autumn calves should be harvested. Separate management of breeding and fattening herds is recommended.

Introduction

State regulation of northern reindeer husbandry was initiated in the Soviet Union in the 1930s and 1940s. At that time, there were 2.2 million domesticated reindeer (Syroechkovsky, 1975). Schools of reindeer management existed (Vasilevisch & Levin, 1951; Baskin, 1970) but many features of reindeer husbandry were determined by ethnic traditions, the capabilities of private owners, and herd sizes. The transition to the socialist economy resulted in the establishment of large cooperative units. Today, an average cooperative supports 12,000 head, though reindeer are managed in herds of 1200–3500 (Koshelev, 1985).

Populations and productivity

Reindeer husbandry is important in the economies of 22 ethnic minorities in the Soviet Union as a source of sustenance and clothing, and

Table 10.1. *Numbers of domestic and wild reindeer in the Soviet Union (× 1000)*[a]

Region	1961		1965		1970		1975		1976		1977		1978		1981		1984	
	dom.	wild	dom.	wild	dom.	wild	dom.	wild	dom.	wild	dom.	wild	dom.	wild	dom.	wild	dom.	wild
Altai	—	—	—	0.1	—	0.1	—	—	—	—	—	1.0	—	—	—	0.4	—	0.4
Amur	12.6	2.0	15.0	2.0	17.3	2.0	16.1	—	14.7	—	14.9	8.0	15.0	—	15.7	4.15	13.5	2.4
Archangelsk	169.9	2.5	174.6	3.6	193.2	9.7	180.0	15.4	183.2	15.4	183.4	20.0	181.8	20.0	192.7	20.0	154.0	6.5
Buryat ASSR	3.9	3.0	4.2	3.0	5.0	5.0	4.8	6.5	4.2	7.0	4.3	7.0	2.6	7.0	1.9	3.7	—	3.5
Irkutsk	7.0	11.0	7.5	11.0	5.4	12.5	4.5	7.0	4.1	20.0	4.0	20.0	4.1	14.5	1.2	20.0	2.6	14.0
Kamchatka	158.8	—	174.8	5.5	178.9	8.0	174.8	7.0	175.6	6.0	176.3	5.5	177.0	5.0	177.8	7.3	166.6	4.5
Karel ASSR	—	1.0	0.2	3.0	0.1	1.1	—	1.0	0.003	8.0	—	3.5	—	6.0	—	6.6	—	5.0
Kemerovo	—	0.9	—	1.0	—	1.0	—	—	—	0.8	—	0.8	—	0.5	—	0.6	—	0.6
Komi ASSR	136.7	4.0	135.9	2.0	147.1	4.3	100.6	5.0	104.9	5.5	107.7	6.0	97.6	6.0	120.6	6.5	65.3	5.0
Krasnoyarsk Territory	136.8	103.0	188.1	120.0	164.8	340.0	142.9	540.0	143.3	543.0	144.2	521.0	126.1	500.0	102.6	523.0	87.0	575.0
Magadan	579.3	3.5	710.3	3.5	738.5	3.5	685.4	3.5	696.7	4.5	710.5	14.0	711.9	14.0	705.4	16.0	626.0	11.0
Murmansk	74.2	7.5	77.1	10.5	81.9	22.0	65.6	6.4	65.2	1.5	65.5	1.5	64.8	1.5	66.4	2.0	62.4	2.7
Omsk	—	0.1	—	0.6	—	0.6	—	0.8	—	0.9	—	0.8	—	0.75	—	0.8	—	0.8
Perm	—	0.4	—	0.4	—	0.6	—	0.22	—	0.55	—	0.6	—	0.55	—	0.25	—	—
Sakhalin	13.1	5.0	15.1	5.0	14.2	3.0	13.4	2.8	13.3	2.8	12.5	3.0	10.7	3.0	3.4	4.4	3.6	5.0
Sverdlovsk	—	—	—	1.0	—	1.0	—	—	—	—	—	1.0	—	—	—	1.0	—	1.0
Tomsk	0.6	7.0	0.3	5.0	0.1	3.5	—	0.7	0.06	2.0	—	2.5	—	2.5	—	1.8	—	5.0
Tuva ASSR	—	10.0	9.1	6.0	11.6	2.5	10.5	3.5	11.1	3.5	11.8	3.5	13.1	4.0	10.0	4.0	9.6	4.0
Tyumen	400.5	—	431.3	10.0	480.9	16.0	440.4	30.0	441.3	28.0	433.8	27.5	411.2	26.0	427.3	21.4	249.6	19.0
Khabarovsk	42.3	5.0	52.4	10.0	50.3	10.5	51.5	11.0	51.6	11.0	49.3	13.0	41.9	13.5	48.4	17.0	44.6	12.0
Chita	11.6	—	14.1	5.6	18.3	6.7	17.3	6.0	15.9	6.0	15.2	5.0	15.6	5.0	14.6	7.5	12.4	7.2
Yakutskaya ASSR	340.6	30.0	359.3	100.0	356.3	160.0	371.9	170.0	377.8	170.0	379.2	170.0	375.9	260.0	371.8	170.0	319.3	250.0
TOTALS	2087.9	195.9	2369.5	308.8	2463.9	613.6	2279.7	836.9	2302.9	836.5	2312.6	835.2	2249.3	889.8	2259.8	875.8	1816.5	943.8

[a]Syroechkovsky (1986).

reindeer meat supplies all northern settlements. In the 1980s, a population of 2.3 million reindeer produced 41,900 tonnes liveweight annually (Mironov, 1982). From 1961 to 1984, the number of domestic reindeer decreased while wild reindeer increased five-fold (Table 10.1).

The Magadan Region, Yakutskaya ASSR, and the Yamal–Nenetsky Autonomous Region account for 60–65% of the total domestic herd (Zabrodin, 1980). The largest-scale reindeer husbandry is in Kamchatka, where typical survival rates are 93% and calving rates exceed 81/100 cows (Khomenko, 1982). On the best ranches, these indices attain 97.4% and 87.4/100 cows. Reindeer ranches are highly profitable; returns on investment were 17–34% at Chukotka in 1976–80 (Garbaretz & Zadorin, 1982).

Murmansk Region

Until the 19th century, reindeer husbandry by the Sami was poorly developed, but wild reindeer were abundant. In the late 19th century, the Komi immigrated from the Pechora River area bringing 5000 domestic reindeer and initiating large-scale husbandry. Labour shortages led to loose or semi-loose systems of management.

Arkhangelsk Region and Komi ASSR

The tundras of the Russian Plain have for ages been used by the Nenetz and Komi for reindeer husbandry. The latter group has its origins not in neighbouring Scandinavia, but more likely in the Altai and the Sayan. In the late 19th and early 20th centuries, most reindeer (up to 150,000) were possessed by the Komi. But as they entered other branches of the economy, reindeer husbandry was left largely to the Nenetz.

Western Siberia

In the 11th century, the Nenetz from the Sayan Mountains settled in Western Siberia, introducing domestic reindeer husbandry. Reindeer husbandry by the Nenetz of the Yamal–Nenetsky Autonomous Region is concentrated in the tundra where a population of 400,000 head has been managed over the last 100–200 years. The Komi also have played an important role contributing their method of managing deer in large herds.

For people living in the taiga (the Mansi–Khanty and Selkupy), reindeer husbandry is less important than hunting. Their reindeer population never exceeded 100,000 head, declining in recent decades to 55,000 partly because of increasing anthropogenic pressures, particularly burning of lichen ranges.

Central Siberia

Reindeer husbandry was developed in Taimyr as late as the 18th century by the Dolgan people. Reindeer husbandry combined elements from the Nenetz (use of dogs for herding, large-scale herd (*taboon*) system), and the Evenki (riding deer in summer and harnessing them to sleds in winter). The Dolgan also practised hunting and fishing. Another nationality of Taimyr, the Nganasan, specialised in hunting and only adopted reindeer husbandry as late as the 19th century, when wild reindeer became scarce. The population of domestic deer has never been large, and despite efforts to rationalise production, it has progressively declined.

The main feature of this region is the abundance of wild reindeer which currently number about 500,000. In Taimyr, a large game management unit was established which harvested about 600,000 animals between 1971–81 at a rate of 60,000–90,000 annually (Syroechkovsky, 1986). Reindeer husbandry remains the second most important branch of the economy. There are 87,000 domestic deer which are important both in the production of meat and hides and in transportation.

Yakutiya

Prior to the 17th century, Yakutiya was rich in wild reindeer. Its Yukaghir people and to a lesser extent the Tungus lived mainly by hunting wild reindeer, procured primarily by spearing at river crossings during migrations. A decline of wild reindeer resulted in persistent famine and decimation of the Yukaghir population by the late 19th century. These people are reluctant to adopt reindeer husbandry. In contrast, the Yakuts, who populated Yakutiya in the early 18th century, took to breeding domestic deer whose numbers increased from 27,500 in 1865 to 380,000 by 1980 (Syroechkovsky, 1986). During the last decade, the wild reindeer population has recovered and now approaches the numbers of domestic reindeer.

Far Northeast

In the 16–18th centuries, the Far Northeast was heavily stocked with wild reindeer which sustained the Chukchi until they turned to reindeer husbandry in the 18th century, taking domestic deer from the Koryaks of Kamchatka as the treasures of war. In a short time, they established a large-scale reindeer husbandry system. Inland Chukchi populations, using both domestic and wild reindeer, increased ten-fold over 150 years to 5000 people.

Until the late 19th and early 20th centuries, the Even people combined reindeer husbandry with hunting. Their taiga husbandry system is character-ised by somewhat smaller herds (dozens of animals) and semi-loose management (the herdsman regularly checks and drives the herd closer to home). From the late 19th century, part of the Even population began to move to Kamchatka and the northern shore of the Sea of Okhotsk.

Mountain taiga regions of Siberia and the Far East

Many nationalities of the mountain taiga regions are engaged in small-scale reindeer husbandry. Reindeer are used for transportation but hunting is the major source of income. The wild reindeer population is small, but along with domestic reindeer and other ungulates (moose [*Alces alces*], red deer [*Cervus elaphus*], musk deer [*Moschus* spp.]) provide the meat and hide requirements of local people.

Range carrying capacity

In the 1930s, the carrying capacity of reindeer range was based on estimates of the lichen resource (Sochava, 1932; Andreev, 1975). In the 1940s and 1950s, a range inventory throughout the entire North indicated that a maximum of 4 million reindeer could be supported by the 450–500 million hectares of available range (Zabrodin, 1980). However, subsequent range assessments revealed a reduction in range to 334 million hectares due to degradation of lichen cover and economic development (Zabrodin, 1980). Lichen resources are being reduced annually by 1.5–2% and, where there are also wild reindeer, by 3–4% (Andreev & Galaktionova, 1983). Moss and lichen tundras are being replaced by herb/shrub tundra and sedge/cotton-grass marshes due to changes in permafrost, runoff, and ground conditions (Utkin, 1980; Polezhaev, 1983).

Nevertheless, in many regions, populations have increased. For instance, carrying capacity on the Taimyr peninsula was estimated in the 1970s at 196,000 head, although ranges supported 450,000 domestic and wild deer (Kolpaschikov, Kuksov & Pavlov, 1983). Syroechkovsky (1975) suggested that the role of subnivean plants was underestimated since domestic reindeer can, of necessity, increase the proportion of undersnow green plants and dry grass in their diet. In the 1950s, lichens constituted 85–90% of the reindeer diet in southern Chukotka, and 60% in northern Chukotka. Currently, the respective values are 40% and 60% (Moryakov & Arefyev, 1982). Reindeer domestication has created new genotypes, such as the Khargin breed of Chukotka.

The balance of seasonal ranges is extremely important. Winter diets provide mainly carbohydrates of varying degrees of digestibility. Spring and summer diets are characterised by green shoots and leaves high in proteins (25% in willows), vitamins and minerals (28–30% in willows). In early autumn, a major dietary item is protein-rich mushrooms. In late autumn, lichens are the principal component in the diet. The seasonal pulse of plant growth in the North is brief. The period when green feeds are available can be extended by following the receding snowline northward, towards the sea, or into the mountains, where spring warming is delayed.

Husbandry
Systems of management

Several husbandry systems are employed in the Soviet Union (Pomishin, 1981). *Close herding* is the main technique in the tundra and forest-tundra zones in Nentsky, Yamal–Nenetsky and Taimyrsky National Regions, Komi ASSR, Yakutiya, Chukotka, and Kamchatka. The *free-camp system*, based on training reindeer to remain in the vicinity of human dwelling, is common in Tofalaria, Tuva, Buryatiya, Mongolia and Sakhalin. *Loose herding*, in which herdsmen periodically gather scattered animals and drive them to new ranges, is practised at Kolmyma, central regions of Yakutiya, northern Chita Region, and the Amur Region. *Seasonal herding* (normally summer and autumn) is practised on the Kola Peninsula, Khanty–Mansiisk Region, and the Krasnoyarsk Territory, Yakutiya. Reindeer are maintained in *fenced enclosures* on the Kola Peninsula, in Yakutiya, and Khabarovsk Territory.

Except for close herding and fenced systems, reindeer are raised under conditions similar to those in the wild. Thus, the detrimental effects of human disturbance, local forage depletion, and disruption of the ethological and spatial population structure are minimised. Although deer are larger and fatter, certain advantages of domestication are lost: the deer may stray, it is more difficult to use them when needed, and the incidence of barrenness is higher.

Free-camp and loose herding systems, long practised in the taiga, are characterised by heavy straying losses and mortality from wolves (up to 15–20% of calves). In the autumn, it is difficult to retain reindeer near camps, because they scatter in search of mushrooms and bulls split the herd into harems. Fences separating winter and summer ranges and blocking passage between river basins have facilitated management in the taiga for a very long time.

Use of fences for the maintenance of reindeer in spring/summer seasons is widely recommended (Syrovatsky, 1979; Mukhachev, 1984; Koshelev, 1985) since it encourages a fuller use of lichens – up to 75–80% of the forage supply, the remainder being green forage. Potential forage supplies are estimated as the annual natural increment of forage plants (10% lichens,

30% green plants) and seasonal availability (65% in spring, 45% in summer, 80% in autumn, and 70% in winter).

The major problems with enclosures include high material and labour requirements during construction, technological difficulties in mountain and remote areas, and rapid destruction of ranges. A possible solution is the use of enclosures during a limited season when reindeer losses are high or when supplemental feeding or range rotation are practised.

Herding

Control of reindeer is based on reinforcing inherent herding instincts. Close cooperation of individuals within the herd and well-developed imitation reactions result in the submission of the will of an individual to the herd. The herding distance, i.e., the distance at which animals respond to one another's behaviour, averages 15–50 m but may be up to 500 m. It is a species-specific space representing the distance at which animals can distinguish conspecifics.

The minimum number of deer which can be grazed as a single herd is about 35 (Baskin, 1970). The normal herds of small private owners in which the deer are not excessively excited is 80–100 head. The optimum herd is 1500 head. However, in tundra regions of Chukotka and Kamchatka, many herds contain 3000–3500 animals.

Western and eastern systems of reindeer husbandry are different. The western system, best represented by the Nenetsky, is characterised by somewhat smaller herds, and the use of dogs and riding reindeer. The eastern system, best represented by the Chukchi and Koryaks, is characterised by managing large herds on foot without dogs.

Population structure

Sex structure of commercial herds is about one male to 18 females with part of the herd comprised of castrated males trained for draught (6% in tundra and 20–30% in taiga regions).

In the 1950s, breeding cows did not exceed 42–45% of the population. Since then, increasing the percentage of females has become an important and highly effective practice. Baskin (1964) demonstrated the possibility of raising the proportion of cows to 54% by harvesting 5-month-old calves. To date, some ranches have up to 60% breeding cows. For a herd of 10,000, up to 2700 are killed each autumn, including 1700 yearlings (Trishin, 1982).

Among other recommended practices are the segregation of females from bulls and young, which corresponds to the natural separation of young before the next calving. It is also useful to maintain separate breeding and fattening herds. Normally, there are three breeding herds and one fattening herd. Each herd requires different management practices for feeding, watering, and regulation of activity rhythms (Baskin, 1970).

Herd health

Only after the transition to the socialist economy did veterinarian services influence the reindeer industry. Such diseases as anthrax and foot-and-mouth disease have been exterminated. The only disease that still presents a problem is necrobacillosis. The causative agent is normally a commensal of the reindeer, but it becomes pathogenic in summer when hoofs are injured or when reindeer graze on clay soils. Despite antibiotics, sulfa preparations and surgery of suppurated hooves, mortality may be considerable in hot years. Warbles were once an important problem. In spring, a single deer can accumulate 300 warble larvae causing extensive damage to the hide. Autumn injections of organophosphate drugs have essentially solved this problem (Solomakha, 1977; Solomakha & Borozdina, 1980).

Genetic selection

Beginning in the 1970s, large-scale work started on inter-population differences of domestic reindeer and establishment of breeds. To date, five breeds have been distinguished: Murmansk, Nenetzy, Evenki, Chukotka, and Sayany (Kinash, 1968; Pomishin, 1981). The Chukotka is bred for meat production, and is short with a barrel-like body, and dark-coated. The contrasting Sayany is bred for riding and is the largest (bulls attain shoulder heights of 115 cm, and weigh up to 150 kg).

The future

Reindeer husbandry is an important branch of livestock breeding in the Soviet Union. Until recently, domestic herds have been the major method for utilising forage resources of the tundra and forest-tundra, as well as the mountain-taiga regions of Siberia and the Far East. However, since the 1960s, increasing populations of wild reindeer have resulted in the economic recovery of hunting. In Taimyr and Yakutiya, hunting wild reindeer equals or exceeds reindeer husbandry in scale, although the economic yield is lower since 30–40% of the domestic reindeer are harvested annually, while hunting yields only 15%.

The choice is not simply wild versus domestic reindeer, or hunting versus husbandry. Domestic reindeer husbandry provides guaranteed meat, hides, and animals for transport. Management of domestic deer in large herds can lead to loss of condition and obviates use of small inaccessible portions of range. The necessity for supervision limits herd mobility. In contrast, wild deer make better use of ranges, including small inaccessible ranges, they are more mobile, use subnivean forages to a greater extent, and are somewhat more adaptable to environmental disturbance.

There is general agreement that both reindeer hunting and husbandry serve the interests of ethnic minorities. After some period in which natives left reindeer husbandry for other branches of economy, the process has reversed. The tradition and harmony between herding or hunting reindeer and folk culture and lifestyle again attract young herdsmen. This trend enhances confidence in further progress of reindeer husbandry.

References

Andreev, V. N. (1975). The state of forage base of reindeer husbandry and the problems of utilisation of ranges by wild reindeer. In *Dikiyi severnyi olen v SSSR*, pp. 68–79. Moscow: Sovetskaya Rossya (in Russian).

Andreev, V. N. & Galaktionova, T. F. (1983). Reindeer ranges in the Yakutskaya ASSR and problems of their use by wild reindeer. In *Dikiyi severnyi olen*, pp. 108–21. Moscow: Nauka (in Russian).

Baskin, L. M. (1964). Regulation of herd structure – the key aspect of zootechnical work. *Magadansky Olenevod*, 13, 16–7 (in Russian).

Baskin, L. M. (1970). *Reindeer Ecology and Behaviour*. Moscow: Nauka (in Russian).

Garbaretz, B. V. & Zadorin, V. I. (1982). Team-shift organization of labour in reindeer husbandry. *Magadansky Olenevod*, 34, 17–20 (in Russian).

Khomenko, N. D. (1982). New in Koryak reindeer husbandry. *Magadansky Olenevod*, 34, 15–7 (in Russian).

Kinash, A. D. (1968). *Morphological Properties and Meat Production of Reindeer of the Polar Urals*. Moscow: Kolos (in Russian).

Kolpaschikov, L. A., Kuksov, V. A. & Pavlov, B. M. (1983). Ecological substantiation of maximum numbers of the Taimyr population and wild reindeer. In *Ecologiya i Ratzionalnoe Ispolzovanie Nazemnykh Pozvonochnykh Severa Srednei Sibiri*, pp. 3–14. Novosibirsk: Vaskhnil (in Russian).

Koshelev, M. (1985). Agrarian complex: state and prospects. *Severnye Prostory*, 1, 20–5 (in Russian).

Mironov, P. E. (1982). Northern reindeer husbandry of the Russian Federation. *Magadansky Olenevod*, 34, 6–10 (in Russian).

Moryakov, V. A. & Arefyev, A. A. (1982). Carrying capacity of reindeer ranges of the Magadan region. *Magadansky Olenevod*, 34, 22–4 (in Russian).

Mukhachev, A. D. (1984). *The Maintenance of Reindeer in Enclosed Pastures in Evenkia*. Novosibirsk: Vaskhnil (in Russian).

Polezhaev, A. N. (1983). The state and prospects of studies on the problems of rational use of ranges in reindeer husbandry (Northern Far East taken as an example). *Nauchno-Tekhnichesky Byulleten SO Vaskhnil*, 47, 22–3 (in Russian).

Pomishin, S. B. (1981). *Problem of Breed and its Refinement in Reindeer Husbandry*. Yakutsk: Yakutsk Publishers (in Russian).

Sochava, V. B. (1932). Tundra studies and reindeer lichen management. In *Sbornik po Olenevodstvu, Tundrovoi Veterinarii i Zootekhnike*, pp. 18–52. Moscow: Vlast sovetov (in Russian).

Solomakha, A. I. (1977). Results of studying baitex as a means of early chemotherapy of oedemogenosis. *Nauchnye Trudy Nauchno-issledovatelskogo Instituta Selskogo Khozyaistva Krainego Severa*, 24, 72–84.

Solomakha, A. I. & Borozdina, N. I. (1980). The biochemical mechanism of toxic action of phosphororganic pesticides on the larvae of warble fly of reindeer. In *Puti Iintensivnogo Razvitiya Olenevodstva na Krainem Severe*. pp. 79–85. Novosibirsk: Vaskhnil (in Russian).

Syroechkovsky, E. E. (1975). The problem of wild reindeer in the USSR at the present state. In *Dikiyi severnyi olen v SSSR*, pp. 14–52. Moscow: Sovietskaya Rosiya (in Russian).

Syroechkovsky, E. E. (1986). *Reindeer*. Moscow: Agropromizdat (in Russian).

Syrovatsky, D. I. (1979). *Technical Recommendation on the Design of Ranches and Size of Fences for Maintenance of Reindeer in Taiga and Montane Taiga Zones*. Moscow: Kolos (in Russian).

Trishin, M. K. (1982). Optimisation of the structure and herd rotation. *Magadansky olenevod*, 34, 29–30 (in Russian).

Utkin, V. V. (1980). Changes of reindeer ranges in the Polar Urals and Polui forest-tundra under the effect of grazing and their rational utilization and protection. In *Selskoye Khozyaistvo Krainego Severa*, 3, Olenevodstvo, pp. 20–2. Magadan: Magadan Publishing House (in Russian).

Vasilevisch, G. M. & Levin, M. G. (1951). Types of reindeer breeding and their origin. *Sovetskaya Ethnografia*, 1, 76–83 (in Russian).

Zabrodin, V. A. (1980). Refinement of the technology of reindeer husbandry. In *Puti Intensivnogo Razvitiya Olenevodstva na Krainem Severe*, pp. 3–9. Novosibirsk: Vaskhnil (in Russian).

11

Reindeer husbandry in Fennoscandia

SVEN SKJENNEBERG

Abstract

Semi-domestic reindeer belong to one subspecies (*Rangifer tarandus tarandus*) but there is some genotypic variation between mountain and forest types, the former being most common. In Norway and Sweden, herds move up to 400 km between summer and winter pastures. In Finland and in some parts of Sweden, herds remain on the same forest ranges year round. In Norway and Sweden, ownership of reindeer is an exclusive right of the Sami people, whereas in Finland this is extended to all residents of the reindeer husbandry area. The total population (winter stock) in Fennoscandia is 900,000 with a calf crop of approximately 315,000. The annual harvest of approximately 235,000 animals provides an average of 6970 tonnes of meat. The reindeer population is increasing, causing heavy pressure on pastures which have been progressively restricted by developments such as roads and hydroelectric impoundments. Modern herding techniques use all-terrain vehicles and fences to control animal movements. To increase output and minimise pressure on winter pastures, the harvest regime is shifting to younger cohorts, including autumn calves.

Introduction

Wild reindeer roamed over much of Fennoscandia for thousands of years and were hunted by Sami, Finnish, and Germanic tribes. The first evidence of reindeer husbandry appears in an account from about AD 890 by the Norwegian chief Ottar who told the court of King Alfred of England that he 'owned' 800 tame 'unbought' reindeer (Vorren & Manker, 1957). Of those, six were used as decoys for catching wild reindeer. More recent primary sources question such an early origin of tame reindeer in Fennoscandia. It is now believed that reindeer husbandry developed among the Sami only after 1600 following the near extermination of the wild reindeer from overhunting. The combined hunting or trapping of reindeer and fur-bearing animals along with fishing in both marine and fresh water encouraged nomadism which gradually changed into herding semi-domestic reindeer (Vorren & Manker, 1957).

There are many theories about the domestication of reindeer (Skjenneberg, 1984). Although animal habituation developed gradually from a hunting culture, many features have been adopted from the husbandry of domestic cattle and horses. This is reflected in the many terms in reindeer husbandry with origins in the prehistoric Nordic/Germanic language (Wiklund, 1918).

Specialisation followed with some Sami continuing to follow their herds to seasonal pastures. Other Sami settled, combining agriculture with fishing, trapping, and keeping of small reindeer herds. Reindeer herds of the nomads gradually increased in size and herding changed from a family occupation to a group concern. Today, a relatively small number of Sami are involved in reindeer husbandry. Most live like their contemporaries, earning a living from 'modern' occupations.

Finnish and Germanic tribes did not adopt nomadic herding. Today in northern Finland, and to a small degree in Sweden and Norway, holders keep small numbers of reindeer as an important supplement to their household income.

Reindeer husbandry is currently practised throughout northern and mountainous parts of the Scandinavian penninsula and Finland (Fig. 11.1). Numbers of reindeer total almost 781,000 (Table 11.1).

Reindeer pastures

Topographic relief in Fennoscandia provides a variety of reindeer pastures. In western Scandinavia, the mountains often rise sharply from the sea and are exposed to an oceanic climate with heavy precipitation. The eastern part of the peninsula is situated in a rainshadow, offering good conditions for ground lichens which are the main winter food of reindeer. Luxuriant summer vegetation is mostly found in western Norway and on valley slopes in mountainous parts of Sweden and northern Finland. The distribution of different plant communities dictates seasonal migrations and consequently the system of herding.

Where reindeer husbandry is practised in northern Finland, the landscape is a fine-grained mixture of bogs, forests, and lower hills and mountains, providing all types of reindeer pasture within a small area.

Productivity and quality of pastures vary with districts, winter weather, and herd size. Lichens in winter pastures are usually the limiting resource, especially in years with difficult snow conditions. In the coniferous forests, beard lichens (*Alectoria, Usnea*) on old trees provide important emergency feed when heavy crusted snow makes ground lichens inaccessible. However, as reindeer grow in summer, lack of green vegetation may severely limit the meat output, as happens in some parts of Norwegian Finnmark. Likewise, withering of herbaceous vegetation towards the end of summer may cause deterioration of animal condition if the herds are kept too long on autumn pastures.

Fig. 11.1. Distribution of reindeer husbandry in Fennoscandia.

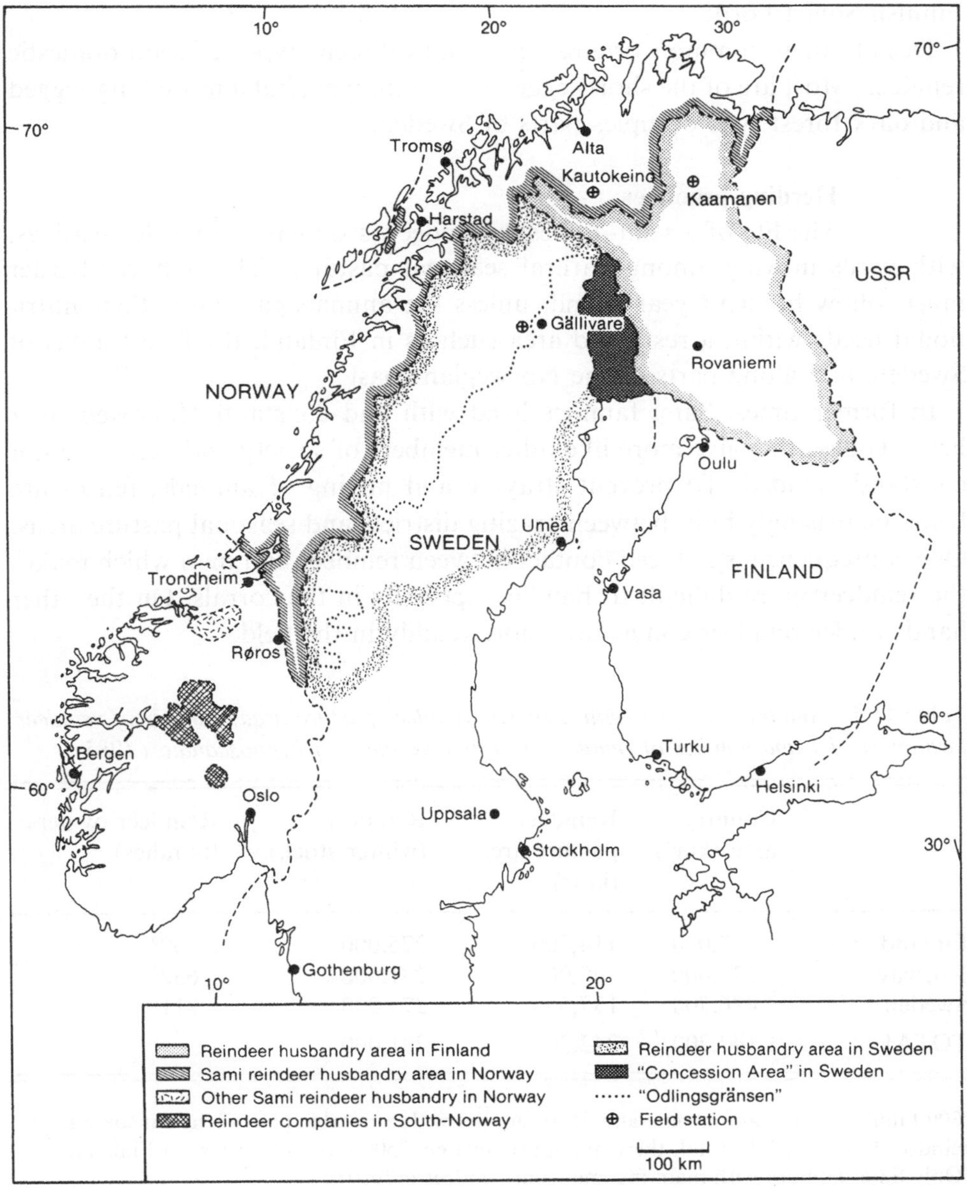

Reindeer husbandry area in Finland
Sami reindeer husbandry area in Norway
Other Sami reindeer husbandry in Norway
Reindeer companies in South-Norway
Reindeer husbandry area in Sweden
"Concession Area" in Sweden
"Odlingsgränsen"
⊕ Field station

100 km

Reindeer types

All semi-domestic reindeer in Fennoscandia belong to the subspecies *Rangifer tarandus tarandus*, and are descendants of wild reindeer which are now reduced to 35,000 in the mountains of southern Norway (Krafft, 1981). However, in eastern Finland, a small population of *R. t. fennicus* has re-established after being extirpated in Finland for about 30 years (Sulkava, 1980). The population immigrated from Soviet Karelia, adjacent to the Finnish/Soviet border.

Despite their common origins, there are different types of semi-domestic reindeer. Most are of the short-legged mountain type, but a more long-legged and dark forest type occupies parts of Sweden.

Herding practices

The life of a semi-domestic reindeer is similar to its wild relatives, with herds moving among natural seasonal pastures. The reindeer herder must follow his herd year-round, unless the animals can cover their nutritional needs within a restricted area such as in Finland, the forest areas of Sweden, and along parts of the Norwegian coast.

In former times, Sami families lived with and constantly followed their herd. Today, they live more like other members of society and herds are not constantly tended. To prevent straying and mixing of animals, fences are being increasingly built between grazing districts and seasonal pasture areas. One consequence is reduced contact between reindeer and man which makes the reindeer more difficult to handle, especially in the corrals. On the other hand, wilder reindeer congregate more readily in the field.

Table 11.1. *Country area, net semi-domestic reindeer pasture area, number of reindeer (winter stock) and number of reindeer owners (families) in Fennoscandia (1985).*

	Country area (km^2)	Reindeer pasture area (km^2)	Reindeer (winter stock)	Reindeer owners (families)
Finland	337,000	114,700	275,000	2,300[a]
Norway	324,000	90,000	231,000	632[b]
Sweden	486,700	137,500	275,000	811
TOTAL	1,147,700	342,200	781,000	—

[a]800 families with main income and 1500 families with a notable part of their income from reindeer husbandry. In total, there are approximately 7500 reindeer owners in Finland.
[b]Only Sami families with main income from reindeer industry.

Winter

Reindeer are tended during winter to prevent straying and mixing with other herds. In heavy winters, herds can mix easily as reindeer range widely to find food. If the pastures are locked by heavy snow or ice, the herd must be moved to better pastures or given emergency feeds. The latter alternative is complicated, laborious, and entails some risk if the herd is not accustomed to concentrates.

Spring

Movement from winter to spring/summer pastures begins in April before snowmelt prevents travel by snowmobile. In parts of northern Norway, a few herders still use sledges pulled by reindeer.

Often herds are split before or during spring migration. Pregnant females may be taken to the calving grounds while the rest of the herd remains on winter pasture for a few more weeks. The success of calving, which starts in early May, has a heavy influence upon the economy of reindeer husbandry. Calving grounds must combine adequate pasture and topography providing good nutrition, shelter, and protection from predators such as foxes (*Vulpes vulpes*), lynx (*Lynx lynx*), wolverine (*Gulo gulo*), and eagles.

Summer

When calves are strong enough in July, roundup and corralling of the herds begin; a stressful time both for herders and reindeer. Calves are often caught with a 'vimpa', which is like a strong fishing pole with a loop at the far end. A new type of corral has been introduced in which calves are separated from their mothers in a 'bridge' system, making it much easier to identify mother–calf relations.

A registered earmark (Fig. 11.2) is cut into the ears of all calves. The same system is used in all three countries, which exchange mutual information among the registrants. Thus, the owner may be found even if the reindeer happens to stray across a border. If the owner is unknown, it may be marked by signs cut into the chest fur. There are special rules for the handling of unmarked calves. More recently, marking with colour-coded plastic buttons has been introduced to facilitate selection of breeding animals.

Autumn

Autumn is the time for three important activities: slaughtering, breeding, and returning to winter ranges. Much of the slaughtering is done before the rut because the meat of stags develops a disagreeable taste. The rut usually starts at the end of September and lasts for about three weeks. This may be followed by a second slaughter before the herds are moved to winter pastures. Nearly all slaughters are conducted in special reindeer slaughter houses or in mobile abattoirs equipped with clean water and other necessities for hygienic handling of meat. All marketed meat is subjected to veterinary inspection.

The main cohort for slaughter is 1.5-year-old males (Table 11.2). Formerly, many of the males were castrated and kept for years, thus using the winter pastures without contributing significantly to meat yields. To improve efficiency, most breeding males are not kept past 2.5 years and the number is minimised to the lowest ratio that will still ensure a high pregnancy rate. In some herds, the male:female ratio is down to 1:17 of the winter stock. The average in Norway is approximately 1:8. Calves also account for a large proportion of the slaughter. In Norway, reindeer owners are financially rewarded by the state for harvesting calves.

Fig. 11.2. Some reindeer earmarks.

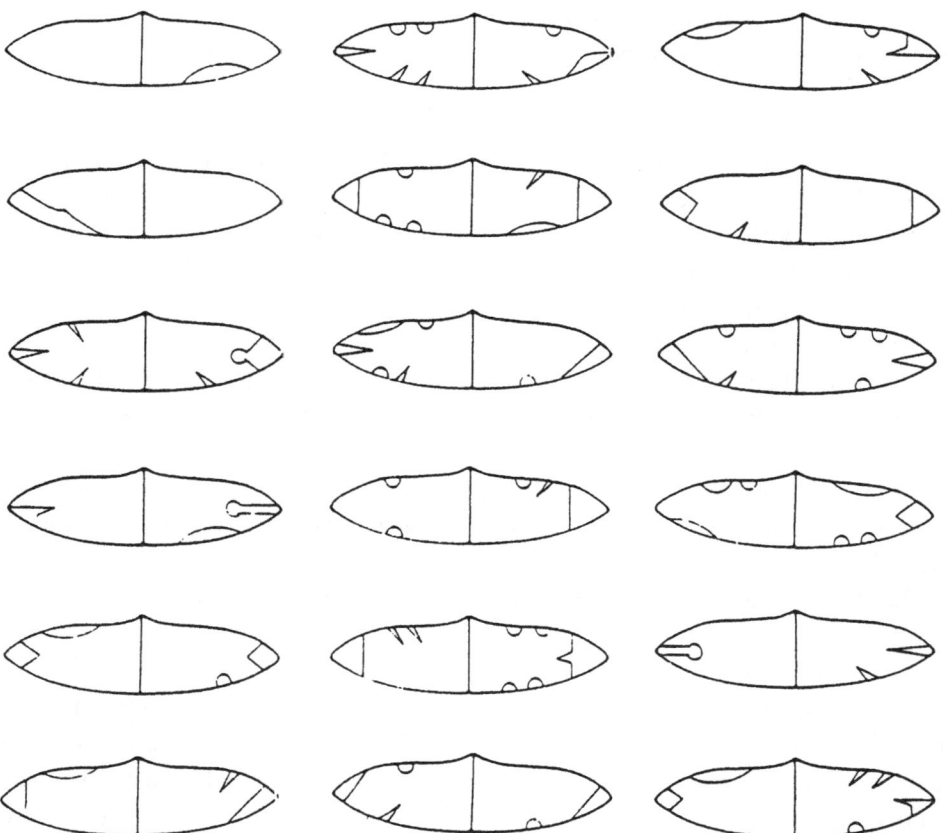

Autumn rain and fog frustrate control of the herds as reindeer stray seeking mushrooms. This increases the likelihood of mixing herds, requiring owners to spend winter months collecting their animals, a very time-consuming task and stressful to both the herders and the animals.

For collecting and moving reindeer, dogs are valuable, although snow-mobiles are used in winter (Fig. 11.3). Airplanes and helicopters sometimes are used to search for reindeer, especially in the autumn when lack of snow makes snowmobiles impractical. Occasionally helicopters are used to move herds, but this has to be done with care, not to stress the animals.

Rights, ownership, and legislation
Finland
Any Finnish citizen who lives where reindeer husbandry is allowed has the right to own reindeer. The reindeer husbandry area is divided into 56 reindeer districts (*Paliskunta*), each with year-round pastures for a single herd; movement between districts is not permitted. The district is an administrative unit, supervised by a board of directors and a reindeer herder boss. The owners must be members of the *Paliskunta*. As prescribed by the Reindeer Law (1868), all districts form a union (*Paliskuntain Yhdistys*) which guides the industry and liaises with the state. Once a year, there is a meeting of reindeer owners called the 'Reindeer Parliament'. The *Paliskuntain Yhdistys* is a semi-official institution partly funded by the state in contrast to Norway and Sweden where administration is a responsibility of the State Department of Agriculture.

Organisation of herding is the collective responsibility of owners, but most

Table 11.2. *Average autumn dressed weight (kg) of different cohorts of reindeer in Fennoscandia.*

Age (years)	Dressed weight (kg)	
	Males	Females
Calves	19	17
1	29	26
2	35	30
3	40	33
5	47	38
> 5	50	40

herding is carried out by herders hired by the district. Infrastructure, such as fences, corrals, and cabins, belong to the district.

Norway

Ownership of reindeer in northern Norway is an ancestral right of the Sami people. This right allows them to graze uncultivated land, whether private or public, without allowance to the landowner. These ancient privileges cover most of the Norwegian area north of Lake Femunden ($62°15'$N). In southern Norway, some reindeer companies are owned by local residents who are licensed and required to pay landowners for use of pastures.

Sami reindeer husbandry is organised as a herding association involving several families, but all family members identify their reindeer with registered earmarks. The legislative authority (The Reindeer Husbandry Law 1978) gives reindeer owners some degree of self-determination. The Sami reindeer area is divided into more than 100 seasonal or year-round grazing districts.

The Agricultural Department is responsible for public administration. There is a Reindeer Husbandry Board for the whole country with subsidiary

Fig. 11.3. Snowmobiles have greatly eased the task of herding reindeer (Photograph by B. Persson.)

boards for the counties and districts. In addition, Sami herders have their own organisation, the Norwegian Society of Sami Reindeer Owners (*Norske Reindriftssamers Landsforbund*).

Sweden

As in Norway, ownership of reindeer is a prerogative of Sami people, except for the 'Concession Area' near the Finnish border where the system is like that in Finland (Fig. 11.1). Civil administration, based on the reindeer law of 1971, is organised in a similar way to that in Norway, except that all reindeer owners must be members of herding cooperatives controlling specific grazing districts, the so-called *Lappby*. It is also an economic unit for the members. The Sami reindeer owners are organised in the *Svenska Samernas Riksförbund*.

The rights of Sami people are rather well protected. Between a line called the 'Cultivation Border' (*Odlingsgränsen*) and Norwegian boundary, they have the right in principle to much of the income from regional exploitation. The income is invested in a special fund (*Samefonden*) for the cultural and economic benefit of the Sami.

International relations

In former times, reindeer herds moved freely across national borders in accordance with natural migrations. On the western coast of Scandinavia, the oceanic climate offers good summer pastures, whereas the best winter pastures are located east of the mountains towards the Baltic Sea.

In 1751, the first convention between Norway and Sweden was signed to allow the Sami to cross the border with their reindeer without limitation. Subsequent agreements have gradually limited those reciprocal rights to certain areas. In 1852, when Finland was still governed by Russia, the Russians closed the border with Norway in some unrelated dispute. This caused great confusion among many Norwegian reindeer owners who lost much of their winter pasture. Many of them emigrated to Sweden with their herds which, in turn, extended the confusion to many Swedish reindeer owners. Today, Norway's boundaries with Finland and the Soviet Union are fenced to prevent straying of reindeer.

Research and development

Organised research on reindeer and reindeer husbandry gained momentum less than 40 years ago. Until recently, reindeer husbandry was very traditional, but has slowly adopted new methods, some of which are based on scientific research. The reluctance was quite natural, as it occurred in remote areas by people separated from the rest of society by geography, culture, and language. Reindeer husbandry has not been the easiest target for quick improvements, as development must be carefully integrated into a balanced ecological and cultural system.

Each country supports research by government agencies, universities, veterinary schools, and other institutions. In Finland, the research group is a division of the Game and Fish Research Institute located in Rovaniemi with the field station at Kaamanen. In Norway, the research division is part of the central administration in Alta with a field station at Kautokeino in Finnmark. In Sweden, there is a research unit under the Swedish Agricultural University with headquarters in Umeå and a field station near Gällivare. At the State Veterinary Institute in Uppsala, there is a special position responsible for registration and control of reindeer diseases.

In 1980, the three countries established a cooperative unit, The Nordic Council for Reindeer Research (NOR), aimed at coordinating limited national resources. The council has a secretariat which is also responsible for editing *Rangifer*, the only scientific journal exclusively publishing papers about reindeer and the reindeer industry. Each year NOR organises a Nordic workshop for those involved in reindeer research.

Growing reindeer numbers have stimulated efforts to evaluate and improve pastures. To date, it has been difficult to develop a good evaluation system and much of the knowledge is still empirically based. Satellite imagery will hopefully improve prospects for control and monitoring of the reindeer pastures.

A related problem is reducing the overwintering breeding herd without impairing productivity. Using mathematical techniques, it is possible to estimate the herd structure which will maximise meat production at each stocking level (Danell, 1985). In some circumstances, it may be profitable to slaughter calves retaining only a few young breeding males. In two herds in Norway, all bulls are slaughtered at 1.5 years old after they have mated. This dependence on young breeding males is possible in districts with good pastures where males reach sexual maturity early. Such changes in slaughter selection result in a much higher meat yield from the same winter stock (Table 11.3).

Artificial feeding to avert starvation has been carried out in many herds, especially in Finland and Sweden. Generally, it is expensive and there is a significant risk of deaths due to gastrointestinal disruptions caused by unsuitable forage and feeding technique. This problem has received much attention, since 20–30% of the calf crop may be lost annually, but the figures differ greatly between years and districts. In Sweden, research based on radio-tracking is elucidating the importance of mortality by predators, diseases, accidents, or starvation.

Handling stress can be a significant source of mortality; attention is now directed to corral designs which minimise stress and improve meat quality (Rehbinder *et al.*, 1982).

Parasitic diseases caused by skin warbles (*Oedemagena tarandi*) and nasal grubs (*Cephenomyia trompe*) are common in Fennoscandia. Many herds are now treated for these parasites with drugs also effective against other internal parasites such as lungworms (mainly *Dictyocaulus*) and the meningeal worms (*Elaphostrongylus rangiferi*). The latter sometimes causes lameness of infested reindeer and may occasionally cause serious losses (Nordkvist, Christensson & Rehbinder, 1984). *Echinococcus* (hydatid cyst) is almost eradicated by combining treatment of herding dogs, sanitary precautions, and veterinary inspection at slaughter (Skjenneberg, 1959, 1983).

Much of the reindeer meat is marketed unprocessed, though development and marketing of processed products is being stressed to increase economic output.

The results of research are communicated to the industry through training programmes, handbooks, and trade journals. Norway has a separate school for reindeer herders, offering a one-year programme. Similar training is offered in Finland as a division of an agricultural school. In Sweden, training is more heavily based on specialised short courses. Manuals and handbooks include Skjenneberg (1965), Anon. (1966), Skjenneberg & Slagsvold (1968), Kurkela (1978), and Kosmo (1985a). In Sweden, there is a series of pamphlets dealing with specific topics. Each country publishes a trade journal providing practical information for reindeer owners: *Poromies* (Finland); *Reindriftsnytt* (Norway); *Rennäringsnytt* (Sweden).

Table 11.3. *Development of the reindeer industry output in a Norwegian county from 1975 to 1985.*[a]

Season	Winter stock	Reindeer slaughtered	Meat (kg)	Average slaughter weight (kg)	Meat output per winter reindeer (kg/head)
1975–78	9,908	1,321	49,700	37.6	5.0
1984–85	10,408	6,487	183,400	27.8	17.6

[a]Kosmo (1985b).

Economy

During the last decade, the economy has improved substantially for several reasons. Rising meat prices is one factor, but new operational techniques and new routines for selection of slaughter animals have been responsible for better output. The contribution of various subsidies differs considerably among the countries.

Industry output

In 1985, total net income from the reindeer industry of Fennoscandia was over US$ 40 million (Table 11.4). The main source of revenue was meat. Export of live animals comprises a very limited part of the sales and only minor amounts of antlers are sold because amputation of soft antlers is prohibited.

Subsidies

In Finland, there is no direct subsidy by the state except for support of the Reindeer Herders' Association. In addition, there is some compensation for reindeer killed by predators or by railway and road traffic.

Since 1976, the Norwegian government has offered a complex set of subsidies aimed at improving harvesting routines, composition of the herds, and marketing of meat. In 1985, the value was approximately NOK 60 million (US$ 8.6 million). Part of the contribution is placed in the *Reindeer Industry Development Fund* which also supports research. Special arrangements are made for free use of marine landing crafts for transporting of herds in spring from the mainland to the many islands with summer pasture. This eliminates risks and losses from swimming when the reindeer are weak after the long winter.

Table 11.4. *Calf crop, harvest and economic profit of reindeer industry in Fennoscandia (1985).*

Country	Calf crop	Harvest	Meat weight (tonnes)	Net income[a] (millions)	Net income (US$ million)
Finland	101,000	95,000	2,800	FIM 74	15.2
Norway	102,000	63,000	1,960	NOK 68	9.3
Sweden	110,000	76,000	2,470	SEK 110	16.0
Total	313,000	234,000	7,230		40.5

[a]Including subsidies and incidental earnings.

In Sweden there is one major support measure; namely, a subsidy for slaughtered reindeer subjected to veterinary inspection (SEK 160/reindeer). The main contribution to the reindeer industry, beside marketing of meat, comes from the Sami Fund. Similar to Finland, the state compensates losses by predation and traffic accidents.

Family income

Differences in working conditions between and within countries prevent proper comparisons of the income from reindeer industry to the average family. We can look upon an example taken from Norwegian statistics (Table 11.5). In Finland, where the reindeer industry is not subsidised, the market price of reindeer meat is US\$ 6.82/kg to the producer (1985). Subsidised prices in Norway and Sweden are US\$ 4.75 and 4.25/kg, respectively. Advanced techniques have increased operational expenses significantly (Table 11.6).

Problems and future prospects

The main problem facing the reindeer industry is competition for pasture both with other herds and from other land uses such as hydroelectric development (construction of dams, river regulation, road construction). These uses cause damage to valuable grazing land and construction may interrupt movements between seasonal pastures. Division of pasture areas by roads and/or railways may limit both natural grazing movements and pasture

Table 11.5. *Total product of reindeer industry in Norway 1985.*[a]

Income	(1,000 NOK)
Sale of meat and by-products	65,994
Compensations	2,265
Other income	830
Gross income	67,429
Expenses	49,202
Net product	18,227
Subsidies	24,995
Tax compensation	8,761
TOTAL	51,983
Interests on rented capital	5,294
NET INCOME	46,689
Net income per family in NOK: 69,685 (US\$: 10,000).	

[a]Anon. (1986).

areas, diminishing the effectiveness of grazing and increasing the risk of overgrazing, especially of lichens which may easily be destroyed for decades.

Improved communications and transportation encourage and increase recreational access to formerly undisturbed reindeer pastures. This human traffic, especially in the sensitive calving time, presents the most serious difficulty for reindeer herders. It frightens the animals and destroys the peace which is so essential for a reindeer to gather enough food to survive the long winter.

Lichens are the most important forage, not only in winter but also in early spring and late autumn. Since lichens absorb most of their nutrition from precipitation, they act as sponges for air pollutants. Some are toxic and hamper lichen growth. Others are a hazard to the reindeer or to people who consume reindeer meat. Radioactive nuclides stay in lichens for decades, contaminating reindeer meat for many years.

The nuclear accident at Chernobyl in late April 1986 deposited radioactive iodine$_{31}$, caesium$_{134}$ and caesium$_{137}$ over central Scandinavia and southern Finland. High levels of radiocaesium in reindeer and game meat required nearly half the harvest of reindeer meat in Sweden and Norway during the first slaughter season after the fallout to be discarded. It was decided to harvest as usual to keep the winter stock down and prevent overgrazing. The

Table 11.6. *Expenses in reindeer husbandry in 1000 NOK, 1985.*[a]

Type of Expense	Amount	%
Transport	1,769	3.8
Commodities, including feeds	744	1.6
Dogs	648	1.4
Travelling	1,205	2.6
Field equipment	5,166	11.0
Terrain vehicles	15,441	33.0
Cars	8,645	18.5
Boats	293	0.6
Cabins, corrals, fences	7,595	16.3
Electricity	912	2.0
Telephone	1,129	2.4
Rental fees	200	0.4
Insurance	539	1.2
Miscellaneous	1,137	2.4
Administration	1,201	2.6

[a]Anon. (1986).

economic loss was compensated by government. The event stimulated a search for means to decontaminate reindeer for radiocaesium. Small amounts of bentonite or ferriferrocyanides on the fodder is effective for artificially fed slaughter animals. A bolus with the latter substance deposited in the numen can be used for reindeer grazing contaminated pasture.

Despite new legislation, reindeer owners still tend to increase their herds and overgraze in winter. One reason is the rising costs of mechanisation. In addition to damaging winter pastures, herds stocked too heavily lose condition, produce less meat and offspring, and become more susceptible to parasites and diseases.

Some antagonism exists between the reindeer industry and forestry in Finland and Sweden where herds graze lichen-rich coniferous forests during winter (Eriksson, 1975). Forestry activities may damage the lichen carpet in a number of ways such as driving on snow-free ground with machines or site preparation for reforestation. Logging slash may act as obstacles to herd movement. Although there seems to be minimal risks of poisoning (Nordkvist & Erne, 1983), fertilising with urea or ammonium nitrate discourages reindeer from grazing upon treated areas for up to two seasons (Eriksson, 1984) and can diminish lichen resources. Clearcutting also exposes snow to wind and sun, creating crusts which prevent reindeer from digging craters for lichens. Removal of old trees with arboreal lichens removes an important nutritional resource in emergencies when deep or hard snow prevents cratering. Occasionally, reindeer herders fell such trees to provide additional winter fodder.

There has always been a conflict between reindeer owners and conservationists about predator control; laws permit predators to survive in the wilderness and compensate reindeer owners for losses. The numbers of reindeer actually killed by predators are uncertain. Numbers of reindeer claimed on compensation programmes in Sweden have varied from 1531 in 1972/3 to 6432 in 1981/2 (Gustavsson, 1985).

Despite many problems, there is optimism and renewed interest among young people to enter the reindeer industry. It offers a healthy lifestyle and reasonable income. In that pasture is limited and cannot be extended, regulations now limit both numbers of reindeer and reindeer owners.

References

Anon. (1966). *Ekonomisk Renskötsel*. Boras: LT'Förlag (in Swedish).
Anon. (1986). Økonomisk utvalg for reindriften. *Reindriftsnæringens totalregnskap*. Alta: Reindriftsadministasjonen (in Norwegian).
Danell, Ö. (1985). Optimal hjordsammansättning – Ett systemanalytiskt problem. *Rangifer*, 2–1984, Appendix 107–11 (in Swedish).

Eriksson, O. (1975). Sylvicultural practices and reindeer grazing in Sweden. In *Proceedings of the First International Reindeer and Caribou Symposium*, ed. J. R. Luick, P. Lent, D. R. Klein & R. G. White, 1972. Biological Papers of the University of Alaska. Special Report Number 1:108–21.

Eriksson, O. (1984). *Effekter av skogsgödsling på renbete och renbetning*. Skogsfakta No. 5. Supplement. Uppsala: Sveriges Lantbruks-universitet, Skogsvetenskapliga Fakultenen. 80–7 (in Swedish).

Gustavsson, K. (1985). *Rennäringen, siffror och diagram*. Lantbruksstyrelsen. Meddelanden 1985: 2. Stockholm: Liber Förlag.

Kosmo, A. J. (1985a). Hvor går samisk reindrift? In *Samer i sör. Årbok* No. 2, ed. S. Fellheim & A. Jåma, pp. 166–70 (in Norwegian).

Kosmo, A. J. (1985b). *Reindrift med Fremtid. Driftsøkonomi med Planlegging*. Tromsø: Reindriftsadministrasjonen & Samisk Utdanningsråd (in Norwegian).

Krafft, A. (1981). *Viltrein i Norge*. Viltrapport 18. Trondheim. Trondheim: Direktoratet for Vilt og Ferskvannsfisk (in Norwegian).

Kurkela, P. (1978). *Miten hoidan poroa*. Kemijärvi: Societas Geographica Fenniae Nordica (in Finnish).

Nordkvist, M., Christensson, D. & Rehbinder, C. (1984). Ett fältavmaskningsförsök med ivermectin (MSD) på renar. *Rangifer*, 4 (2), 2–9 (in Norwegian).

Nordkvist, M. & Erne, K. (1983). The toxicity of forest fertilizers (ammonium nitrate) to reindeer. *Acta Zoologica Fennica*, 175, 101–5.

Rehbinder, C., Edquist, L. E., Lundström, K. & Villafañe, F. (1982). A field study of management stress in reindeer (*Rangifer tarandus* L.). *Rangifer*, 2 (2), 2–21.

Skjenneberg, S. (1959). Ekinokokkose hos rein i Kautokeino. *Nordisk Veterinärmedisin*, 11, 110–23 (in Norwegian).

Skjenneberg, S. (1965). *Rein og Reindrift*. Lesjaskog: A/S Fjell–Nytt (in Norwegian).

Skjenneberg, S. (1983). Advance in the study of veterinary medicine in the reindeer/caribou since 1978. *Acta Zoologica Fennica*, 175, 91–8.

Skjenneberg, S. (1984). Reindeer. In *Evolution of Domesticated Animals*, ed. I. L. Mason, pp. 128–38. London and New York: Longman.

Skjenneberg, S. & Slagsvold, L. (1968). *Reindriften og dens Naturgrunnlag*. Oslo, Bergen, Tromsø: Universitetsforlaget (in Norwegian).

Sulkava, S. (1980). Population of the wild forest reindeer, *Rangifer tarandus fennicus* Lönnberg 1909, in Finland. In *Proceedings of the Second Reindeer/Caribou Symposium*, ed. E. Reimers, E. Gaare & S. Skjenneberg, pp. 681–4. Trondheim: Direktoratet for Vilt og Ferskvannsfisk.

Vorren, Ø. & Manker, E. (1957). *Samekulturen*. Tromso Museum Skrifter, 5. Tromsø: Tromsø Museum (in Norwegian).

Wiklund, K. B. (1918). Om renskötselns uppkomst. *Ymer* 3, 249–73 (in Swedish).

12

Reindeer husbandry in North America

GEORGE W. SCOTTER

Abstract

Reindeer (*Rangifer tarandus*) were first imported into North America from Siberia in 1891. Those reindeer increased in Alaska to 641,000 by 1932 and then crashed. During the past decade, the number of reindeer in Alaska has been relatively stable at about 20,000 animals. The early success of the Alaskan reindeer stimulated similar interests in Canada. Of six attempts to develop reindeer husbandry in Canada, the Mackenzie Delta herd, established in 1935, and the Belcher Islands herd, established in 1978, are the only ones still in operation. The Mackenzie Delta herd was established with 2382 reindeer to supplement regional wildlife resources and to improve the Inuit's (Eskimo) economic condition by creating a viable industry with native-owned herds. All the native herds, established from the nucleus herd, eventually reverted back to government ownership. The reindeer business was sold to private owners in 1974 and became a financial success. The reindeer herd introduced on the Belcher Islands has expanded exponentially and has received community support. Environmental and animal welfare concerns associated with reindeer husbandry are modest. If land-use problems can be resolved and modern technologies are applied to reindeer husbandry in Alaska and northern Canada, the future appears moderately promising.

Introduction

Domestication of reindeer dates back about 2000 years and is one of the oldest known livelihoods in the arctic and subarctic regions. Reindeer husbandry was confined to Eurasia until the late 1800s. This article reviews the history of reindeer husbandry in North America with emphasis on the Canadian experiments.

Alaska

Between 1891 and 1902, 1280 reindeer were imported into Alaska from Siberia. The introduction was directed by Dr Sheldon Jackson, first

Superintendent of Education in Alaska, to supplement the food supply of coastal indigenous people. Their food was then derived mainly from wild, free-ranging sea and land mammals. Reindeer numbers reached 641,000 by 1932, but were reduced to about 25,000 by 1950 (Fig. 12.1). Factors that led to herd declines included predation, disease, mixing and loss to caribou herds, poaching, adverse weather conditions, overstocking of ranges, and reduced range quality (Lantis, 1950; Hanson, 1952; Sonnenfeld, 1959; Brady, 1968; Klein, 1980). At present, there are about 23,000 reindeer in Alaska and the number of animals has been relatively stable for the last decade. Since 1940, herd ownership has been restricted to indigenous people.

Stern *et al.* (1980) published a comprehensive historical summary and analysis of the reindeer industry in Alaska that covers the period from 1891 through 1977. There is little need to repeat most of their report. However, a review of production and its value since 1960 may assist with predicting the future of the industry. From 1960 through 1977, the annual production of reindeer meat (dressed carcass weight) in Alaska averaged 225,000 kg, with

Fig. 12.1. Estimated total reindeer numbers in Alaska, 1891 to 1985 (data from Stern *et al.* (1980) and E. L. Arobio, pers. comm.).

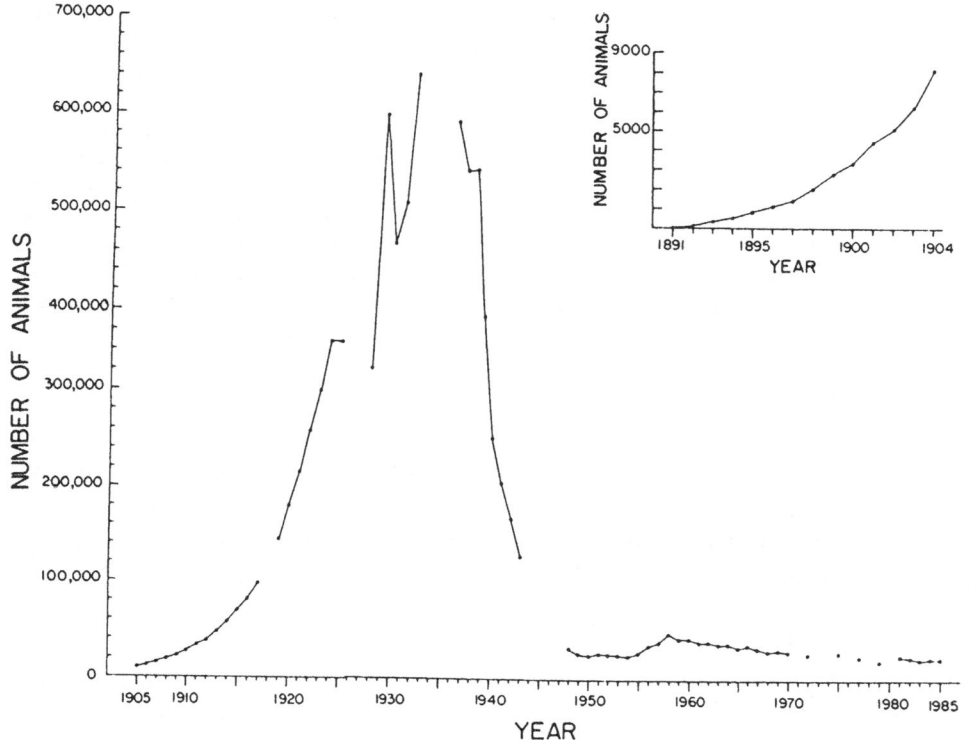

the lowest production of 130,000 kg in 1976 and the highest of 342,000 kg in 1968 (Stern *et al.*, 1980). Value of the meat produced averaged US$ 236,000 and ranged from a low of US$ 166,000 in 1972 to a high of US$ 324,000 in 1968 (Stern *et al.*, 1980).

The price for velvet antler increased steadily from 1969 to 1977. Owners received US$ 2.20/kg in 1969 and as much as US$ 52.27/kg in 1977 (Stern *et al.*, 1980). Adding the value of antlers, hides and meat by-products to those of the meat, the total annual value of reindeer products from 1972 through 1977 averaged US$ 343,000/year and ranged from a low of US$ 215,000 in 1972 to a high of US$ 501,000 in 1976 (Stern *et al.*, 1980).

Recently, state and federal agencies have expended considerable effort to develop a profitable operation. For example, Arobio, Thomas & Workman (1979), Arobio (1981), and Thomas *et al.* (1983) examined the optimum reindeer herd structure using mathematical programming techniques. The management option that provided the largest returns was to maximise velvet antler production. Based on meat and antler prices for 1977, they estimated the potential net returns for herds of 1000, 2000, and 3000 animals at US$ 30,000, US$ 62,000, and US$ 106,500, respectively. Herds with less than 1000 reindeer do not appear economical. Naylor *et al.* (1980) discussed possible impacts of the proposed federal land management policy in association with the Alaska Native Claims Settlement Act (ANCSA) and the combined effects on the future economics of reindeer herding. Apparently, there are economic advantages of more intensive management over operation of the herds as subsistence activities (Naylor *et al.*, 1980).

Newfoundland experiment

The apparent success of early reindeer introductions in Alaska stimulated interest in similar introductions into Canada. Dr W. T. Grenfell, supported by the Boston Transcript newspaper and Canada Department of Agriculture, purchased 300 reindeer in Norway. They were brought to Cremeliere, near St Anthony, Newfoundland, in 1908, along with three Sami families to herd them (Grenfell, 1919). The herd increased to about 1300 by 1912, but its Sami herders, discouraged by the unfavourable climate and low pay, returned home. Poaching, indifferent care, and a lack of understanding by the local people contributed to the herd's decline.

Grenfell was in France during World War I and upon his return only 230 reindeer were found. The experiment had failed, but Grenfell remained enthusiastic about establishing a viable reindeer industry in Canada. In 1918, Grenfell, with assistance from the Canadian government, took about 150 of the remaining reindeer to Rocky Bay, Quebec, on the north shore of the St

Lawrence River. Those reindeer were subsequently moved to Anticosti Island in the Gulf of St Lawrence and allowed to run wild. However, they did not thrive, probably because suitable forage was lacking or possibly because of transmission of a parasite (*Parelaphostrongylus tenuis*) from white-tailed deer (*Odocoileus virginianus*). That parasite is highly pathogenic to both moose (*Alces*) and reindeer (Anderson, 1972; Anderson & Prestwood, 1981). Reindeer were last seen on the island in 1949 (Cameron, 1958).

In 1911, the Department of the Interior purchased 50 reindeer from Grenfell. The animals were driven and transported by ship, rail, horse-drawn wagons, and scows from St Anthony, Newfoundland, toward their destination in the Great Slave Lake region of the Northwest Territories (Hedlin, 1961; Inglis, 1969). Deaths and straying throughout the journey caused heavy losses. By 1916, the herd was reduced to a single reindeer which the herder ate, thus ending an experiment costing more than Can$ 60,000 – probably the most expensive steak dinner in history.

Hudson's Bay Reindeer Company

In May 1919, the Canadian government appointed a Royal Commission to investigate the possibilities of reindeer and muskox (*Ovibos moschatus*) herding in the arctic and subarctic regions of Canada. Before the Commission's report was completed, V. Stefansson, after resigning his position on the Commission, applied for and was granted grazing privileges on more than 260,000 km^2 of southern Baffin Island. He then persuaded the Hudson's Bay Company to set up a subsidiary, Hudson's Bay Reindeer Company, of which he was a director and the technical adviser (Stefansson, 1964). S. T. Storkerson was hired to study the grazing prospects on the leasehold and reported enthusiastically on the vegetation and its suitability for reindeer.

Representatives of the Hudson's Bay Reindeer Company went to Norway to purchase reindeer, and on 13 October 1921 the *Nascopie* sailed with 627 reindeer and six Sami herders. Seventy-seven reindeer died before arrival in Amadjuak Bay, Baffin Island, on 1 November. The released reindeer scattered in all directions in search of food. The herders were able to round up only 260 of the 550 animals released.

The reindeer did not find the quantity and quality of forage Storkerson so optimistically reported, for he had evidently mistaken mosses or unpalatable lichens for desirable ones. The reindeer lichens, so important in winter, were very scarce, and the reindeer had to forage over large areas in search of suitable food. By the summer of 1923, only 181 reindeer remained. That autumn, the last of the Sami herders returned to Norway. During the winter of 1924/25, most of the herd disappeared. On 27 May 1927, the grazing permit of the Hudson's Bay Reindeer Company was cancelled and the enterprise officially concluded. Begun with high hopes, the Baffin Island experiment ended a total failure and cost Can$ 200,000 (Stefansson, 1964).

Officials of the Hudson's Bay Company blamed the lack of feeding grounds while Stefansson blamed inefficient herding and management. However, the originators of this scheme apparently wrongly assumed that good range for free-ranging caribou equated to good range for domesticated reindeer. They forgot that while free-ranging and domesticated forms of *Rangifer* eat the same kinds of forage, wild individuals can roam free in search of food while domesticated ones must be kept on a limited range, rich enough to allow herding, if they are to be of any use to their owners. Even the best management practice for reindeer would probably not have forestalled failure of the Baffin Island experiment because of the lack of suitable forage to support the herd in a limited area.

Mackenzie Delta experiment

The Royal Commission report (Rutherford, McLean & Harkin, 1922) recommended that small experimental reindeer herds be established in several places. Thus in April 1926, A. E. Porsild was appointed to make a general botanical reconnaissance of northwestern Canada with special reference to reindeer pasture and other general conditions that would be important to future reindeer husbandry. Porsild (1929) concluded that the arctic coast and Eskimo Lakes region of the District of Mackenzie had a carrying capacity of at least 250,000 reindeer while the Great Bear Lake basin could support 300,000 more.

In 1929, the Canadian government and the Lomen Reindeer Company of Alaska agreed to deliver 3000 reindeer to the Reindeer Preserve near the Mackenzie River Delta. The delivery was completed in 1935, after a drive which took six winters and five summers (Scotter, 1978). The final tally was 2382 reindeer (Fig. 12. 2), comprising 1498 does, 611 bucks, and 273 steers. Only 10% of the animals were from the original herd; 90% had been born on the 2600 km trek. This was fewer than the 3000 animals agreed on, but the birth of 800 fawns a few weeks after arrival made up for the short delivery.

This reindeer venture was intended to improve the economic plight of the native people by providing a stable supplement to their diet and an opportunity for employment (Fig. 12.3). The region's herds of native caribou (*R. t. groenlandicus*) and other wild mammals had seriously dwindled. This followed the arrival of traders in the Arctic and the introduction of firearms to the natives. The decline of wild mammals was further exacerbated when whalers engaged Inuit people to hunt for meat to feed their crews. Wild

Fig. 12.2. Estimated number of fawns, slaughtered animals, and total reindeer numbers at the end of the year at the Reindeer Preserve, 1935 to 1985 (data from Krebs (1961), Hill (1967), and files of Canadian Reindeer Ltd.).

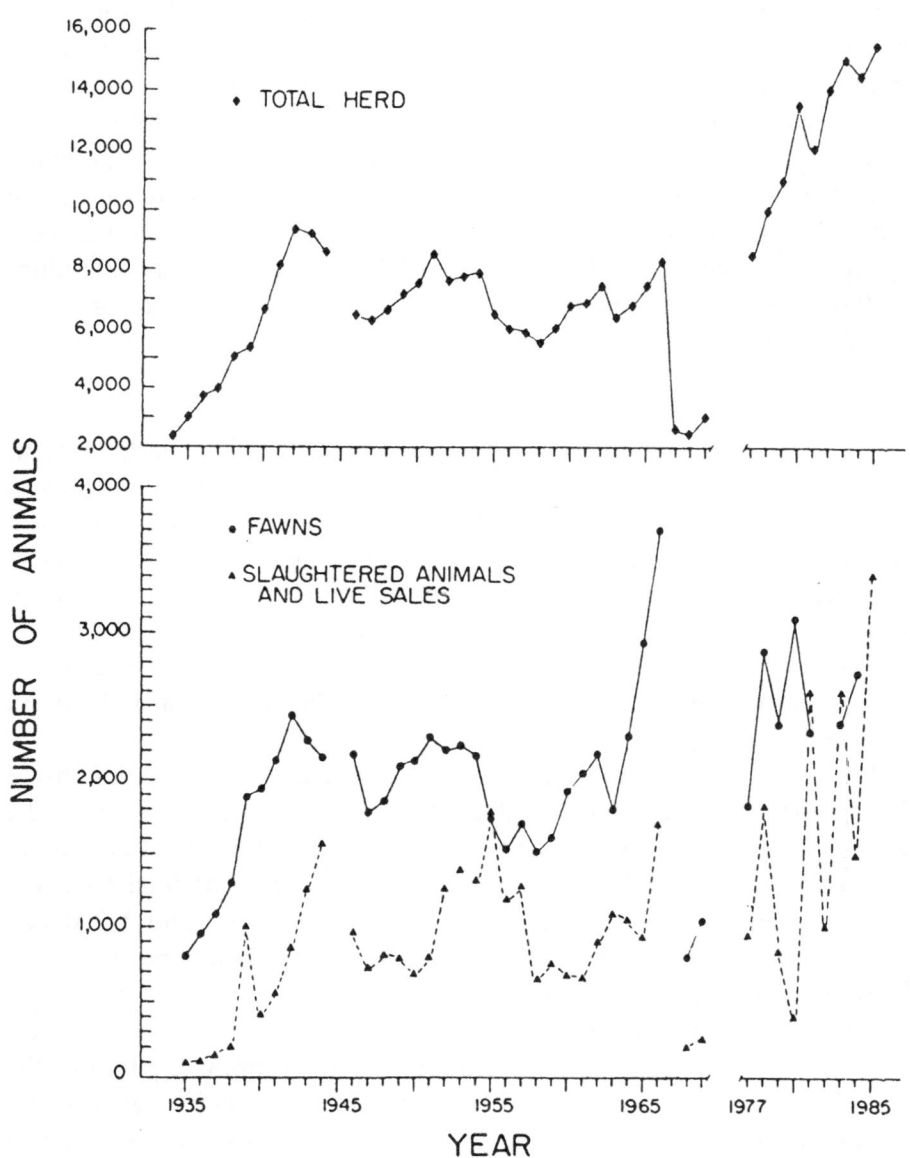

mammals had been the principal livelihood of the Inuit, therefore, a shortage of game forced them into commercial trapping in order to obtain money to buy food and other necessities. Thus, their economic plight varied with the fluctuating supply of fur-bearing animals and the price of pelts.

The Reindeer Preserve

The Reindeer Preserve (Fig. 12.4), near Inuvik, Northwest Territories, is bound by the Beaufort Sea on the north, the Mackenzie River on the west, and the Anderson River on the east. The preserve of 17,000 km² was established in 1933 and was enlarged to 47,000 km² in 1952.

The coastal area of the Reindeer Preserve lies in the arctic climate zone while the southern part of the Preserve lies in the subarctic climate zone. The mean annual total precipitation is low, ranging from 15 cm at Tuktoyaktuk to 28 cm at Inuvik. Fog is common in summer, especially along the coast. Mean temperatures in July and August are only about 10°C. Winters are long, cold, and dark with more snowfall inland than on coastal sites.

Fig. 12.3. A Sami herder with his Inuit apprentices at the Reindeer Preserve in 1937 (Photograph by A. E. Porsild).

There are several vegetation types on the Reindeer Preserve. The northern portion including Richards Island, the Pleistocene Coastlands, the northern portion of the Fluted Plains, and the northern part of the Anderson River Uplands, lies within the tundra zone. The southern portion is in the boreal forest zone. The general vegetation sequence from north to south is tundra; tundra with scrub willow (*Salix* spp.) and ground birch (*Betula glandulosa*);

Fig. 12.4. MacKenzie Delta Reindeer Preserve showing the suggested areas for summer, winter, and spring-autumn grazing.

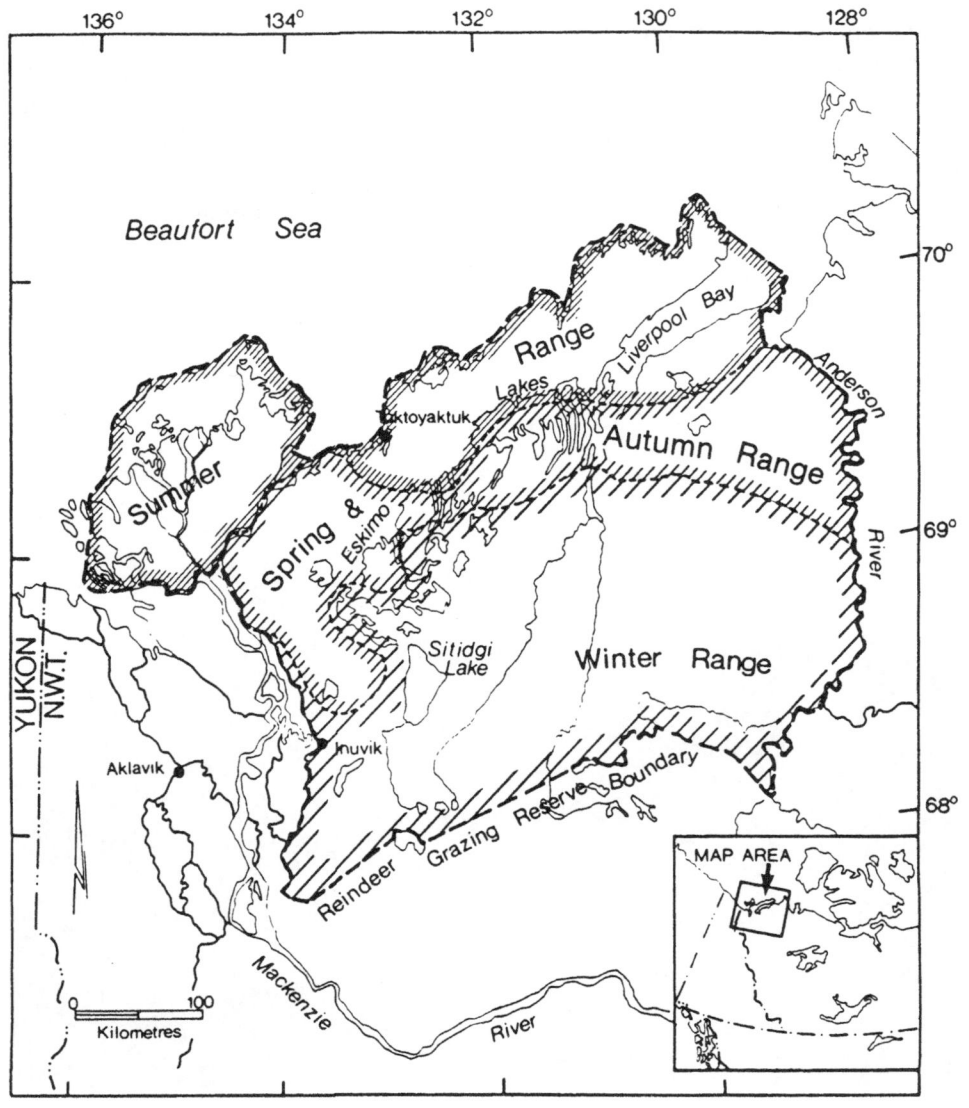

scrub willow and ground birch; woodland and tundra with much scrub willow and ground birch; open woodland (Fig. 12.5); and continuous woodland (Mackay, 1963).

Without the benefit of maps or aerial photographs, Porsild estimated the Arctic Coast in northwestern Canada would support 250,000 reindeer. The Reindeer Preserve now encompasses about 30% of the coastal area; therefore, based on Porsild's estimate, it should support approximately 85,000 reindeer (Hill, 1967). Porsild (1936) estimated the original Preserve (17,000 km²) would indefinitely support 25,000 reindeer.

However, in 1965, S. B Johansson, a former manager, estimated that the enlarged Preserve (47,000 km²) would carry only 30,000 reindeer. Scotter (1968) studied range conditions and trends on the Preserve in 1965 and 1966 and agreed with Johansson's estimate, provided that good range management practices were employed and all 47,000 km² used. Only about one-quarter of the Preserve has been under intensive use. The forested portion is often unused because herding is difficult there.

In general, early estimates of carrying capacity for reindeer were too high, partly because the time required by lichens (Fig. 12.6) for recovery from

Fig. 12.5. Reindeer feeding in an open woodland during the winter (Photograph by R. F. Nowosad).

grazing was never considered in the calculations. In Sweden, Skuncke (1969) concluded that reindeer need 8–10 times more range than previously calculated. Skogland (1986) demonstrated a decrease in adult female body size and lower recruitment rates in wild reindeer, caused mainly by winter food limitations, in response to greater population densities.

Fires have burned over many square kilometres of open woodland and scrub vegetation between the Caribou Hills and Kugaluk River. Fires on the winter range must be controlled to maintain the Preserve's carrying capacity. A major fire could destroy the whole reindeer operation. The risk of fire is increasing as the area is becoming more accessible to local people and tourists.

Oil exploration on the Preserve expanded rapidly in the 1960s and 1970s but its effects on carrying capacity and seasonal movements of reindeer have been, in general, minimal and temporary.

Management of the herds

The original plan was to set up a government-owned main herd from which smaller herds would be formed. Once established, the smaller herds would be turned over to suitable herdsmen, each assigned to particular

Fig. 12.6. Lichens in the genus Cladonia are an important winter forage for reindeer (Photograph by G. W. Scotter).

winter and summer ranges. As these herds increased in size, their holders would repay the number of animals they had been given so that herds would eventually become self-supporting units. These objectives were never reached.

Between 1938 and 1954, six Inuit-owned herds of about 1000 animals each were established. Most of these herds increased in numbers for a few years, declined, and finally reverted to government ownership. The owners of the first two herds were drowned at sea in a boating accident in 1944 and their herds were amalgamated and placed under government supervision. The last Inuit-owned herd was returned to the government in 1964.

The reasons for this failure were many and complex. Lantis (1954) suggested that the Inuit resisted reindeer herding because they were not enthusiastic about the monotonous herding tasks. To change a hunter into a herder would mean changing not only his life style but also his whole psychology. At that time, most Inuit were unwilling to turn from hunting and trapping and a settled community life to become mere followers of reindeer. In the late 1950s, more rewarding employment, such as wage labour and more remunerative trapping, became available. Such tasks were seasonal and did not have to be carried on throughout the year to be profitable.

Sonnenfeld (1959) suggested that the inland Eurasian herders, who originally hunted wild reindeer, could better make the transition to herding domesticated reindeer because they were accustomed to following animals. He speculated that reindeer ranching might have been more successful among inland Inuit in Alaska than among the coastal Inuit who lived in permanent settlements for a good part of the year. The same theory could perhaps be applied to the Inuit of the Mackenzie Delta.

Predation, mixing with resident wild caribou, poaching, and diseases such as foot rot and an undetermined weak-bone ailment that occasionally afflicts the herds, were among other problems.

The Mackenzie reindeer operation also suffered from attempts to apply the less modern Sami practices. Some unfavourable practices included close herding and unsatisfactory sex structures with a high ratio of males to females. The assumption that Sami pastoral practices were suited to the Canadian reindeer operation is questionable. The Fennoscandian reindeer operation has itself been plagued with problems (Scotter, 1965; Chapter 11), although the adoption of modern management practices was improving the outlook for their industry, at least until meat sales were affected by radionuclide contamination from the Chernobyl disaster in 1986.

The early Mackenzie Delta reindeer operation lacked effective direction partly because some of its managers did not have experience with livestock or wildlife management, partly because a coherent, practical policy was not set, and partly because of a lack of biological data. The lack of funds and the great distance to the decision-making body in Ottawa were also major problems.

From 1960 to 1968, the reindeer project was in the hands of a private contractor who proposed to make it self-supporting. However, during that period costs exceeded revenues by a ratio of three to one.

Herding management was casual from 1963 until 1968, with the animals occasionally being observed from the air. Between 1960 and 1968, the reindeer population decreased from 8400 to about 2800 animals (Fig. 12.2) when the contract was terminated. Preobrazhenskii (1968) suggested that free and semi-free grazing should not be used in Russia. However, in Finland, Maki (1966) noted that intensive reindeer husbandry was giving way to more extensive methods. Based on the results obtained in Canada, casual herding with only sporadic supervision was ineffective.

The Canadian Wildlife Service was asked to take charge of the project in 1968 so that the future of reindeer herding in Canada could be assessed. The Canadian Wildlife Service conducted scientific studies on the animals and their ranges (Scotter, 1968, 1972; Scotter & Miltimore, 1973; Nowosad, 1975). Management techniques that would ensure a higher yield of meat at reasonable cost were attempted and a study on native ownership and economic viability of the reindeer herd was completed (Stager & Denike, 1972). A number of alternative management strategies were considered by the Federal Government. It concluded that the reindeer industry could be operated economically, independent of government assistance. Consistent with the programme of transferring control and ownership of commercial projects to native people, the government sold the entire reindeer operation for Can$ 50,000 in 1974 to a new company called Canadian Reindeer Ltd which was owned by a former chief herder, Silas Kanagegana. At the time of the purchase, the herd numbered about 5200 (Nasogaluak & Billingsley, 1981).

Kanagegana operated the herd essentially without any financial assistance from government until his deteriorating health forced him to sell the business in 1978. The herd was then sold to William Nasogaluak, the current owner of Canadian Reindeer for Can$ 250,000.

Lack of sufficient cash flow severely limited development of the reindeer industry at the start of the private enterprise period. A small local market and the inability to sell reindeer meat in the rest of Canada because of government meat inspection regulations were two major reasons causing insufficient cash flow. With the help of Agriculture Canada, a portable slaughter facility was developed, and later improved, which met meat inspection standards. Then it was possible to sell meat that was surplus to northern requirements to markets in southern Canada. The development of southern markets was assisted by the building of landing facilities at Inuvik for large aircraft. Cargo aircraft, which generally returned south empty, provided attractive backhaul rates for reindeer meat.

The next major development influencing the Canadian reindeer industry was the opening of the Dempster Highway, linking the Mackenzie Delta with southern Canada via the Yukon. That road permitted vehicular transport-ation of meat to southern markets more economically than by air freight and facilitated the sale of live reindeer as well, which now provide substantial revenues. Most of the exported meat is sold to the restaurant trade in North America.

The economics of the reindeer industry improved further with the sale of velvet antlers to the Oriental pharmaceutical trade. Antlers were harvested for the first time in 1977 and were processed locally. Thereafter, antlers became a major new source of income. In fact, the income from the sale of antlers exceeded that from meat in some years.

The improvement in the company's cash-flow position as a result of access to southern markets for meat and the new cash crop for antlers permitted changes in the management strategy for the herd. Helicopters were used for summer roundups, to assist with ear tagging calves, to carry out castrations so that steers would be available for subsequent slaughtering, to closely examine the health of animals, and to select and mark animals for slaughter during winter. Better equipment, such as snowmobiles, was provided for the herders and a fixed-wing aircraft was purchased to improve surveillance for stray animals.

Under the management of Canadian Reindeer Ltd, the percentage of females in the herd has ranged from 42–48%, fawns as a percentage of the herd varied from 21–35%, and fawns as a percentage of females ranged from 51–78%. An unprecedented finding was the occurrence of twin fetuses in up to 25% of the females (Godkin, 1986a). Breeding bulls constitute about 30% of the herd; however, the present management goal is to reduce that to about 15%. Pregnancy rates were high with over 99% of the adult females pregnant each year and herd recruitment has averaged 28% yearly (Godkin, 1986b). The total number of animals slaughtered combined with live sales has varied annually from 400 to 2600. Herd size has increased from about 10,000 in 1978 to the present estimate of 15,000 (Fig. 12.2).

Some improvements in management by Canadian Reindeer Ltd may still be possible. In northern Russia, for example, Baskin (Chapter 10) reported greater than 60% breeding females, a rate of reproduction of 87 fawns per 100 females, a 1:18 male to female ratio, and the harvest of 2700 animals per year from a herd of 10,000 reindeer. Prime adult males produce up to 7 kg of antler while older females average only about 2 kg. Therefore, the higher ratio of older males in the Canadian herd may be justified because of heavier antler production in males and the high price for velvet.

Disease and parasites have been only minor problems to the Mackenzie Delta herd (Godkin, 1986b). Foot rot is the most severe disease while the nostril fly (*Cephenomyia trompe*) and the warble fly (*Oedemagena tarandi*) are the major parasites. Ivermectin is highly effective in controlling those parasites (Dieterich, 1986; Godkin, 1986b).

Predation is a minor problem, also. Wolves (*Canis lupis*) and barren-ground grizzlies (*Ursus arctos*) are the major predators. Red foxes (*Vulpes vulpes*), Arctic foxes (*Alopex lagopus*), and ravens (*Corvus corax*) kill some newborn fawns (Nowosad, 1975).

The expansion of the herd from 5200 animals in 1974 to approximately 13,500 by 1980 (Fig. 12.2) caused some concern about potential overgrazing, particularly of the lichen wintering range. Sims (1983) developed a ground-truth and remote sensing programme to assess the winter rangeland on the tundra. Baseline data supported with photographs will provide future monitoring capabilities to assess range condition, trend, and use. Computer-aided digital classification of landsat data was applied to reindeer range in Alaska (George, Stringer & Baldridge, 1977).

Current estimates of gross income from the reindeer herd are about Can$ 1 million annually. The reindeer industry generates local employment, makes meat available for local consumption at lower prices than imported meat, and earns foreign currency from international sales.

Despite the marked increase in income and improved management of the herd, Canadian Reindeer Ltd's future is jeopardised by two new problems – land claims settlement and changes to the hunting regulations in the Northwest Territories that permit hunting of caribou on the Preserve at any time.

Nasogaluak purchased the reindeer herd with the understanding that grazing rights of the Preserve would be granted to him in accordance with the Reindeer Regulations established to control the industry. However, in 1984, the Government of Canada signed a land claims settlement with the Inuvialuit of the Western Arctic. The land transferred to the Inuvialuit, under the Committee for Original People's Entitlement (COPE), includes a substantial portion of the Reindeer Preserve. The major implication of the land transfer is that Canadian Reindeer Ltd can no longer expect free and unconditional use of this portion of the Preserve which is now COPE land. In addition, the Reindeer Regulations prohibiting the hunting of reindeer on the Preserve are now viewed as having no force on Inuvialuit lands. Currently, considerable killing of reindeer is alleged to be taking place. The current owners believe that they can no longer operate a viable business because the recent land claims settlement did not provide for reindeer grazing.

A potential buyer for the reindeer herd is the Inuvialuit Regional Corporation (IRC) which now controls most of the land crucial to the commercial success of the enterprise. IRC is prepared to buy the business but not at a price acceptable to Canadian Reindeer Ltd. IRC offered about Can\$ 1,300,000 for the operation. Their offer assumes that the reindeer project could not operate as a 'family' business operation but would be faced with more costly management practices including expenditures for grazing permits, public relations, and other compensation. An independent assessment, based on cash flow, values the operation at between Can\$ 4,900,000 and Can\$ 5,200,000. Despite their best efforts, the parties have been unable to agree on a final price; therefore, Canadian Reindeer Ltd has negotiated with China and Korea for the sale of live reindeer. However, an export permit for the sale of live animals outside Canada has been revoked by the territorial government for technical reasons. This action is being challenged in the courts by Canadian Reindeer Ltd on the grounds that the territorial government has no jurisdiction over the export of reindeer from the Northwest Territories.

Belcher Islands

The most recent attempt to establish a reindeer herd in Canada is on the Belcher Islands, Northwest Territories, in the southeastern part of Hudson Bay. Sixty reindeer were obtained from Canadian Reindeer Ltd and transported by air to Flaherty Island on 9 March 1978. The group consisted of 50 females, of which 45 were assumed to be pregnant, and 10 males.

An increase from 60 to between 500 and 700 animals in eight years indicates that the Belcher Islands reindeer herd has become successfully established. Local residents have made limited harvests since 1983. During 1983–86, about 165 reindeer have been harvested and 18 are known to have died accidentally. Another 70 males and 30 female adults are to be harvested during 1986/87 (M. Ferguson, personal communication). In addition, the animals have expanded their range to at least two other islands. The local community owns and manages the reindeer herd. As a result, there has been no poaching by the local people and negligible predation.

Environmental and animal welfare concerns

The conflicts between domestic reindeer and their wild counterparts in Eurasia and North America were reviewed by Klein (1980). The major conflicts are: the loss of domestic reindeer to wild herds; food competition, primarily for winter forage; and transmission of diseases and parasites between the populations. Most domestic reindeer that join with wild herds are killed by predators, hunters, or the rigours of migration and winter foraging. The surviving reindeer appear to have low breeding success and little genetic influence on wild populations. Herded reindeer feed more intensively than free-ranging herds; therefore, the probability of range overgrazing is much greater for them. Caribou or wild reindeer are much more selective in their feeding, partly by choice and partly because they are not restricted by herding activities. Several diseases and parasites of reindeer and caribou could conceivably be transmitted when the animals come into contact or make use of the same ranges. However, the transmission of most diseases and parasites is often overemphasised because they are common to both reindeer and caribou (Klein, 1980).

Reindeer husbandry sometimes conflicts with the subsistence needs of local people. For example, residents of Tuktoyaktuk could not, in the past, legally hunt caribou on the Reindeer Preserve. That caused a great deal of animosity towards the reindeer operation.

Also, there are impacts on wildlife other than caribou. Indirect effects on wildlife can be substantial because reindeer herders use snowmobiles in pursuit of predators (wolves, bears, wolverines (*Gulo gulo*), and foxes), and because of subsistence hunting and gathering incidental to herding (primarily moose, muskox, and waterfowl hunting, and egg gathering). Reindeer sometimes eat muskrat (*Ondatra zibethicus*) houses in the late autumn or winter exposing the muskrats to freezing conditions which has caused local trappers to protest a loss of income from such destruction.

Reindeer herding is often more politically attractive than subsistence hunting and gathering yet fewer people usually benefit from it. Our society is biased toward cash economies.

Animal welfare concerns have not yet focused on reindeer herding, velvet antler harvesting, and related activities. However, reindeer owners are now becoming aware of this potential problem and are emphasising humane treatment of animals and humane slaughter for meat production. The public concern for reindeer husbandry appears to be no greater than that for range livestock in western North America.

Future of reindeer husbandry

Reindeer husbandry has not contributed its full potential to the economy of northern North America. The potential to increase production of meat and other products and to increase income through improved management practices and greater entrepreneurship is evident. The reindeer industry is an intelligent and biologically sound use of a renewable resource. Given the demonstrated potential for local employment, the profit to native people, and the priority that governments attach to viable renewable resource development in the North, the reindeer industry may finally reach its potential after many years.

The evolution of reindeer husbandry in North America, however, is not likely to continue without problems. Political issues such as the settlement of native land claims and the subsequent management of those lands will continue to challenge reindeer owners, given the divergent philosophies of other groups about the best uses of such lands. ANCSA in Alaska and COPE in Canada are forcing reindeer owners to come to grips with political realities. Failure to do so will threaten the survival of the industry. In addition, many of the improved management practices have been possible because of the additional income from the sale of velvet antlers, which has been volatile and may be ephemeral. Should that market weaken and prices fall, the economic outlook for the reindeer industry could be less promising. The development of speciality products such as reindeer sausage and jerky (dried meat) could help compensate for possible losses from antler sales.

Several factors are currently improving the outlook for the industry. In addition to the recent high prices for velvet antlers, the Chernobyl disaster in the USSR has caused sharply increased prices for reindeer meat. In Canada, for example, the price of venison has increased from Can$ 5.00/kg to nearly Can$ 11.00/kg since the Chernobyl nuclear disaster contaminated some Scandinavian reindeer. The possibilities for export of meat to European gourmet markets should be pursued. The downturn in the gas and oil industry in Alaska and the Beaufort Sea region should stimulate greater interest in renewable resource industries such as reindeer ranching. Finally, the application of modern technologies, such as remote sensing of rangelands and profit-maximising programming, could increase the efficiency and effectiveness of reindeer husbandry. There is every indication that reindeer husbandry in North America can become a profitable industry of moderate size in the future.

References

Anderson, R. C. (1972). The ecological relationships of meningeal worm and native cervids in North America. *Journal of Wildlife Diseases*, 8, 304–10.

Anderson, R. C. & Prestwood, A. K. (1981). Lungworms. In *Diseases and Parasites of White-Tailed Deer*, ed. W. R. Davidson, Tall Timbers Research Station Miscellaneous Publication, 7, 266–317.

Arobio, E. L. (1981). Optimum herd structure in Alaska reindeer herds. *Agroborealis*, 13, 32–7.

Arobio, E. L., Thomas, W. C. & Workman, W. G. (1979). Mathematical programming for considering management options in Alaska reindeer herding. In *Proceedings of the Second International Reindeer/Caribou Symposium*, ed. E. Reimers, E. Gaare & S. Skjenneberg, pp. 690–9. Trondheim: Direktoratet for Vilt og Ferskvannsfisk.

Brady, J. (1968). The reindeer industry in Alaska. *Alaska Review of Business and Economic Conditions*, 5, 1–20.

Cameron, A. W. (1958). *Mammals of the Islands in the Gulf of St. Lawrence*. Bulletin 154. Ottawa: National Museum of Canada.

Dieterich, R. A. (1986). Some herding, record keeping and treatment methods used in Alaskan reindeer herds. *Rangifer*, Special Issue 1, 111–13.

George, T. H., Stringer, W. J. & Baldridge, J. W. (1977). Reindeer range inventory in western Alaska from computer-aided digital classification of landsat data. In *Proceedings of the Eleventh International Symposium on Remote Sensing of Environment*, vol. 1, pp. 671–82. Ann Arbor: Environmental Institute of Michigan.

Godkin, G. F. (1986a). Fertility and twinning in Canadian reindeer. *Rangifer*, Special Issue 1, 145–50.

Godkin, G. F. (1986b). The reindeer industry in Canada. *Canadian Veterinary Journal*, 27, 488–90.

Grenfell, W. T. (1919). *Forty Years for Labrador*. Boston and New York: Houghton Mifflin.

Hanson, H. C. (1952). Importance and development of the reindeer industry in Alaska. *Journal of Range Management*, 5, 243–51.

Hedlin, R. (1961). Reindeer for the North. *The Beaver*, 291, 48–54.

Hill, R. M. (1967). *Mackenzie Reindeer Operations*. Northern Coordination and Research Centre Report 67–1. Ottawa: Department of Indian Affairs and Northern Development.

Inglis, G. (1969). 'And then there were none ...'. *North*, 16(2), 6–11.

Klein, D. R. (1980). Conflicts between domestic reindeer and their wild counterparts: a review of Eurasian and North American experience. *Arctic*, 33, 739–56.

Krebs, C. J. (1961). Population dynamics of the Mackenzie Delta reindeer herd, 1938–1958. *Arctic*, 14, 91–100.

Lantis, M. (1950). The reindeer industry in Alaska. *Arctic*, 3, 27–44.

Lantis, M. (1954). Problems of human ecology in the North American Arctic. *Arctic*, 7, 307–20.

Mackay, J. R. (1963). *The Mackenzie Delta Area, N.W.T.* Memoir 8. Ottawa: Geographical Branch, Mines and Technical Services.

Maki, T. V. (1966). Reindeer husbandry as an example of game ranching. In *Proceedings of the Sixth World Forestry Congress*, Vol. 3, pp. 3688–93.

Nasogaluak, W. & Billingsley, D. (1981). The reindeer industry in the western Canadian Arctic: problems and potential. In *Proceedings of the First International Symposium on Renewable Resources and the Economy of the North*, ed. M. M. R. Freeman, pp. 89–95. Ottawa: Association of Canadian Universities for Northern Studies, Canadian MAB Program.

Naylor, L. L., Stern, R. O., Thomas, W. C. & Arobio, E. L. (1980). Socio-economic evaluation of reindeer herding in northwestern Alaska. *Arctic*, 33, 246–72.

Nowosad, R. F. (1975). Reindeer survival in the Mackenzie Delta herd, birth to four months. In *Proceedings of the First International Reindeer/Caribou Symposium*, ed. J. R. Luick, P. C. Lent, D. R. Klein & R. G. White, pp. 199–208. Fairbanks: University of Alaska.

Porsild, A. E. (1929). *Reindeer Grazing in Northwest Canada*. Ottawa: Department of the Interior.

Porsild, A. E. (1936). The reindeer industry and the Canadian Eskimo. *Geographical Journal*, 88, 1–19.

Preobrazhenskii, B. V. (1968). Management and breeding of reindeer. In *Reindeer Husbandry*, ed. P. S. Zhigunov, pp. 78–128. Jerusalem: Israel Program of Science Translation.

Rutherford, J. G., McLean, J. S. & Harkin, J. B. (1922). *Report of the Royal Commission to Investigate the Possibilities of the Reindeer and Musk-ox Industries in the Arctic and Sub-Arctic Regions of Canada*. Ottawa: Department of Interior.

Scotter, G. W. (1965). Reindeer ranching in Fennoscandia. *Journal of Range Management*, 18, 301–5.

Scotter, G. W. (1968). *Study of the Range Resources and Management of the Canadian Reindeer Operation*. Edmonton: Canadian Wildlife Service.

Scotter, G. W. (1972). Chemical composition of forage plants from the Reindeer Preserve, Northwest Territories. *Arctic*, 25, 21–7.

Scotter, G. W. (1978). How Andy Bahr led the great reindeer herd from western Alaska to the Mackenzie Delta. *Canadian Geographic*, 97, 12–19.

Scotter, G. W. & Miltimore, J. E. (1973). Mineral content of forage plants from the Reindeer Preserve, Northwest Territories. *Canadian Journal of Plant Science*, 53, 263–8.

Sims, R. A. (1983). *Ground-Truth and Large-Scale 70 mm Study of Reindeer Winter Rangeland, Tuktoyaktuk Peninsula Area, N.W.T.*. Ph.D. Thesis. Vancouver: University of British Columbia.

Skogland, T. (1986). Density dependent food limitation and maximal production in wild reindeer herds. *Journal of Wildlife Management*, 50, 314–19.

Skuncke, F. (1969). *Reindeer Ecology and Management in Sweden*. Biological Paper 8. Fairbanks: University of Alaska.

Sonnenfeld, J. (1959). An arctic reindeer industry: growth and decline. *Geographical Review*, 49, 76–94.

Stager, J. K. & Denike, K. G. (1972). *Reindeer Herding in the Mackenzie Delta Region: A Social and Economic Study of a Northern Resource Industry*. Vancouver: University of British Columbia.

Stefansson, V. (1964). *Discovery*. New York: McGraw-Hill.

Stern, R. O., Arobio, E. L., Naylor, L. L. & Thomas, W. C. (1980). *Eskimos, Reindeer and Land*. Agriculture Experiment Station Bulletin 59. Fairbanks: University of Alaska.

Thomas, W. C., Arobio, E. L., Naylor, L. L. & Stern, R. O. (1983). An alternative management system for Alaska reindeer herds. *Agricultural Systems*, 11, 1–16.

SECTION E

Extensive containment systems: game ranching

NEIL FAIRALL

This section deals with production of essentially wild ungulates held in large enclosures. Because of its extensive nature, this form of utilisation is restricted to continents with extensive tracts of less intensively utilised land. Among countries there are differences in emphasis based on available resources both in animals and land, cultural or aesthetic values placed on the animals, and legislation authorising use of the resource.

Africa with its variety of species and relatively large numbers has been in the vanguard of this development. The first studies of wildlife utilisation were undertaken in Zimbabwe by the American biologists Dasmann & Mossman (1960). These studies were initiated in containment systems but were translated to the use of game in large free-living herds. This form of game utilisation was seen as an extension of the subsistence use that had been a part of the African lifestyle since time immemorial. It was envisaged that a managed community of wild ungulates could meet the protein needs of an African population where this commodity was in short supply. However, it soon became clear that some of the expectations were unrealistic and the greatest success has since been in a variety of containment systems where the required level of management expertise was available (Fairall, 1984).

In North America, game has also been used for subsistence since prehistoric times, but later this resource was decimated to such an extent that protective legislation was introduced. This legislation has been counterproductive to the commercial use of indigenous game meat (Chapter 14). However, in the United States it has led to the introduction of exotic species (White, 1987), a facet that has not yet emerged in other areas. Canadian interest focused initially on crossbreeding bison with domestic cattle, an unsuccessful enterprise. However, interest has since broadened to other species and has followed the basic premise of the African situation (Chapter 13).

The Canadian and American contributions to this section deal with recent developments in the field and give adequate coverage of the subject in these regions. The complexity of tropical wildlife production systems cannot be captured in a single chapter. Additional aspects are discussed by Blankenship & Qvortrup (1974), Berry (1975), Johnstone (1975), and Hudson, Stelfox & Hopcraft (1984). The earlier work is mainly of historic interest as subsequent developments and political changes have modified many factors although the concepts remain valid. Young (1984) and Bothma (1986) have produced handbooks that comprehensively cover present knowledge of African game ranching in a form that can be used by the practitioner.

Advantages of ranching

As a production mode, game ranching capitalises on the ecological adaptations of indigenous species while satisfying the requirement of establishing ownership and managing a closed system. Many species are more efficient than domestic stock in climatically stressful environments or where certain endemic diseases are prevalent. Game ranching also offers ample potential for using the specialised and complementary feeding habits of wild ungulates to balance range utilisation, presumably achieving higher production than possible with a single species (Dasmann, 1964; Talbot *et al.*, 1965; Walker, 1979).

Nevertheless, not all wild animals are suitable for game ranching (de Vos, 1973). In Southern Africa with more than 26 ungulate species, only eight are serious contenders for game ranching and, of these, only springbok (*Antidorcas marsupialis*), impala (*Aepyceros melampus*), blesbok (*Damaliscus dorcas*), and kudu (*Tragelaphus strepsiceros*) are presently of real economic potential (Conroy & Gaigher, 1982).

Constraints

Despite these opportunities for ecologically sound land use, game ranching has several constraints in relation to property, disease control, and hygienic slaughter.

Under extensive management systems, small properties often do not provide an adequate economic return, and on extremely large properties the erection of a game fence becomes prohibitively expensive, particularly in rough or wet terrain. One answer to this problem lies in flexible fencing requirements. On large properties, modest fencing standards might be acceptable as the movement across the fence would either be small relative to the total population or balanced by reciprocal movement from adjoining properties. Ordinary stock fencing is also an adequate barrier for some species such as blesbok and springbok.

Whereas disease resistance is an advantage, the fact that game can be carriers limits options for exporting breeding stock and unprocessed products. In South Africa, foot-and-mouth disease and corridor disease (*Theileriosis*) effectively rule out buffalo on game ranches whereas malignant catarrhal fever and swine fever do the same for wildebeest and warthog, respectively. The extensive nature of such enterprises makes it difficult to establish and maintain disease-free herds.

Although slaughter requirements for local markets usually can be attained, the availability of game meats carefully standardised for size and fat cover and slaughered with *ante-* and *post-mortem* inspection at licensed abattoirs erodes the competitiveness of ranch venison on international markets. Although traditional markets accept field-killed carcases, new gourmet markets are less likely to be as forgiving.

Future needs

Game ranching faces a confusing future especially with regard to the relative roles of sport and meat production. Benson (1986a,b), in a survey of game utilisation in South Africa, clearly indicates that hunting is the most important single source of income. In trophy hunting, the most lucrative form, the hunter retains only the trophy and meat marketing forms a part of the enterprise. Hunting also only removes part of the crop and for management purposes cropping is essential. Whatever the eventual emphasis, productivity will be the determining factor in the long term and this implies a better understanding of optimal stocking rates and population structures.

Optimal stocking rates are determined empirically in domestic ranching situations. Knowledge has been accumulated over many years and can be readily evaluated, the more so because relatively uncomplicated species mixtures are the rule. For wildlife communities, very little of this information exists and each practitioner manages quite subjectively. Scientific study of this complex ecological problem has not progressed appreciably and biologists presently despair of finding a valid answer in the case of complex species mixtures (Mentis & Duke, 1976; Mentis, 1977). In the interim, the application of better animal mass units scaled on the basis of metabolic growth factors for each species and age group (Meissner, 1982) is preferred to some general power function of body weight. Use of available scientific data and the empirical knowledge of experienced game farmers in knowledge-based expert systems models (Starfield, Owen-Smith & Bleloch, 1985) also holds promise.

In the past, game populations have been harvested by controlling numbers, with little attention given to the structure of herds in terms of sex and age. This contrasts with the domestic animal enterprise where the sex and age classes are carefully manipulated to achieve the greatest efficiency. The question of optimal harvests is treated in standard texts (Caughley, 1977; Eltringham, 1984; Bothma, 1986). These are, however, only used by biologists and very little of this information is available to the practitioner. The basic data needed to apply these relatively sophisticated techniques are also not available for most species. Determining sex and age structures are, however, within the capabilities of most managers, and this data can be evaluated by simulation techniques to show which production strategies are likely to be most efficient (Fairall, 1985).

References

Benson, D. E. (1986a). Game farming survey 3: The part played by hunters. *Farmers Weekly*, 18 April, pp. 32–3.

Benson, D. E. (1986b). Game farming survey 5: What of the future? *Farmers Weekly*, 2 May, pp. 43–5.

Berry, M. P. S. (1975). Game ranching in Natal. *Journal of the South African Wildlife Management Association*, 5, 33–7.

Blakenship, L. H. & Qvortrup, S. A. (1974). Resource management on a Kenya ranch. *Journal of the South African Wildlife Management Association*, 4, 185–90.

Bothma, J. du P. (1986). *Wildplaasbestuur*. Pretoria: van Skaik.

Caughley, G. (1977). *Analysis of Vertebrate Populations*, Chichester: John Wiley, 234 pp.

Conroy, A. M. & Gaigher, I. (1982). Venison, aquaculture and ostrich meat production. *South African Journal of Animal Science*, 12, 219–22.

Dasmann, R. F. (1964). *African Game Ranching*. Oxford: Pergamon Press.

Dasmann, R. F. & Mossman, A. S. (1960). The economic value of Rhodesian game. *Rhodesian Farmer*, 30, 17–20.

de Vos, A. (1973). Ecological conditions affecting the production of wild herbivorous mammals on grasslands. *Advances in Ecological Research, 6, 137–83.*

Eltringham, S. K. 1984. *Wildlife Resources and Economic Development*. Chichester: John Wiley.

Fairall, N. (1984). The use of non-domesticated African mammals for game farming. *Acta Zoologica Fennica*, 172, 215–8.

Fairall, N. (1985). Manipulation of age and sex ratios to optimize production from impala *Aepyceros melampus* populations. *South African Journal of Wildlife Research*, 15, 85–8.

Hudson, R. J., Stelfox, J. B. & Hopcraft, D. (1984). Wildlife production systems and programmes in Kenya. *Acta Zoologica Fennica*, 172, 225–6.

Johnstone, P. (1975). Evaluation of a Rhodesian game ranch. *Journal of the South African Wildlife Management Association*, 5, 43–51.

Meissner, H. H. (1982). Theory and application of a method to calculate forage intake of wild African ungulates for purposes of estimating carrying capacity. *South African Journal of Wildlife Research*, 12, 41–7.

Mentis, M. T. (1977). Stocking rates and carrying capacities for ungulates on African rangelands. *South African Journal of Wildlife Research*, 7, 89–98.

Mentis, M. T. & Duke, R. R. (1976). Carrying capacities of natural veld in Natal for large herbivores. *South African Journal of Wildlife Research*, 6, 67–74.

Starfield, A. M., Owen-Smith, N. & Bleloch, A. L. (1985). A rule based population model for adaptive management. *South African Journal of Wildlife Research*, 15, 59–62.

Talbot, L. M., Payne, W. J. A., Ledger, H. P., Verdcourt, L. D. & Talbot, M. H. (1965). The meat production potential of wild animals in Africa. *Technical Communications* 16, Commonwealth Agricultural Bureaux.

Walker, B. H. (1979). Game ranching in Africa. In *Management of Semi-Arid Ecosystems*, ed. B. H. Walker, pp. 55–81. Amsterdam: Elsevier.

White, R. J. (1987). *Big Game Ranching in the United States*. Mesilla (New Mexico): Wild Sheep and Goat International.

Young, E. (1984). *Wildboerdery en Natuurreservaatbestuur*, Swartklip: Eddie Young Publication.

13

Game production in western Canada

LYLE A. RENECKER, CHARLES B. BLYTH &
CORMACK C. GATES

Abstract

Since 1970, interest in commercial game production has increased as conventional agriculture searched for innovative ways to diversify. Initially, Elk Island National Park, with its productive populations of bison (*Bison bison*), moose (*Alces alces*), and wapiti (*Cervus elaphus*) served as a model for the emerging game industry. However, several legal and economic forces now favour more intensive systems. Development of commercial bison and wapiti operations was motivated largely by attractive returns for breeding stock (US$ 2500), meat (US$ 8/kg), and velvet antlers (US$ 100/kg). In early 1987, there were 251 commercial operations in the prairie provinces holding 5100 bison and 2900 wapiti. Despite several obstacles including inimical legislation and limited supplies of breeding stock, these numbers are increasing rapidly. This new industry provides an incentive for landscape conservation on private lands and offers a culturally consistent livelihood for native people.

Introduction

The western interior of Canada once supported a highly productive large herbivore community which included bison, wapiti, moose, mule deer (*Odocoileus hemionus*), white-tailed deer (*O. virginianus*), and pronghorn antelope (*Antilocapra americana*). These ungulates have provided a subsistence and economic base for native peoples since glaciers retreated some 12,000 years ago (Ray, 1974). But, by the end of the 19th century, this rich grazing system was largely replaced by conventional agriculture. Today, declining profitability of crop and livestock production has encouraged farm diversification. Though in its infancy, game ranching in western Canada is being actively considered as one of several alternative agricultural technologies.

Emergence of game production

During prehistoric and early historic times, subsistence hunters followed two different hunting patterns. Plains Indians depended primarily on plains bison (*B. b. bison*) for food, clothing, and shelter. Boreal Indians subsisted on a more diverse harvest of furbearing animals, waterfowl, deer, moose, wapiti, wood bison (*B. b. athabascae*), and caribou (*Rangifer tarandus caribou*). To both cultures, the parkland (ecotone between the boreal forest and the great plains) was an important part of their seasonal subsistence pattern. With the onset of winter, bison from the adjacent plains moved northward to the parkland (Roe, 1970) followed by Plains Indians who relied upon them for survival and the wood and habitat afforded by this habitat (Ray, 1974). Concomitantly, some Boreal Indians moved southward to winter in the parkland.

By 1790, the Northwest and Hudson Bay fur trading companies had penetrated the western prairie provinces of Manitoba, Saskatchewan, and Alberta. Trade in pemmican (a mixture of dried pounded meat, fat, and often berries) soon exceeded that of furs. The journals of traders and explorers provide numerous references to the provisioning of large quantities of game meat. During the mid 1800s, Fort Edmonton required approximately 500 bison annually representing a daily consumption of over 300 kg (DeSmet, 1847; Webb, 1967). Most bison were procured from September to May whereas moose, deer, wapiti, and smaller animals provided an alternative food source during summer when bison moved southward to the plains (Losey, 1978).

By the late 1860s, heavy demands on wild herds had taken their toll. The depletion of ungulates did not go unnoticed. Early conservation efforts included legal proclamations such as 'The Buffalo Ordinance' of 1877. This and other legislation was enacted to prevent further depletion of wildlife from uncontrolled hunting, but was largely ineffective. As a result, plains bison were eliminated from the parkland and prairies, and only a few wood bison were left in the northern boreal forest (Soper, 1941). Wapiti were reduced to a few remnant populations.

During the next 30 years, forest succession proceeded rapidly with vastly reduced grazing by native herbivores and fewer fires (Lewis, 1976). With legal land title and rail transport systems in place by the late 1800s, agricultural development and timber harvesting proceeded at an uncontrolled pace. Agricultural development initially concentrated in the most productive landscapes, extending more slowly into marginal northern areas.

Early interest in game production was in response to dwindling numbers of wild animals in the late 1800s (Sifton, 1916). The adaptability of wapiti to intensive management was soon recognised (Lantz, 1910). In 1915, the Dominion Department of Agriculture established an experimental bison/cattalo (bison x cattle hybrid) herd to evaluate prospects of developing a new breed for western agriculture (Rorabacher, 1970).

Renewed interest in game ranching in western Canada stems largely from recognition of the high productivity of wild herbivores, a need for landscape conservation, declining profitability of conventional agriculture, and growing consumer interest in leaner meats. Initial attention focused on ranching natural assemblages of large herbivores including bison (grazer), moose (browser), and wapiti (mixed feeder) (Fig. 13.1). The model for this mixed-species production system was Elk Island National Park (EINP), which seemed to demonstrate that native grazing systems were reasonably productive and practical to manage (Telfer & Scotter, 1975).

Extensive game ranching is a low-input strategy whose economy of scale derives from the relationship between perimeter (fencing costs) and area (production capacity). However, few private landowners control enough land (over 25 km^2) to exploit this extensive option (Hudson, 1981; Hudson & Blyth, 1986) although lands controlled communally by native people are suitable. Despite the advantages, the experience of the first extensive commercial ranch in Canada (the Kikino Metis Settlement in north-central Alberta) was that difficulty controlling stock on large properties results in a loss of economic opportunity, such as harvesting velvet antlers.

The alternative is to dilute investment costs by increasing stocking rate on a smaller land base through seeding tame pastures, fertilising, and winter feeding. These operations exploit the advantages of subdivisional fences to manage grazing intensity, maximise stock productivity through supplementary feeding, and capitalise fully on commercial markets.

Current status

Although the concept has been discussed and practised for nearly two decades, the industry in western Canada is still in its infancy and enabling legislation is just now being developed. Only the Kikino Metis Settlement in central Alberta has established a large-scale ranch although interest among Indian and Metis Settlements is high. In Alberta, at least 19 of 39 settlements are embarking on game ranching ventures (G. Lynch, Alberta Fish and Wildlife Division, personal communication). Because of the financial constraints of development and limited availability of breeding stock, most private entrepreneurs are interested in intensive game farming on smaller properties.

In early 1987, there were 251 commercial game ranches and farms in

Fig. 13.1. Primary species considered in a mixed-species grazing system in the boreal forest. Top, bison (grazer); middle, moose (browser); bottom, wapiti (mixed feeder) (reprinted with permission from Renecker & Kozak, 1987).

Alberta, Saskatchewan, and Manitoba holding 5100 bison and 2900 wapiti (Table 13.1). Most of these operations were small (64–128 hectares) with few operations raising more than 200 head of either species.

The plains bison is legally defined as a domestic animal in most provinces (Chapter 19). The wood bison is officially endangered and therefore cannot be used commercially. Wapiti ranching is permitted in Alberta and Saskatchewan, but slaughter is only allowed in the latter. Currently, meat supplies are absorbed by local markets in Saskatchewan but a small number of carcases serve restaurants in British Columbia, Ontario, Quebec, and several maritime provinces.

Trade of indigenous wildlife has been controlled by stringent provincial regulations except in the Northwest Territories where there is a traditional commercial/subsistence trade among native people. Although game meat is sold in several eastern provinces, most western provinces do not permit the sale of game meat from indigenous wildlife except plains bison. Therefore, the main current commodities are breeding stock and velvet antlers.

Exports of wapiti from Canada to New Zealand from 1984–86 totalled 286 head (J. Parliament, Agriculture Canada, personal communication). Annual sales are predicted to exceed 200 animals in the near future, representing a gross revenue of approximately US$ 540,000. Records of production and a guarantee of temperament are now standard requirements by New Zealand buyers. Within Canada, domestic trade exceeded 400 head during 1986 at a value of approximately US$ 2.5 million. Records indicate that 1092 bison were imported from 1980–86 both for slaughter and breeding purposes (A. Dagenais, Agriculture Canada, personal communication).

Table 13.1. *Bison and wapiti on private ranches in Alberta, Saskatchewan, and Manitoba, May 1987.*

Province	Number of ranches	Number of animals
Alberta		
Bison	113	3,500
Wapiti	85	1,800
Saskatchewan		
Bison	18	800
Wapiti	20	1,000
Manitoba		
Bison	13	800
Wapiti	2	100

Canada's contribution of velvet antler to the primary market in Korea is small compared to the volume exported by New Zealand, China, and the Soviet Union. Canada produced approximately 3200 kg of wapiti velvet in 1986, valued at more than US$ 204,000. Although the major outlet for wapiti velvet has been Korea, buyers from Hong Kong and ethnic markets within Canada will easily absorb velvet antler produced during the early development of the industry. The major constraint is the lack of a 'marketing desk' where velvet can be pooled and sold in bulk to buyers who normally require quantities of at least one tonne.

Land base and range productivity

Most of northern Alberta, Saskatchewan, northern British Columbia, Manitoba, and the subarctic of the Northwest Territories and Yukon are covered by aspen-dominated parkland and boreal forest, some of which has served for livestock grazing, marginal agriculture, fur trapping, sport hunting, and forestry. Although northern landscapes could be used for wildlife production, most jurisdictions restrict commercial operations to private lands in the agricultural zone. The agricultural area which holds the greatest promise for game production covers about 820,000 km^2 (Fig. 13.2).

The boreal forest, aspen parkland, and prairie are best described as a continuum. In the southern parkland, prairie grasslands predominate. Aspen (*Populus tremuloides*) and balsam poplar (*P. balsamifera*) become increasing important components of the canopy as the proportion of meadows decreases with increasing latitude. Coniferous trees, primarily white spruce (*Picea glauca*), black spruce (*P. mariana*), and jackpine (*Pinus banksiana*), increase with latitude, dominating boreal forests in the subarctic. Meadows are scattered throughout the boreal forest and are particularly well developed in the Slave River Lowlands and the Peace/Athabascae Delta. Without perturbations, such as flooding, clearing, and fire, meadows are colonised by woody species and commonly succeed to forest (Rowe & Scotter, 1973).

Production of native forage in the parkland and boreal forest is generally high (Table 13.2). However, production varies widely in response to a number of environmental conditions such as thermal radiation, moisture, fertility, and plant community composition. Nutritional quality of forage is an important criterion of range carrying capacity. During winter, forage digestibility is marginally adequate for maintenance as a result of lower protein (5–8%) and dry matter digestibility (27–45%) (Renecker, 1987). However, the growth pulse during spring and summer, although brief, is sufficiently high (protein content of 20–30%; dry matter digestibility of 70–77%) to support rapid growth and fattening of native ungulates (Renecker, 1987).

Productivity

Records of early fur traders and explorers describe an abundance
of ungulates in the western interior. The journals of Alexander Mackenzie
gave an impression of the productivity of the plains in 1793:

> All this country, to the south branch of the Saskatchewan,
> abounds in beaver, moose-deer, fallow deer, elks, bears, buffaloes,

In 1869, Issac Cowie (Ogilvie, 1979) recorded an encounter with one of the
last large herds of bison:

Fig. 13.2. Vegetation zones of western Canada, according to Zoltai (1975).

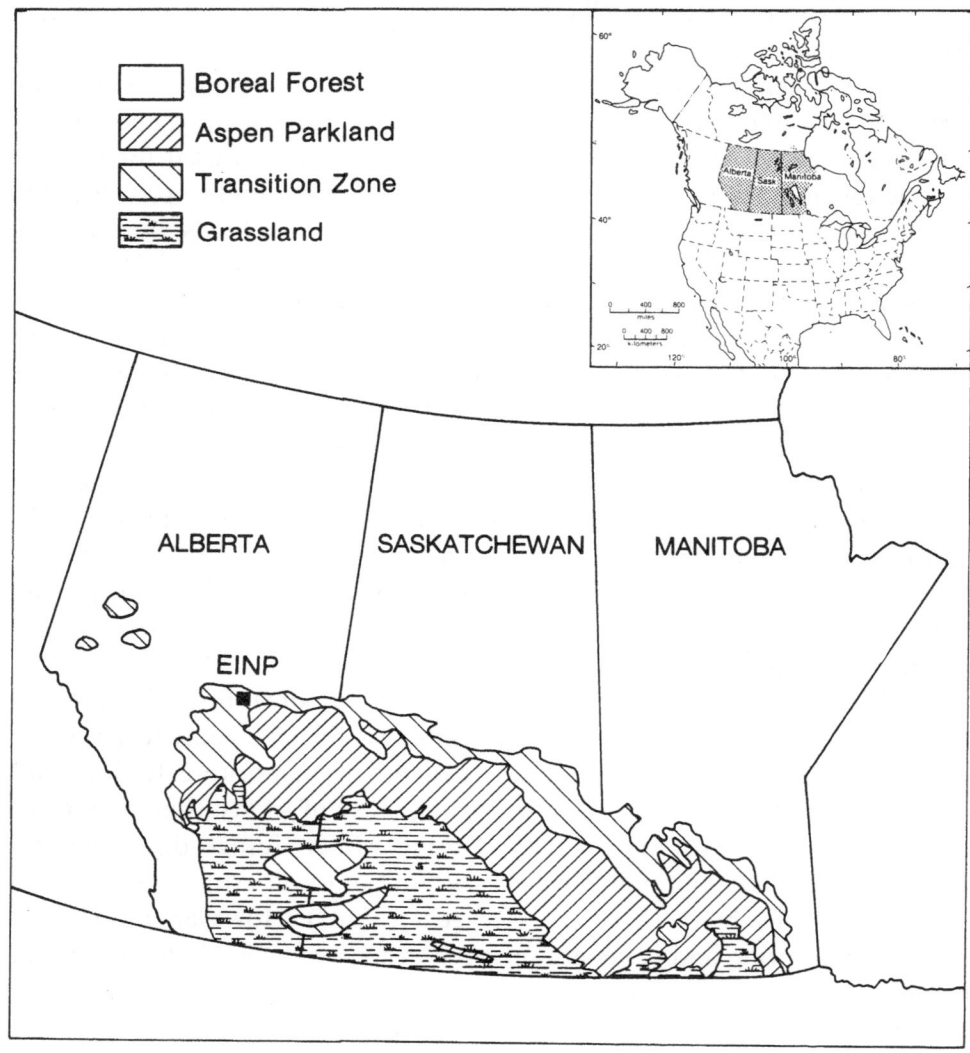

> Our route took us in the midst of the herd which opened in front
> and closed behind the train carts like water round a ship.

Based on range areas and carrying capacities, pristine populations of perhaps
60 million bison and 10 million wapiti have been estimated (Seton, 1974). The
best information on the dynamics of this grazing ecosystem came from the
80-year history of Elk Island National Park.

Stocking and harvests

Since their introduction, in 1907, plains bison have remained the
dominant herbivore at Elk Island followed by wapiti, moose, and deer (Fig.
13.3, Table 13.3). The live ungulate biomass has varied widely around a long-
term average of 58 kg/hectare. Although the carrying capacity for bison is
high at Elk Island ($8/km^2$, 36 kg/hectare), it can be increased ten-fold with
supplemental feeding and pasture management. Similarly, stocking rates of
wapiti under intense management on private farms can exceed 375 kg/hectare
($250/km^2$).

Table 13.2. *Available biomass of forage in habitats utilised by native herbivores.*

Habitat	Peak biomass (kg/hectare)	Reference
Boreal–Parkland transition		
Wet meadows	948-2,640	Telfer & Scotter (1975)
Mesic meadows	516-1,298	Telfer & Scotter (1975)
Upland meadows	861	Telfer & Scotter (1975)
Grassland-snowberry	1,082	Renecker & Hudson (1986a)
Aspen forest	1,590	Renecker & Hudson (1986a)
Willow-sedge	3,523	Renecker & Hudson (1986a)
Scrub-poplar grassland	2,322	Renecker & Hudson (1986a)
Open parkland		
Rough fescue grassland	1,207	Bailey & Wroe (1974)
Mixed-wood boreal forest		
Aspen-cranberry	390	Renecker (personal communication)
Aspen-spruce-bearberry	570	Renecker (personal communication)
Boreal forest		
Wet sedge meadow	2,090	Bailey & Penner (1973)
Dry sedge meadow	516-1,316	Gates (personal communication)
Willow sedge savanna	278-650	Gates (personal communication)
Spruce forest	199	Gates (personal communication)
Dwarf birch-willow bog	336	Gates (personal communication)

Removals by slaughter and live capture has totalled 17,679 animals since 1907. Although the first harvest occurred in 1928, regular annual harvests to control populations did not begin until 1935. Average harvests from this time until 1960 were slightly over 14 kg/hectare. After major reduction campaigns

Table 13.3. *Average carrying capacity and productivity in the northern portion of Elk Island National Park, 1907-85.*[a]

Species	Herd size (animals)	Density (animals/ hectare)	Biomass (kg/hectare)	Herd increment (%/year)	Annual harvest (kg/hectare)
Plains bison	840	0.082	35.6	17.4	5.5
Moose	127	0.029	8.8	—	0.9
Wapiti	511	0.05	12.2	19.1	1.5
Mule deer	118	0.016	0.8	3.8	0.0
White-tailed deer	63	0.005	0.3	46.7	0.02
Total	1,659	0.18	57.7	—	7.92

[a]Current area 11,000 hectares.

Fig. 13.3. Liveweight biomass of plains bison, wapiti, and moose at Elk Island National Park (drawn as proportion of total biomass).

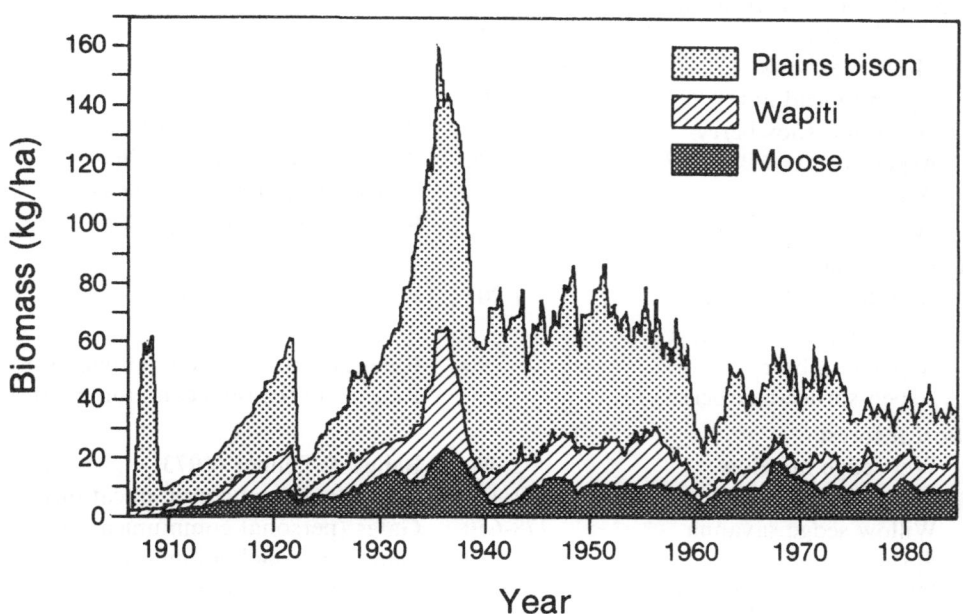

in 1959/60, the annual harvest required to control the new lower population was almost half that required during the previous 25 years, averaging 7.3 kg/hectare on a liveweight basis.

These empirical results are close to predictions by conventional stocking calculations and linear programming. Telfer & Scotter (1975) estimated the carrying capacity to be 54 kg/hectare with a sustained yield of 11 kg/hectare/ year. Hudson & Blyth (1986) used linear programming to explore optimal stocking patterns which would maximise forage utilisation and sustained yield. The analysis indicated that annual sustained productivity could be increased from 9.4 to 14 kg/hectare by increasing the proportion of grassland and maintaining populations of 14 bison, 8 wapiti, and 2.8 moose/km^2.

The problem with such calculations is that they do not account for density-dependence. To correct this deficiency, Blyth (1989) used time series analysis to obtain empirical estimates of K (density at which population growth $= 0$) and more importantly MSY (equilibrium density at which maximum sustained yield is achieved) based on 80 years of data from Elk Island (Table 13.4). The high stocking rates can be attributed to the ecological adaptations and dietary specialisation of these ungulates. Although maximum stocking rate and production can be reached at ecological carrying capacity where plant succession reaches a balance with animal density, because of public pressure for pristine systems, EINP has been managed at or below MSY.

Growth and meat yields

Routine winter slaughters and roundups at Elk Island have provided a great deal of information on growth (Fig. 13.4). Sex differences in the weights of plains bison become pronounced after 3.5 years of age. With pasture management and supplemental feeding, growth rates are higher with

Table 13.4. *Maximum sustained yields (MSY), equilibrium densities (K), and ecological carrying capacities for Elk Island National Park.*[a]

Species	Maximum sustained yield (no./km^2)	Equilibrium density (no./km^2)	Ecological carrying capacity (no./km^2)
Plains bison	2.5	12.0	19.0
Wood bison	1.0	5.0	6.0
Wapiti	2.4	7.0	10.0
Moose	0.8	2.8	5.3

[a]Data for most recent representative time period (Blyth, 1989).

males and females reaching 730 kg and 635 kg, respectively, at 2.5 years of age (S. Biewald, Alexco Foods Ltd, personal communication).

At maturity, female wood bison are larger than female plains bison, but mature weights of males are similar. Wood bison from the Mackenzie Bison Sanctuary in the Northwest Territories had dressed weights which were 54% of liveweights (Gates, personal communication) and 18% higher than carcase weights reported by Halloran (1960) for plains bison.

Wapiti at Elk Island are large relative to other populations of the Rocky Mountain subspecies (Peek, 1982; Blyth, 1989). Dressed weights of males approach 60% of liveweight (Renecker, personal communication). Smith (1973) found that yield of retail meat cuts was 54% of the field dressed weight for wapiti shot by hunters. However, when animals were slaughtered under controlled conditions, the total yield of commercial meat cuts (European style) is 78–81% of the dressed carcase (Renecker, personal communication).

Records of moose weights at Elk Island are incomplete and do not follow a smooth curve, which may reflect some environmental influence upon indi-

Fig. 13.4. Growth curves for ungulates at Elk Island National Park.

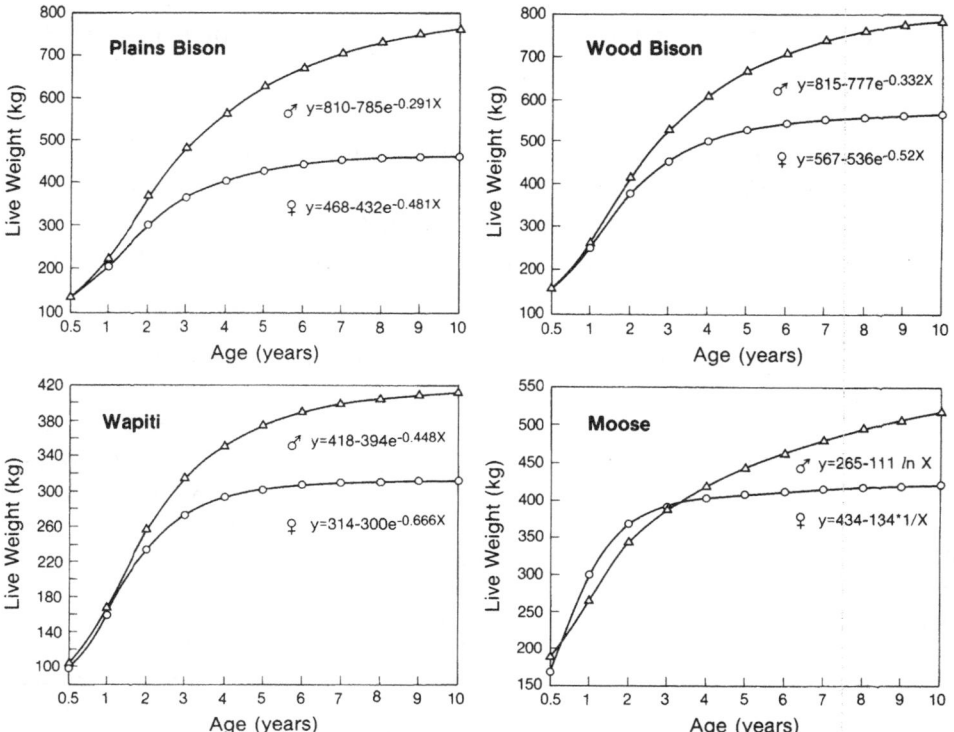

vidual cohorts. Dressing percentages have varied from 47% (Blyth, 1981) to 50% (Blood, McGillis & Lovass, 1967).

Technology and management
Infrastructure
Legislation governing game ranching has evolved as new technology has developed. Initially, the Alberta government stipulated that perimeter fences must be of 9-gauge wire netting at least 2.1 m high, with 15 cm mesh, and with wooden posts spaced at 4 m intervals. Such high fencing standards were considered necessary to withstand challenges from bison, wapiti, and moose, especially when animals are first introduced or during the rut. However, with adequate feed and appropriate management of breeding groups, fences are seldom tested by farm-bred stock and standards can be relaxed with substantial reductions in development costs (Table 13.5). Provincial governments in western Canada have since relaxed these requirements acknowledging the effectiveness of lighter high tensile netting.

Adequate yard construction is costly, but the benefits are justified. Handling facilities for bison consist of a working alley of several stanchions separated by sliding gates and with one side hinged for quick release of excited animals. Squeezes constructed for bison are necessarily larger and stronger than those for cattle and are equipped with a stop gate.

Handling facilities for wapiti can be of lighter construction. Animals are sorted using a round drafting corral into small holding pens with 2.4 m walls of vertical boards. They can be moved singly into the crush which has a drop-floor and side panels made of either foam or plywood.

Range management
Knowledge of habitat and forage preferences is fundamental for evaluating stocking rates on extensive properties. For plains bison, sedges and grasses comprise the majority of the diet throughout the year. In spring and summer, browse (primarily willow) is a major component of wood bison diets in the Mackenzie Sanctuary suggesting diet diversification when forage quality is high. However, wood bison depend largely on sedge during winter periods (96% of wood bison diet in the Mackenzie Bison Santuary) (Gates, personal communication).

Wapiti are mixed feeders and show greater versatility than bison. In winter, wapiti may include more than 60% browse in their diet; however, during early spring foraging shifts to green growth in sedge meadows and finally forbs and leaves predominate in diets of late spring and summer (Gates & Hudson, 1981, 1983; Nietfeld, 1983). In autumn when forage quality declines with plant senescence, wapiti concentrate on leaf litter which can comprise almost 60% of their diet (Nietfeld, 1983). When snow conditions during winter permit, wapiti crater for fine grasses, legumes, and leaves. As snow deepens and available biomass is reduced, woody twigs become increasingly important.

The staple winter foods of moose are woody twigs but dependence on bark increases during early spring (Cairns, 1976; Renecker, 1987). During spring and summer, moose increase intake to about 10 kg dry matter/day by stripping leaves from trees and shrubs (Renecker & Hudson, 1985) and increase their selectivity for green leaves and forbs during autumn when variation in forage quality is greatest (Renecker & Hudson, 1986a). Although moose are extremely cold tolerant they are easily heat stressed and rely

Table 13.5. *Specifications and costs of game fencing.*

Materials	Specifications			Cost (US$/km)
	Mesh dimensions (cm)	Smooth wire spacings (cm)	Height (m)	
Perimeter fence				
9 gauge wire netting (Watchman)	15	—	2.1	6,800
12 gauge netting (1.9-2.0 m high) plus 2-wire electric	5-20	20	2.4	3,800
15-wire electric	—	5-20	2.4	2,500
Race fence				
12 gauge netting (2.0 m high) plus 4-wire electric	5-20	20	2.8	3,400
Subdivisional fence				
12 gauge netting (1.2 m high) plus 4-wire electric	5-20	20	2.0	2,300
12 gauge netting plus 1-wire electric offset 30 cm on outrigger	5-20	—	1.9	2,700
15-wire electrified (wood posts every 12 m and 2.4 m fibreglass stays every 3.7 m)	—	5-20	2.4	2,300

heavily upon wetlands to mitigate thermal imbalances (Renecker & Hudson, 1986b).

There are benefits to hastening the flush of plant growth in spring. Fire can be used as a tool to remove litter from meadows, topkill shrubs and trees, remove decadent growth, maintain the browse supply, and to recycle nutrients (Wright & Bailey, 1982). Similarly, fire can be combined with herbicide spray programme to reduce forest cover and expand grasslands. However, use of fire is controlled both provincially and regionally in western Canada and may be restricted when forest fire hazards are high.

Husbandry

The cycle of management of a representative commercial game enterprise based on wapiti and bison is illustrated in Fig. 13.5. Calving occurs between mid May and early June for both bison and wapiti. Government regulations in Alberta require that wapiti calves be tagged and registered within 30 days of birth. Hinds are usually placed in small pastures prior to parturition for close observation, protection from predators, and habituation to humans.

Velvet antlers are removed from wapiti stags in early June using methods described by Denholm (1979). Until recently veterinarians have been constrained by lack of a suitable antagonist for xylazine, the general sedative-

Fig. 13.5. Management calendar for a bison–wapiti game ranch or farm.

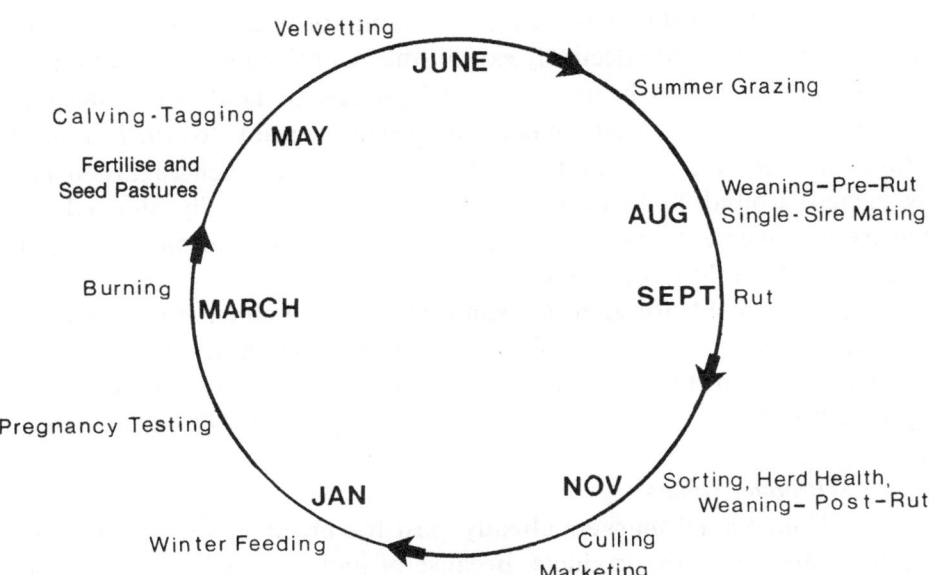

analgesic used in the procedure. Recent studies have shown that a combination of yohimbine and 4-aminopyridine is an effective antagonist for xylazine, eliminating problems of prolonged recumbency (Renecker & Olsen, 1986a,b).

There is a controversy of whether to wean wapiti before or after the rut. The advantages of pre-rut weaning in early September are that high-quality forage is readily available. Climatic conditions are usually less favourable when animals are weaned in November which can result in static body growth. On the other hand, if calves remain with the hinds during the autumn rut they learn the location of supplementary winter feed. For bison, the rut begins in mid July when forage production and quality are high. Generally, bison calves are not weaned until the autumn roundup.

As stocking rates increase on intensive operations, supplementary winter feeding is necessary. A feeding programe usually begins in November and can be used to train stock to move into corrals for handling. For wapiti, a combination of alfalfa and whole oats provides an adequate winter ration. Generally, mature wapiti hinds consume 4.5–6.6 kg dry matter/day of alfalfa during winter (Renecker, personal communication). Bison are less selective grazers and can best utilise grass hay supplements. Overwinter loss of perhaps 10% of bodyweight is advisable to capitalise on compensatory gains in spring and to minimise calving difficulties.

Production and economics
Investment and production costs
Only government agencies, community projects, and native groups possess the land base to effectively exploit the extensive management option. Most private operators begin with small parcels of land which meet the minimum government requirements (65 hectares prior to 1987, now 45 hectares in Alberta). To dilute financial investments, fences are usually constructed around 8–20 hectare parcels, the area is fully stocked, and intensive management practices are employed. These operations can expand as a positive cash flow is realised.

Capital investments for game are similar to those of conventional livestock enterprises, but variable costs of production are comparatively low (Table 13.6). Under circumstances where beef is produced at US$ 1.24/kg, cash costs of production may be US$ 0.94/kg for bison and US$ 0.86/kg for wapiti.

Product values
Commercial markets already exist for breeding stock, velvet antlers, meat, and other by-products. Because of limited supply, current prices

Table 13.6. *Representative budget for a wapiti ranch which would encompass 240 hectares when the maximum breeding population is reached.*

Item	Cost (US$)
Capital investment	
Over 8-year period	
Land (@US$ 400/hectare)	104,000
Boundary fence	38,000
Yards	8,800
Internal fence	19,000
In year 1 only	
Machinery	
(tractor, front end loader,	
ATC, truck, trailer, etc.)	30,000
Crush	2,200
Electronic scale	3,000
Breeding stock	
5 mature stags (@ US$ 3,330)	16,650
45 mature hinds (@ US$ 2,590)	116,550
Annual variable costs at enterprise maturity	
Velvet antler removal (@ US$ 22/animal)	3,030
Herd health (@ US$ 7.4/animals)	4,820
Sorting, culling, etc. (@ US$ 7.4/animal)	4,880
Supplemental feed	22,940
Fuel, oil, grease, equipment maintenance	
utilities, taxes	2,370
Pasture management (@ US$ 14.8/hectare)	3,850
Annual fixed costs at enterprise maturity	
Principal plus interest (20% down	
payment; loan for 30 year period	
@ 10.75% interest)	24,617
Depreciation (enterprise maturity)	6,000
Total fixed plus variable costs	72,507
Annual returns at enterprise maturity	
Velvet antler sales (400 kg Super A Grade	
@ US$ 92/kg)	36,800
Live animal sales (112 yearlings @ US$ 2,220)	246,400
Meat sales (culled and surplus stock;	
18,485 kg @ US$ 4.70/kg)	86,880
Total returns	370,080
Annual balance sheet summary at enterprise maturity	
Cash inflow	370,080
Cash outflow (fixed plus variable costs)	72,507
Net profit	297,553

for breeding stock are high, ranging from US\$ 3330–8600 for mature wapiti stags and US\$ 2590–4700 for mature hinds. Bison bulls are valued at about US\$ 1440 and cows between US\$ 1100–1800.

Both export and domestic markets are strong. New Zealand has been the primary buyer of Canadian wapiti but increasingly animals are being redistributed to meet a growing local demand. A strong domestic market also exists for bison. With renewed interest and a higher price for leaner meats, producers are scrambling to obtain breeding stock, inevitably inflating its value.

Velvet antlers from wapiti stags provide a major source of revenue. However, yields vary widely with both age and plane of nutrition. Stags pastured on native range and offered no supplementary feed will produce approximately one kilogram of velvet antlers at one year of age. Production tends to increase at a rate of one additional kilogram/year until the stag is 8 years old when production remains static or declines marginally. However, with supplemental feeding, production can be increased to at least 1.5–1.75 kg as a yearling with an additional 1.5–1.75 kg of production each year until the animal is 8 years old when production decreases. If harvested at precisely the right time in early June, a mature stag can produce velvet antler valued at US\$ 740.

Ultimately, the industry depends upon the sale of meat. In Saskatchewan, Quebec, and Ontario, the only provinces permitting slaughter of farmed wapiti, the live weight and carcase values of wapiti are US\$ 4.75/kg and US\$ 11.10/kg, respectively. Currently, bison are sold on domestic markets for twice the price of beef (US\$ 2.10/kg live weight and US\$ 4.90/kg carcase weight). Export opportunities exist but the product must be in stable supply and consistent in quality if interest is to be maintained. Because of constraints on meat quality and hygiene required by the European Economic Community, it is doubtful whether field slaughter could be permitted for meat destined for export.

Although hunting of ranched game generally is prohibited, on several Indian and Metis Settlements hunters can purchase a bison hunt. The head and cape are valued at US\$ 3500. Traditional activities of Indian and Metis people, such as horse-back riding, guiding, and camping can be combined to provide spin-off benefits of US\$ 2400/package.

Implications

The main goals of commercial game production are to diversify animal agriculture, to offer a culturally consistent livelihood for native people, and to encourage landscape conservation. At a time when the profitability of conventional crop and livestock production in Canada is declining, the combination of low operating costs and high value of bison and wapiti presents a new opportunity for farm diversification. Game ranching offers more efficient use of marginal lands and opens new market opportunities.

Native people are important potential beneficiaries. In a recent report (DIAND, 1984), the Canadian Department of Internal Affairs and Northern Development expressed concern about rising costs of access to fish and game resources for many native people in the Northwest Territories. The report noted a decline in reliance on game and fish resources related to the high cost of harvesting in terms of both capital outlay and operating costs. One of the suggested remedial measures was to develop alternative food production technologies suited to the northern environment. Game production ranks high among alternative technologies which could contribute to regional self-sufficiency and which would be consistent with traditional native lifestyles.

The most controversial aspect of game production is its environmental impact. Hard-fought debates have raged between those who perceive game ranching as an environmentally sensible agricultural alternative, and those who predict the destruction of wilderness values and loss of hunting opportunities. Although game ranching may provide a tangible benefit for private landowners who apply conservation principles to their properties, there is fear that this will be counterbalanced by poaching if the sale of game meat is permitted (Geist, 1985). On the other hand, the constant supply of venison into the marketplace and stringent processing regulations may reduce the incentive to market poached meat (Renecker & Kozak, 1987).

As this new industry develops, the welfare of wild species on commercial operations will come under greater public scrutiny. Procedures such as capture, mechanical restraint, chemical immobilisation, and velvet antler removal will be questioned. For the industry to develop with strong public support, considerable emphasis must be placed on animal welfare. For this reason, provincial game producer associations have developed Codes of Ethics which function as guidelines for the new industry.

References

Bailey, A. W. & Penner, D. F. (1973). *The Potential of Rangelands in the Fort Providence, NWT Area to Support Muskox: An Interim Report on Research Studies, 1973*. Report to Government of the North West Territories. Contract No. 73-1-252.

Bailey, A. W. & Wroe, R. A. (1974). Aspen invasion in a portion of the Alberta parklands. *Journal of Range Management*, 27, 263-6.

Blood, D. A., McGillis, J. R. & Lovass, A. L. (1967). Weights and measurements of moose in Elk Island National Park, Alberta. *Canadian Field Naturalist*, 81, 263-9.

Blyth, C. B. (1981). *Moose Reduction Program, Elk Island National Park, December 1, 1980 to December 12, 1980.* Parks Canada Report on Resource Conservation, Elk Island National Park, Alberta.

Blyth, C. B. (1989). *The Dynamics of the Parkland Grazing System at Elk Island National Park, Central Alberta.* M.Sc. Thesis. Edmonton: University of Alberta.

Cairns, A. L. (1976). *Distribution and food habits of moose, wapiti, deer, bison and snowshoe hare in Elk Island National Park, Alberta.* M.Sc. Thesis. Calgary: University of Calgary.

Denholm, L. J. (1979). Veterinary aspects of antlerogenesis and commercial velvet antler production. In *Proceedings No. 49 Deer Refresher Course*, pp. 253–9. Sydney: The Australian Museum.

Department of Indian Affairs and Northern Development (DIAND). (1984). *Northern Food Costs – Overview.* Ottawa: Department of Indian Affairs and Northern Development, Economic Strategy Division.

DeSmet, P. (1847). *Oregon Missions and Travels over the Rocky Mountains in 1845–46.* New York: E. Dunigan.

Gates, C. C. & Hudson, R. J. (1981). Habitat selection by wapiti in a boreal forest enclosure. *Naturaliste canadien*, 108, 153–66.

Gates, C. C. & Hudson, R. J. (1983). Foraging behaviour of wapiti in a boreal forest enclosure. *Naturaliste canadien*, 110, 197–206.

Geist, V. (1985). Game ranching: threat to wildlife conservation in North America. *Wildlife Society Bulletin*, 13, 594–8.

Halloran, A. F. (1960). American bison weights and measurements from the Witchita Mountains Wildlife Refuge. *Proceedings of the Oklahoma Academy Science*, 41, 214–8.

Hudson, R. J. (1981). *Agricultural Potential of Wapiti.* 60th Annual Feeders' Day Report, pp. 80–6. Edmonton: Department of Animal Science, University of Alberta.

Hudson, R. J. & Blyth, C. (1986). Mixed grazing systems of the aspen boreal forest. In *Rangelands: A Resource Under Siege*, ed. P. J. Joss, P. W. Lynch & O. B. Williams, pp. 380–3. Canberra: Australian Academy of Science.

Lantz, D. E. (1910). *Raising Deer and Other Large Game Animals in the United States.* Bulletin Number 36. Washington: United States Department of Agriculture, Biological Survey.

Lewis, H. T. (1976). Maskuta: The ecology of Indian fires in northern Alberta. *Western Canadian Journal of Anthropology*, 7, 15–53.

Losey, T. C. (1978). *Prehistoric Cultural Ecology of the Western Prairie-Forest Transition Zone, Alberta, Canada.* M.Sc. Thesis. Edmonton: University of Alberta.

Nietfeld, M. T. (1983). *Foraging behavior of wapiti in the boreal mixed-wood forest, central Alberta.* M.Sc. Thesis. Edmonton: University of Alberta.

Ogilvie, S. C. (1979). *The Park Buffalo.* Calgary: National and Provincial Parks Association of Canada.

Peek, J. M. (1982). Elk (*Cervus elaphus*). In *Wild Mammals of North America: Biology, Management and Economics*, ed. J. A. Chapman & G. A. Feldhamer, pp. 851–61. Baltimore: Johns Hopkins University Press.

Ray, A. J. (1974). *Indians in the Fur Trade: Their Role as Hunters, Trappers and Middlemen in the Lands Southwest of Hudson's Bay.* Toronto: University of Toronto Press.

Renecker, L. A. (1987). *Bioenergetics and Behavior of Moose (*Alces alces*) in the Aspen-Dominated Boreal Forest*, Ph.D. Thesis. Edmonton: University of Alberta.

Renecker, L. A. & Hudson, R. J. (1985). Estimation of dry matter intake of free-ranging moose. *Journal of Wildlife Management*, 49, 785–92.

Renecker, L. A. & Hudson, R. J. (1986a). Seasonal foraging rates of free-ranging moose. *Journal of Wildlife Management*, 50, 143–7.

Renecker, L. A. & Hudson, R. J. (1986b). Seasonal energy expenditures and thermal regulatory responses of moose. *Canadian Journal of Zoology*, 64, 322–7.

Renecker, L. A. & Kozak, H. M. (1987). Game ranching in western Canada. *Rangelands*, 5, 213-6.

Renecker, L. A. & Olsen, C. D. (1986a). Antagonism of xylazine hydrochloride with yohimbine hydrochloride and 4-aminopyridine in captive wapiti. *Journal of Wildlife Diseases*, 22, 91–6.

Renecker, L. A. & Olsen, C. D. (1986b). Reversing Rompun: The North American experience. *The Deer Farmer*, 28 (Dec–Jan), 31.

Roe, F. G. (1970). *The North American Buffalo: A Critical Study of the Species in its Wild State*. Toronto: University of Toronto.

Rorabacher, J. A. (1970). *The American Buffalo in Transition: A Historic and Economic Survey of the Bison in America*. St Cloud: North Star Press.

Rowe, J. S. & Scotter, G. W. (1973). Fire in the boreal forest. *Quaternary Research*, 3, 444–64.

Seton, E. T. (1974). *Life-Histories of Northern Animals. An Account of the Mammals of Manitoba. Grass Eaters*. New York: Arno Press.

Sifton, C. (1916). *Conservation of Fish, Birds and Game. Proceedings of a meeting of the Committee on Fisheries, Game and Fur-bearing Animals, Commission of Conservation Canada*. Ottawa, 1 and 2 November, 1915. Toronto: Methodist Book and Publishing House.

Smith, F. C. (1973). *Quality and Quantity of Meat from Big Game Carcasses of Wyoming*. M.Sc. Thesis. Laramie: University of Wyoming.

Soper, J. S. (1941). History, range and home life of the northern Bison. *Ecological Monographs*, 11, 347–412.

Telfer, E. S. & Scotter, G. W. (1975). Potential for game ranching in boreal aspen forests of Western Canada. *Journal of Range Management*, 28, 171–80.

Webb, R. (1967). *The Range of White-Tailed Deer in Alberta*. Edmonton: Alberta Department of Lands and Forests.

Wright, H. A. & Bailey, A. W. (1982). *Fire Ecology, United States and Southern Canada*. New York: John Wiley.

Zoltai, S. C. (1975). *Southern Limit of Coniferous Trees on the Canadian Plain*. Information Report NOR–X–128. Edmonton: Environment Canada, Forestry Service.

14

Ranching native and exotic ungulates in the United States

TERENCE P. YORKS

Abstract

Laws pertaining to the management of game animals in the United States are built on two basic facts. First, the history of European settlement is replete with local extinction or near extinction of wild ungulates through excessive harvests. Second, wildlife is legally considered a public rather than private resource. However, both native and exotic animals are being increasingly seen as economically valuable. Through extension of trespass laws, fee hunting is becoming a primary management objective on private lands. Trade in game meat and by-products, despite growing recognition of health, culinary, and other benefits, remains limited by the legal prohibitions assumed necessary to protect the public resource, and by an inadequately developed marketing infrastructure. Intensive game farming in fenced enclosures is restricted by old laws, the animal welfare lobby, and the belief that properly managed free-ranging combinations of native animals are more productive.

Introduction

Native ungulates raised for commercial purposes in the continental United States include most of the species present at the time of Columbus's arrival. Of particular note are bison (*Bison bison*), wapiti (*Cervus elaphus*), deer (*Odocoileus virginianus* or *O. hemionus*), pronghorn antelope (*Antilocapra americana*), and occasionally bighorn sheep (*Ovis canadensis*). 'Commercial production' ranges from highly organised, tightly controlled efforts on small acreages for bison, wapiti, and deer to purely sport activities on wholly unfenced, public lands (White, 1986).

In this chapter, the interaction between public game wardens and private managers provides a dominant theme. The resultant desire to overcome the limitations of a well-entrenched bureaucracy in control of all aspects of native game utilisation has played a significant role in bringing several dozen species of 'exotic' ungulates from other continents into commercial production enterprises, first for sport hunting and later for meat production. Some of the principal exotics now utilised for commercial purposes include red deer (*Cervus elaphus*), axis deer (*Axis axis*), blackbuck (*Antilope cervicapra*), mouflon (*Ovis musimon*), aoudad (*Ammotragus lervia*), sika deer (*Cervus nippon*), fallow deer (*Dama dama*), nilgai (*Boselaphus tragocamelus*), kudu (*Tragelaphus imberbis*), and Russian boar (*Sus scrofa*).

Management of game animals in the United States as an economic resource is by no means a new idea. Tribal territorialism on the part of pre-Columbian populations was centred to an important degree around known ranges of native ungulates. Meat, hides, and by-products of these animals were simply critical to these people for survival, comfort, and trade (Josephy, 1968). Management practices may have included selective harvesting to maintain desired population levels and the use of fire to maintain vegetation quality (Martin, 1978).

Yet, good management by native Americans of their animal resource was not universal. It may even be possible that post-Pleistocene megafauna (e.g., mastodons, mammoths, and ground sloths) were pushed into final extinction by overharvesting, and both individuals and tribes wasted meat and/or joined in the post-Columbian excess harvests to supply the fur trade (Martin, 1978). Despite such contradictory examples, knowledge by pre-Columbian tribes of animal habits was almost certainly superior to our own (Jacobs, 1980). There remains a reservoir of knowledge for contemporary management in the yet not too distant past.

With the arrival of Europeans, commercial harvests of native game began, and took place with essentially no concern for their effects on animal populations. By the early 1800s, the severe impacts of game harvests and subsequent land clearing were already being felt. For example, the last wapiti in Indiana was apparently killed in 1818 (Butler, 1934). James Fenimore Cooper's (1841) then well-known *Leatherstocking Tales* have frequent comments on habitat degradation, the consequences of which are more familiar to most people from the writings and work of modern-day environmentalists.

Among native ungulates, only the deer have maintained a range or numbers even remotely like that of their ancestors. One reason may be seen in the records of Lewis and Clark, who maintained their expedition's health through the daily harvest of approximately 15 kg/person. The larger harvest of game meat by commercial hunters in the early days of America would have been even more prone to loss by spoilage with an even greater impact on animal numbers.

Today, with refrigeration, and a higher value placed on a limited resource, beef consumption averages only about 0.25 kg/day on a live-weight basis.

Given the disparity in effective consumption between past and present, and a domestic animal herd equal in magnitude to the total pre-Columbian native animal population to meet this relatively modest current demand, the reason for reductions in game populations with the movements of the pioneers should be obvious.

To underline the quantitative impact of this early market hunting, 600,000 white-tailed deer hides were shipped from just the port of Savannah, Georgia, from 1755–73 (Rue, 1978), despite rather primitive weapons and distribution systems. The subsequent and better publicised demise of the massive bison herd can be seen as but the rapid culmination of two centuries of rather heedless exploitation which had occurred regardless of attempted legal regulation. Enactment of restrictions on game utilisation in the United States began in 1646 in Rhode Island (Rue, 1978).

The rapid decline of wild ungulate numbers, along with the extirpation of many populations, eventually resulted in more effective laws in all states, including absolute prohibitions on trading game meat. These long-standing laws have the unintended impact of making ranching of native animals much more difficult. Where native animals were able to maintain even part of their original range, the corollary tradition of public ownership of game animals and consequent free access to them has persisted strongly, even on lands otherwise perceived as private. While harvesting regulations have become universal, user ('trespass') fees to compensate landowners are often seen as grossly undemocratic (Anon., 1986a). However, the bison, through its near extinction and subsequent movements controlled by private individuals, has rather curiously moved from legal consideration as a public resource to strictly private.

Meanwhile, human tastes turned away from lean, distinctly flavoured game to relatively bland, inseparably fatty, 'civilised' domestic meats. In a practical vein, poor handling of venison by individual hunters exacerbated the public image of native meats as strongly flavoured and tough. The 'macho' image of animal carcasses brought home from the field on the vehicle fender does little for the quality of the meat, and offends public sensibilities.

In contrast to the commonly held view that European livestock must somehow be superior producers to native animals, visionaries have made sporadic attempts to harness the potential of native ungulates. True domestication was tried most often with wapiti; whereas the animals bred quite successfully on private estates, they proved less tractable as draft animals (Kellogg, 1887).

By the 1950s, there was considerable interest, particularly in Texas, in the potential for sport fee hunting on private lands, especially since local economics had become such that deer hunting leases had a gross return at least equal to domestic livestock operations (Ramsey, 1965). A concurrent surge in importation and herd development took place for a variety of exotic ungulates. This began in the 1920s, first as a curiosity, then as a sport hunting resource, and more recently as a source of meat. The interest in exotics has a less commercial correlate in the raising of species that are endangered or threatened on other continents, which includes restocking efforts of animals like Pere David's deer (*Elaphurus davidianus*) to China and the preservation of these and other unique gene pools.

The overall size of the commercial game industry remains quite small compared to that for domestic livestock. Pat Hoctor of the *Animal Finder's Guide*, Prairie Creek, Indiana, suggested (personal communication) that the annual gross value of all trade in native and exotic mammals in the United States for zoos, private herds, and other commercial production amounted to about US$ 300 million distributed among 17,000 active animal raisers. While this is a large number, the domestic cattle industry grossed an equal amount every *three days* in slaughter animal value alone.

There is but one non-domestic meat distributor in Texas, the state with the largest production (M. Schultz, Texas Agricultural Extension Service, personal communication). While other states have operations such as Durham Ranch in California, the Wymans in Colorado, and several fallow deer farms in New York, their yearly total meat handling is less than the hourly output of a major domestic meat processor like Monfort of Colorado. The total volume of corporate game meat processing is small enough that wildlife management texts tend to look at commercial production as a trivial enterprise in comparison with non-commercial production on lands managed for recreation. For comparison, the calculated recreational value of hunting in the sparsely populated state of Wyoming alone is estimated to exceed US$ 200 million (Anon., 1985a). Fortunately for the game industry, the conceptual boundaries between hunting and commercial production are becoming increasingly blurred.

There is a surge of interest in a culinary variety of meats, as well as in the avoidance of fats, cholesterol, steroids, and antibiotics which are associated with more conventional meats. Native and exotic ungulates appear to have advantages in all of these categories. For example, choice beef averages 49% water, 380 kcal/100 g, 14% crude protein, and 41% fat (Watt & Merrill, 1963), in contrast to wapiti which are 70% water, 175 kcal/100 g, 22% crude protein, and 6% fat, while deer and antelope are similar in general composition to the wapiti (Field, Smith & Hepworth, 1973). Purchasable game meat has a strong appeal for those disinterested in hunting and/or butchering their own supply.

There also has been a 'back to the land' movement, with a search for traditional values. Because of human population increases, people are less able to find places where they can 'get away from it all', and game animals represent an aspect of land perceived to be 'wild'. There has been a decline in the quality of hunting on public lands and access to private properties.

Accordingly, there has been increasing interest in paying to hunt on private land to assure limited competition as well as for better trophy quality. Hunter lease fees, which could either be classified under recreation or called a part of commercial game production, amounted to US$ 110 million for native deer in Texas in 1974 (Teer, 1984). When this quantified hunter interest is combined with an unsaturated market for restaurant or packaged game meat, there is every reason to expect explosive growth of commercial game ranching in the United States in the immediate future.

Size and structure of commercial production

The overall extent of the commercial industry is approximately US$ 300 million when rated as the value of all 'wild' animals in trade. This is a considerable sum, and one that is only part of the total. However, only a fraction directly applies to ungulates, and each operation trades only US$ 20,000 on average per year, so most must be very small indeed. On the other hand, this statistic is for actual sale of animals. The overall structure of commercial game related activities could be divided into two classes, each with a subclass, i.e., fee hunting of native game or exotics, and native game or exotics for meat and by-products.

Fee hunting of game on individual property means payment to a land-owner for access to the animals (also called a 'trespass fee'). This can range from less than US$ 100 to enter the land and remove a female antelope, to over US$ 5000 for a guided hunt with provision of a vehicle, meals, lodging, and other services to harvest a trophy wapiti or a large exotic. Generally speaking, the price varies with the attendant services, trophy size, and the relative rarity of the animal. Animals in these operations may be either contained within game fences or free-ranging. A slight variant on the fee hunt is a lease, where an individual or a group rents a piece of property for a specified period for the purpose of hunting.

Production of meat and by-products, including leather and velvet antlers, must by law be accomplished within game-proof fences, since the American tradition of public ownership of wildlife requires that meat sold comes from animals originating outside the local area. This requirement precludes use of locally adapted game animals.

An immediate problem with examining the specifics of the game industry also relates to the historical character and perception of individual independence. Game operations have often developed with an adversarial relationship to the law. They fall outside common practice, are widely scattered, and are too variable to be easily regulated. Further, game ranches tend to occur in more rural areas, distanced from the concerns of centralised authority. The laws were developed simplistically and with the assumption that the only way to stop poaching was to stop any trade at all. The legal power of game wardens is counterbalanced by a willingness to ignore small operations. Many operators perceive a danger in providing statistical information about their enterprises. They feel embarrassed if it is small, or fear taxation and further regulation if it is large. Only the bison industry nationally, through the National Buffalo Association, and the exotics industry, with the Texas Exotic Association, have so far formed producers' organisations which have consistently profiled the statistics of their industry.

Table 14.1 presents the most careful documentation of the commercial production of ungulates in the United States, by following the growth of the Texas exotics industry. To put this in proper perspective, less than 100,000 exotic ungulates of all species need to be compared with the more than 2.5 million white-tailed deer (Cook, 1984) and 14 million cattle (Anon., 1986b) that exist in Texas. In economic terms, however, the impact of this relatively small number of exotics is considerably greater than might be expected. For most prospective hunters, photographers, or meat consumers, a substantial cost to fulfil a desire to obtain such animals locally becomes modest when compared to a journey to other continents to find these animals in native habitats.

Mike Hughes of the Texas Wild Game Cooperative, which is the sole marketer of exotic meat in that state, reports (unpublished data) that the five largest exotic ungulate ranches supply half of the state's annual total of 600 trophy and 800 live axis deer, as well as 200 trophy and 600 live blackbuck. Roughly 400 axis deer are harvested for meat annually, although Hughes estimated current demand is 20 times that number.

Of the 500,000 hunters of native deer in Texas, about half lease or own the land on which the animals are procured (Teer, 1985). This percentage reflects a ten-fold increase in leasing in the 50 years such activity has been surveyed. The percentage of land leased and/or owned outright where wild ungulate production is of significant importance is commonly assumed to be particularly high in Texas, and due to the small percentage of public lands in that state.

Yet, in a recent survey in Wyoming, where half of the state is publicly owned, more than half of the landowners indicated that fee hunting formed

Table 14.1. *Population trends of introduced artiodactyls in Texas.*[a]

Common Name(s)	Scientific Name	1966	1971	1974	1979	1984
Axis deer	*Axis axis*	6,450	11,171	19,581	22,799	38,035
Blackbuck antelope	*Antilope cervicapra*	4,125	5,470	7,339	9,639	18,789
Mouflon	*Ovis musimon*	10,000	16,169	15,254	9,536	15,394
Aoudad/Barbary sheep	*Ammotragus lervia*	1,300	3,217	3,531	8,451	14,651
Fallow deer	*Dama dama*	445	2,617	4,483	7,922	10,507
Sika deer	*Cervus nippon*	875	2,036	3,042	6,217	7,956
Nilgai	*Boselaphus tragocamelus*	4,000	4,120	2,786	2,241	
Russian boar	*Sus scrofa*	10,000			1,123	
Wapiti	*Cervus elaphus*				969[b]	
American bison	*Bison bison*				761[b]	
Ibex/Ibex cross	*Capra* sp.		6	25	485	
Red deer	*Cervus elaphus*	95	307	407	434	
Addax	*Addax nasomaculatus*		2	22	172	
Eland	*Taurotragus oryx*	25	66	78	150	
Red deer/wapiti hybrid	*Cervus* sp.			88	126	
Lotham sheep	*Ovis* sp.		150	200	126	
Llama	*Lama glama*	8	22	116	107	
Argali/mouflon hybrid	*Ovis* sp.		1		104	
Barasingha	*Cervus duvauceli*	40	49	51	79	
Zebra	*Equus zebra*	10			76	
Markhor goat	*Capra falconeri*			7	55	
Springbok	*Antidorcas marsupialis*		4	6	53	
Arabian gazelle	*Gazella arabica*		9		52	
Oryx	*Oryx* sp.	9	38	42	82	
Beisa oryx	*Oryx beisa*				45	
Gemsbok	*Oryx gazella*			43	41	
Scimitar horned oryx	*Oryx tao*				32	
Sable antelope	*Hippotragus niger*	4	7	13	28	
Pigmy goat	*Capra* sp.		5	42	25	
Defassa waterbuck	*Kobus defassa*				24	
Common waterbuck	*Kobus ellipsiprymnus*			1	1	
Camel (one hump)	*Camelus dromedarius*			12	23	
Sambar	*Cervus unicolor*	60	85	50	22	
Black wildebeest	*Connochaetes gnu*	5				
Blue wildebeest	*Connochaetes taurinus*				19	
Giraffe	*Giraffa camelopardalis*	4	8		17	
Impala	*Aepyceros melampus*	1	24	9	16	
Roan antelope	*Hippotragus equinus*				12	
Greater kudu	*Tragelaphus capensis*	3	19		12	
Blesbok	*Damaliscus dorcas*	1	10	8	11	
Muntjac	*Muntiacus muntjac*				10	
Sitatunga	*Tragelaphus spekii*	5	14	31	10	
Dama gazelle	*Gazella dama*				9	
Lesser kudu	*Tragelaphus imberbis*			25	8	
Guanaco	*Lama guanacoe*	1		4	5	
Nyala	*Tragelaphus angasi*		1	1	4	
Banteng	*Bibos javanicus*				3	

Table 14.1. (*cont.*)

Common Name(s)	Scientific Name	1966	1971	1974	1979	1984
Water buffalo	*Bubalus bubalis*				1	
Topi	*Damaliscus korrigum*		3	2	1	
Grant's gazelle	*Gazella granti*				1	
Brocket deer	*Mazama* sp.	25	12			
White kob	*Adenota kob*			1		
Lechwe/waterbuck	*Kobus* sp.		8			
Nile lechwe	*Kobus megaceros*			1	1	
Hartebeests	*Alcelaphus* sp.			1		
Suni antelope	*Nesotragus moschatus*		2			
Thompson gazelle	*Gazella thomsoni*		3			
Himalayan tahr	*Hemitragus jemlahicus*	4	7	5		
Hawaiian black sheep	*Ammotragus* sp.			6		
Iranian red sheep	*Ammotragus* sp.			6		

[a]Temple (1982) and Jones (1986).
[b]First time included in survey.

an important part of their income (Powell & Bahr, 1987). This suggests that trespass fee hunting, even in states with widely dissimilar land ownership patterns, may be more prevalent than is widely thought. The 190 ranches included in the Wyoming survey averaged 10,000 hectares. Each hosted 75 hunters per species present (deer, antelope, and/or wapiti). A 1979 resource assessment (Yorks & McMullen, 1980) suggests that 0.7 animal-unit-months/hectare of domestic grazing are typical in Wyoming. This, when combined with the average ranch size for the state, gives a figure of 600 domestic animals per ranch, which interestingly enough is the traditional viable economic unit for livestock. Wyoming's Game and Fish Department figures (Anon., 1985b) show a 15–20% harvest of game populations each year. Combining these two figures suggests that game populations on fee-charging Wyoming ranches are similar in magnitude to the number of domestic animals.

Wyoming's Game and Fish Department gives a figure of 400,000 mule deer for the state, with 120,000 hunters achieving a 50% success rate (Anon., 1985b). The level of emotional and financial investment typical of American hunters should indicate that this relatively low overall success rate on no-fee land is a direct inducement to interest them in hunting with a trespass fee.

The Vermejo Ranch in New Mexico has been one of the leaders in the movement towards fee hunting on fenced properties (Wolfe, 1980). This 200,000 hectare territory was purchased in 1973 by the Pennzoil Corporation for US$ 27 million, and 20% of the land was donated to the US Forest

Service for tax purposes. Wapiti were restocked in 1911 from Wyoming, because this species had been hunted out in New Mexico. From 1926–33, the ranch was operated by Henry Chandler as the Vermejo Club, an elite hunting, fishing, and recreation camp. Cattle were added only during the Depression, while big game remained part of the enterprise in a reorganisation begun in 1945 by W. J. Gourley. In 1957, the wapiti herd was augmented by the purchase of several hundred animals from Yellowstone Park. Current guest-related operations gross 150% of the ranch's million dollar annual yield from cattle.

Hunter success at Vermejo exceeds 60% for wapiti, averages 85% for deer, and is typically 100% for antelope. Wapiti antlers, the prime consideration for trophy hunters, are consistently larger than six points, in comparison to only 14% of racks reaching this size in the nearby public White River National Forest. Vermejo hosts 500 wapiti hunters per year, who account for 20% of New Mexico's harvest of this species. More than two-thirds of the actual harvest is antlerless, and part of a management strategy is to augment antler development. Vermejo currently charges US$ 6500 for a bull wapiti during trophy hunts and US$ 4500 thereafter, US$ 1000 for antlerless wapiti, and US$ 1300 for antelope.

Dana, Baden & Blood (1985) report that for a much smaller enterprise on 3000 hectares in Montana, hunters achieved an 80% success rate for mule deer bucks. This Eastern Slope Land Management Association is of particular importance because it represents smaller ranchers banding together to take advantage of the roaming characteristics of native ungulates. Cooperation among ranchers – like the 100,000 hectare Elk Mountain Safari operation at Saratoga, Wyoming – and its profitability could mark the beginnings of a trend.

The Deseret Ranch in Utah represents cooperation of another sort. The Latter Day Saints (Mormon) Church purchased 100,000 hectares of deeded and 200,000 hectares of federal lease property in 1980 for evangelistic use. The ranch has diverse ecosystems, from salt desert shrub to alpine tundra. According to ranch general manager Gregg Simmonds (personal communication), the number of hunters was reduced from 3000 to 350 during a quality improvement phase in ranch development. Success returns exceed 80% for deer and wapiti, and the US$ 100,000 annual investment now made for wildlife is doubled in its return from hunter fees, as opposed to a 2% net return on investment from their domestic cattle herd. The ranch managers plan to massively expand their emphasis on native ungulates, and are currently restocking bison and antelope. Harvests of this private herd for packaged meat are still on the drawing board. Current Deseret hunting fees are US$ 1000 for an 11-day non-guided deer hunt, and US$ 600 for a six-day hunt, US$ 3000 for a six-day guided hunt, US$ 5000 for guided antlered wapiti hunts, and US$ 25/day for antlerless wapiti hunts without guides (Anon., 1986a).

The number of meat-oriented ranches, other than for bison or domesticated animals, is presently limited, especially outside Texas. The Wyman Ranch is one of only two in Colorado, and is representative of operations in Rocky Mountain states where raising native game for meat is legal. It stocks 200–300 animals bought initially from private sources (Petersen, 1986). Twenty wapiti are taken by trophy hunts each year, and 80 for meat. Half the resident bulls are cropped for antler velvet, and exporting the velvet currently provides about half of the overall return. In comparison, Wyoming, the state probably best known for game animals, allows no raising of game animals of any kind for meat, or the erection of game proof fencing.

The Lucky Star Ranch in northern New York State initially imported red deer and fallow deer, and it now directly supplies New York City restaurants with 100–150 animals as whole carcases from its 1000 animal resident herd (J. von Kerckerinck, personal communication). Whole carcase sales are utilised to circumvent US Department of Agriculture inspection required for the sale of individual cuts. Three hundred other red deer are in residence near Auburn, New York, as the exotic meat supply for the Culinary Institute of America in Hyde Park (P. Duckelsbuehler, personal communication).

Technology and productivity

In comparison with the commercial game industry in countries such as New Zealand and Scotland, the US industry must be considered rather primitive. The normal source of extension information is still state Fish and Game or Agricultural Extension personnel, whose training and experience are with extensive rather than intensive management. Information networks or on-site experts are rare for any commercially focused management of either native or exotic ungulates. Hence, productivity tends to parallel that of strictly 'natural' herds, with very few exceptions.

Woven wire fences remain the primary choice for separating properties. Ungulates are mainly fed supplemental hay or pelleted grain/legume combinations in times of resource stress, such as drought or extreme snow depths. Veterinary care is similarly basic. The extensive nature of the animals' distribution plays a major role in this situation, as does the legal status of the animals as a quasi-public resource. The adoption of European techniques for genetic improvement is strictly experimental (Cook, 1984).

Transport and slaughter laws are based on domestic animal requirements, resulting in considerable difficulties for owners and traders. Deer transport regulations, from interpretations of what is humane for cattle, require that animals have sufficient space in which to turn around, which results in many broken vertebrae in these more temperamental creatures. Inconsistent county, state, and interstate laws seriously restrict meat harvest and sales.

At the Vermejo Ranch in New Mexico, a full-time game biologist oversees winter helicopter surveys of the wapiti herd, mapping their distribution and age and sex ratios (Wolfe, 1980). During the winter months, box traps are employed to monitor herd health. Necropsies are performed on animals which die from natural causes, and hunter kills are surveyed for antler size, tooth wear, and kidney fat, as well as for conception in female animals. Wolfe indicates that their conception rate ranges from 65–95%. The wapiti herd is harvested to maximise the number of 6–9-year-old bulls, because of the economic desirability of their racks.

For the five Texas exotic ungulate ranches which provide such statistics, M. Hughes (unpublished data) calculated that axis deer with a ten-to-one doe to buck ratio produce one fawn per doe per year, and that 80% of these fawns survive, along with 87% of the adults. Through the year, 700 axis deer and 200 blackbuck were harvested from the 50 ranches participating in the Texas Wild Game Cooperative. Harvesters used silenced 30 calibre rifles to minimise stress on the herd.

Economics

Both secrecy within the game industry and its still undeveloped status limit assessment of investment requirements and profitability. Game meat prices in markets and restaurants, estimated through a personal survey, tend to be 50–100% higher than for comparable domestic meat cuts. Because management tends to be as extensive as possible, the primary investment limitations for the commercial production of ungulates remain land, fencing, and purchase of breeding stock. Other requirements are minimal, especially given the native animals' longer adaptation to North American plants and climates and their superior self-distributing patterns.

P. Hoctor (personal communication) estimated that for part-time family commercial game operations, sales averaged US$ 15,000, amortised annual equipment investments were US$ 5000, veterinary care/medicines totalled US$ 1000, and supplemental feed was US$ 5000. These small operators average less than 50 animals and utilise less than 50 hectares. M. Hughes (unpublished data) calculated that returns on investment for exotic ungulates raised for meat in Texas, based strictly on animal purchase prices, vary from

3% to 60% annually over time and with market values for the meat. Typical carcase values were US$ 3–5/kg for venison.

Trophy hunts, especially if they include ancillary services (guides, food, and lodging), traditionally have been thought to provide the highest return for fenced ungulate production in the United States, although the statistics quoted by Petersen (1986) for meat production indicate a change where such operations are feasible. Powell & Bahr (1987) reported average lease charges of US$ 1000 per wapiti, US$ 300 per deer, and US$ 150 per antelope in Wyoming, where the raising of game animals for meat sales remains illegal. These lease rates, interestingly enough, approximate the game meat carcase values quoted by Hughes for retail sales, which are 50% to 100% higher than the carcases of similarly sized domestic livestock. Where fine trophies are found, or other services are provided, the cost of a deer and an antelope can rise to US$ 2500 (M. Scott, P Cross Bar Outfitters, Gillette, Wyoming, personal communication), a deer at another site to US$ 3500 (Guynn, 1985), or a wapiti to US$ 5000 (Wolfe, 1980; Simmonds, personal communication). The key is the additional services and/or the perceived uniqueness of trophy antler size.

Data of Yorks & McMullen (1980) indicate average domestic grazing in the Texas Hill Country to be about two animal-unit-months/hectare annually. This gives a lease value of US$ 3/hectare/year at government rates or about US$ 25/hectare/year at private rates. This already puts the average annual lease of US$ 25/hectare/year for Hill Country game (Teer, 1985) – up from US$ 3/hectare in 1960 – at the upper end of the value of land for domestic livestock. The lease value is rapidly rising in economic importance. On coastal land near Corpus Christi, Texas the Welder Wildlife Foundation has a 7800-hectare experimental ranch which returns a net of US$ 30,000 annually on domestic cattle, while the lease value of hunting for the 1000 head deer herd has exceeded US$ 50,000. The total for quail, deer, and javelina (*Tayassu tajacu*) gross return is over US$ 75,000 (Teer, 1985).

Considerations for the future

Native ungulates are still treated as if publicly owned, even when wholly resident on private property. Animal management and harvests remain strictly under the jurisdiction of state Game and Fish Department officials constrained by inflexible laws. In turn, state legislatures are bound by conflicting public interest groups. Game ranchers who wish to include trophy hunting as part of their income find the ability to decide seasons, and especially to spread out hunts, severely restricted. The type of animal and the kinds of people allowed to hunt may run counter to the operator's desire. For example, in Wyoming permits for non-resident hunters are allocated among districts by random computer selection. An individual rancher cannot assure the hunters who would be welcome to come back that they may indeed be able to obtain the necessary permit to do so.

The concept of public ownership of all game animals also means that meat operations cannot begin with animals initially resident on any given individual's property. They must purchase their breeding stock elsewhere. Such a limitation not only makes the animals expensive, but also reduces the potential for utilisation of animals best adapted to local climate and vegetation.

The often conflicting and confusing patterns of federal, state, and local regulations covering transport, slaughter, and meat sales from undomesticated animals further hampers practical game ranching. The present laws were enacted to counter a history of gross overharvests and were particularly oriented to preclude the sale of poached meat. Law-makers and regulators had only the transport and slaughter considerations for domestic animals as a model. As a sign of recent progress, a field slaughter procedure for various ungulates has recently been approved by the US Department of Agriculture for meat slated for interstate trade. However, an official inspector must be on site during the slaughter operation. For a ranch which has a limited number of animals and is many miles from an inspector's home base, the cost of the inspection may well be prohibitive.

On the other hand, technological advances may alter the potential of game meat marketing in the United States. Serotyping of tissues is being touted in public health journals for tracing illicit meats. It now seems feasible to assure the public of the legality of game meat, but this important concept remains undeveloped, and poaching remains a serious concern. Nevertheless, it has been argued that an industry can more effectively police itself than is possible by overstressed government agencies. More immediately, commercially available tapes affixed to meats can indicate by their colour whether or not proper storage conditions have been met.

Native game animals have superior weight and hides in the late autumn, and so only by frozen distribution of their meat is it possible to have optimised harvests with the highest overall value for each animal and still provide product availability year-round. Freezing is by no means a new technology, but it has not been widely successful with domestic red meat marketing because of decided consumer preferences for chilled meat. However, freezing is widely used in homes and is similarly adopted by the food service industry to assure product consistency and obviate inventory problems. Further, consumers seem to have little difficulty with purchase of frozen fish or poultry. An essentially new product being brought into the distribution system could be sold only in frozen form, thus bypassing consumer conditioning for refrigerated domestic red meats, and gourmet reluctance to accept frozen products.

Game animals have the advantage of lower fat content, higher protein, and lower cholesterol than domestic animals, as well as less intramuscular fat which allows easier and more effective trimming. This is a critical practical issue in marketing since oxidative rancidity of fat is the primary cause for quality deterioration of food during frozen storage. The augmented shelf-life potential for game meat as a frozen product is doubly important during the growth phase of new products. For restaurants, spotty initial demand can be met with a minimum of overhead, and the same is true for retail meat markets. The advent of microwave defrosting obviates a common problem with frozen products in the past, i.e., the time it takes to ready them for final preparation.

Food preparation is an area in need of research because techniques need to be developed to maintain the low fat advantage of game all the way to the table. Smothering the product in hollandaise sauce, despite its delightful taste, or the use of any other high calorific treatment will quickly negate the suggested health benefits for game meat.

In the United States, two approaches to the commercial production of game animals appear to be taking shape. One utilises intensive management, taking a leaf from domestic animal practices, building increasingly complex fencing to more carefully control the location, timing, and duration of grazing. The other seeks to employ more fully the adaptational superiority of undomesticated ungulates. This latter approach assumes that given large areas in which to graze, native animals will essentially manage themselves. The commercial producer's responsibility is to manage the land itself for optimal habitat qualities.

The quantitative advantage of the free-ranging approach is most difficult to prove, but it makes sense. In natural circumstances, native ungulates move according to seasonal forage availability, range much further from water than conventional livestock, utilise steeper slopes, and operate without herders. This random movement of often large herds within a given range gives the advantages touted by Alan Savory (1985) – such as even distribution over long periods of time of localised hoof action and short duration heavy use – without the necessity of complex fence building. To be truly effective, this free-movement strategy does require cooperation among neighbours. The Elk Mountain Safari effort near Saratoga, Wyoming, where four large ranches pool fee hunting privileges while letting herds move freely, is a major step forward in combining diverse grazing resources to maximise the natural advantage.

Probably the largest single issue concerning free-ranging herds is finding an equitable means of sharing the gain in commercial production through the use of habitat improvement practices. The economic return from game animals which move among properties has traditionally gone only to the owner of the land on which the animal is harvested in the autumn. This limited distribution allows no benefits for the manager of the game's summer or winter ranges. This is critical if the other seasonal ranges are, as usual, on a different property, unless those other individuals can benefit from non-harvest recreation, like photography. Otherwise, the owners who do not realise revenue reasonably perceive the game animals as destructive of the forage resource. However, if the pecuniary benefits from actual meat or trophy harvest could be distributed to all land managers in proportion to the time each animal spends grazing on individual property, a more direct incentive for habitat improvement would be created.

This is a sociological as well as a technical issue, but it is central to the whole question of the commercial game production. The formation of cooperatives like the few described above seems to bode well, but much more remains to be done. The difficulty of knowing where and what the animals have been doing has been a serious impediment to schemes for distribution of returns in the past. The decreasing cost and increasing sophistication of animal monitoring techniques, whether based on ultralight aircraft or satellites, is of particular importance. Such monitoring makes possible the accurate knowledge of animal movements which are necessary for truly equitable distribution of return to those owners on whose lands the animals have grazed.

Ecologically, multiple species grazing can have a profoundly positive impact. The differing dietary preferences among ungulates allow more even distribution of grazing pressure. One species may concentrate on shrubs, another on forbs, while others have specialised preferences among the grasses. The propensity of native animals to forage further from water than domestic species also encourages balanced grazing pressure. Winter foraging characteristics are augmented as well. Native ungulates can survive without that most expensive of ranch operations, winter feeding. In the historical past, however, occasional serious losses did occur, and such losses would be difficult for contemporary economic practice to sustain, but the extent of loss may be no greater than for the domestic herds over time. The use of forage plants when they are not actively growing is as a rule of thumb the least damaging. Hence, winter grazing by native ungulates is not only cheaper than the confined feeding typical for domestic animals, but it can also be easier on the rangelands. Supplemental winter hay feeding, seen as necessary for native wapiti herds in Wyoming, is primarily a result of winter rangeland being withdrawn for human-oriented purposes (Toman, 1986). Generally, free-ranging grazing by multiple species offers the potential for increased vegetative diversity and productivity, with the concurrent advantages for soil and water conservation.

Economically, beyond the direct benefits of lower costs to produce more animals, the advantages of game animals are manifold. The hides tend to be more valuable. Their hair-on insulation value, despite problems with hollow hair breakage, is one not yet commercially explored. The specific utility implicit in the various leathers is well known in specialty markets. The range of flexibility to strength ratio, including softness maintained, for game leather is by no means fully exploited in broader venues. Antler velvet, because of its presumed aphrodisiac and other folk medicine qualities, now commands high prices in the international market. The potential advantages of the meat have already been described.

What is needed is an effective marketing organisation to match product characteristics with these growing demands and to assure product quality. Current small operations have serious problems with consistent product availability, as well as with contamination with bone chips and rumen fluid, generally poor butchering practices, and variable tenderness through inadequate age selection. Poor slaughter, butchering, and cooking practices can result in an unsatisfactory product.

Even when product quality is adequate, there has been a paucity of recipes. Such may seem a trivial area, but preparation techniques now in use in restaurants and homes assume a meat with a bland and fatty starting point. The uniquely desirable characteristics of game meats actually act as a hindrance if recipes which were appropriate for domestic meat products are employed. For sales to grow at the rate they should, serious attention must be given to this sector so that the inherent quality of the meat is preserved and enhanced during the final stages of preparation before consumption, but at the same time taking care to preserve the meat's potential health advantages. Numerous recipes need to be brought forward from the past, and from Europe and New Zealand, when and where game has been more popular, as well as creating entirely new ones.

For animals, there is an important long-term consideration for commercial production, especially in a climate of strong interest in animal welfare. A popular perception is that there is mythical gentleness in wild ungulates and killing them for any purpose is unethical. Animal welfare organisations are mounting multi-million dollar campaigns against hunting and any game trading. Activists attempting to stem this trend have experienced difficulties in being heard or supported. Yet, having essentially wild animals which are dispatched quickly and unexpectedly in the field instead of an abattoir would seem more humane than practices applied to domestic stock. Frank Waters (1941) has extensively addressed this issue in his novel, *The Man Who Killed the Deer*.

The National Cattlemen's Association tends to respond to public attention on slaughter or growth condition issues by ignoring them completely, assuming that any discussion may lead to undesirable thoughts. Animal welfare may be an uncomfortable issue but it is one which very well could stop any distribution of game animals or their products in the United States unless it is effectively confronted.

References

Anon. (1985a). *State of Wyoming, 1985 Annual Report*, Vol. 7, Resources. Cheyenne: Department of Administration and Fiscal Control, Research and Statistics Division.

Anon. (1985b). *The Mule Deer of Wyoming*. 2nd edn. Cheyenne: Wyoming Game and Fish Department.

Anon. (1986a). *Commercialization/Privatization of Wildlife*. Cheyenne: Biological Services, Wyoming Game and Fish Department. 43pp.

Anon. (1986b). *Western Livestock Roundup, March*. Agricultural Extension Service, University of Wyoming. Laramie: University of Wyoming.

Butler, A. W. (1934). Wild and domesticated elk in the early days of Franklin County, Indiana. *Journal of Mammalogy*, 15, 246–8.

Cook, R. L. (1984). Texas. In *White-Tailed Deer: Ecology and Management*, ed. L. K. Halls, pp. 457–74. Harrisburg: Stackpole Books.

Cooper, J. F. (1841, reprinted 1979). *The Deerslayer*. New York: Dodd, Mead. 573pp.

Dana, A., Baden, J. & Blood, T. (1985). Ranching and recreation: covering the costs of wildlife production. In *Holistic Ranch Management Workshop Proceedings*: 28–30 May, Casper, Wyoming, ed. J. Powell, pp. 61–76. Laramie: Wyoming Agricultural Extension Service.

Field, R. A., Smith, F. C. & Hepworth, W. G. (1973). *The Elk Carcass, The Mule Deer Carcass, The Pronghorn Antelope Carcass*. Agricultural Experiment Station Bulletins 594, B–589, B–575. Laramie: University of Wyoming.

Guynn, D. (1985). Wildlife as a ranch enterprise. In *Holistic Ranch Management Workshop Proceedings*: 28–30 May, Casper, Wyoming, ed. J. Powell, pp. 77–99. Laramie: Wyoming Agricultural Extension Service.

Jacobs, W. R. (1980). Indians and ecologists and other environmental themes in American frontier history. In *American Indian Environments: Ecological Issues in Native American History*, ed. C. Vecsey & R. W. Venables, pp. 46–64. Syracuse (NY): Syracuse University Press.

Jones, R. F. (1986). Where the deer and greater kudu play. *Sports Illustrated*, 65, 56–67.

Josephy, A. M. Jr (1968). *The Indian Heritage of America*. New York: Bantam. 397pp.

Kellogg, T. D. (1887). Domestication of the elk. *American Field*, 28, 126–7.

Martin, C. (1978). *Keepers of the Game*. Berkeley: University of California Press.

Petersen, D. (1986). The Wyman elk ranch. *Mother Earth News*, March/April, 62–4.

Powell, J. & Bahr, S. (1987). Preliminary Survey: Wyoming Farm and Ranch Outdoor Recreation Enterprises. Laramie: Wyoming Cooperative Extension Service.

Ramsey, C. W. (1965). Potential economic returns from deer as compared with livestock in the Edwards Plateau region of Texas. *Journal of Range Management*, 18, 247–50.

Rue, L. L. (1978). *The Deer of North America*. New York: Crown Publishers.

Savory, A. (1985). Holistic resource managment. In *Holistic Ranch Management Workshop Proceedings*: 28–30 May, Casper, Wyoming, ed. J. Powell, pp. 1–10. Laramie: Wyoming Agricultural Extension Service.

Teer, J. (1984). Lessons from the Llano Basin. In *White-Tailed Deer: Ecology and Management*, ed. L. K. Halls, pp. 261–90. Harrisburg: Stackpole Books.

Teer, J. (1985). Strategies and techniques for production of wildlife and livestock on western ranches. In *Holistic Ranch Management Workshop Proceedings*: 28–30 May, Casper, Wyoming, ed. J. Powell, pp. 91–100. Laramie: Wyoming Agricultural Extension Service.

Temple, T. (1982). *Records of Exotics*, Vol. 3. Kerrville: Herring Printing Company.

Toman, T. (1986). *Elk feeding programs of the Wyoming Game and Fish Department and the Elk Refuge*. Jackson (Wyoming): Society for Range Management, Summer Meeting, 21 July.

Waters, F. (1941). *The Man Who Killed the Deer*. New York: Pocket Books.

Watt, B. K. & Merrill, A. L. (1963). *Composition of Foods*: Agriculture Handbook No. 8. Washington: US Department of Agriculture.

White, R. J. (1986). *Big Game Ranching in the United States*. Mesilla (New Mexico): Wild Sheep and Goat International.

Wolfe, G. (1980). Goals and procedures on a large western ranch. In *Forty-Second North American Wildlife Conference*, pp. 271–7. Washington: Wildlife Management Institute.

Yorks, T. P. & McMullen, C. (1980). *The Central and Eastern Grasslands* (preprint). Natural Resource Ecology Laboratory. Fort Collins: Colorado State University.

15

Game ranching in southern Africa

JOHN D. SKINNER

Abstract

In the 1960s, game ranching was proposed as a more productive form of land use than domestic stock husbandry. Since then, there has been a steady increase in the numbers of wild ungulates on farms but these still only number hundreds of thousands compared to millions of domestic ungulates. When correct expertise is applied, higher productivity can be attained by combining wild and domestic ungulates. Game do not compete with domestic stock for all resources and can complement domestic livestock to improve productivity. On the other hand, game have been implicated in the spread of livestock diseases and are not well suited to manipulation to facilitate veld management. Life history features, growth, carcase yield, and meat quality for several species are discussed. Particulars are provided for planning and running a game ranch, including stocking rates, herd management, and economic considerations.

Introduction

Africa is considered the birthplace of the modern concept of game ranching (Dasmann & Mossman, 1960; Dasmann, 1962, 1964), particularly southern Africa south of the Kunene and Zambezi rivers. It can broadly be divided into three zones: arid zone, savanna zone (which contains a grassland zone), and the South West Cape zone (Fig. 15.1).

Before settlement by the Bantu and particularly Europeans who introduced firearms, vast herds of some 25 ungulate species roamed the savanna and migrated across the arid zone. These ungulates were overexploited; some such as the blue antelope (*Hippotragus leucophaeus*) and quagga (*Equus quagga*) were exterminated while many others were hunted to the point of extinction. From the mid 1800s, trade in game skins provided an important source of revenue. One firm exported 157,000 wildebeest and blesbok skins in 1866, another nearly 500,000 wildebeest, blesbok and zebra skins in 1870. This hide trade led to the decimation of plains ungulates in the grassland savanna and arid zones, and restricted the distributions of most species.

Game had little commercial value during the first half of this century despite its importance as a source of food. Conservation departments controlling game on private lands only came into existence by 1950. The profitability of game utilisation never reached a level comparable with that of conventional livestock husbandry (Joubert, 1968). The result was a continual decline in numbers and distributions of ungulates.

During the late 1950s, an increasing number of expatriate biologists noted that much of the soil was not suited to crop production and that there was a

Fig. 15.1. The main biotic zones in southern Africa (after Smithers (1983)).

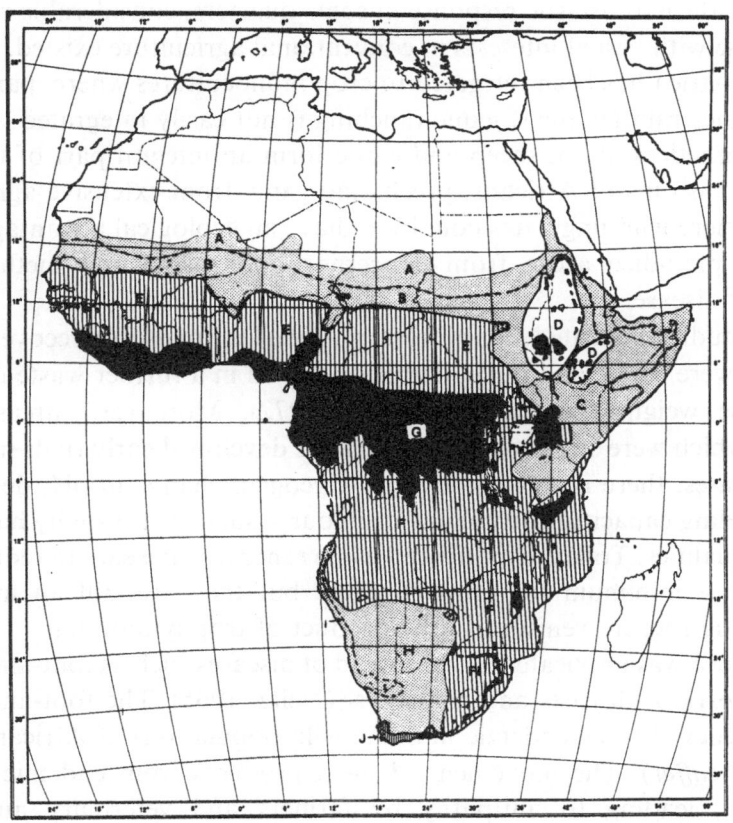

Key to Symbols

Arid zones ||||| Savanna zones   ■ Forest zones ||||| South West Cape zone (Cape Macchia or *fynbos*)   ■ Isolated patches of montane or evergreen forest

concomitant chronic shortage of protein for human consumption. This focused attention on the possibilities of exploiting native ungulates for meat. Although the value of game for the tourist industry was recognised, wildlife for commercial meat production was a novel concept.

Idealists promulgated the view that wild ungulates were disease-resistant, independent of drinking water, or otherwise adapted to local conditions. Above all, they argued that ranching a community of wild browsers and grazers resulted in greater biomass densities, because of limited interspecific dietary competition. Unlike wildlife, monocultures of domestic livestock were alleged to seriously overgraze vegetation (Dasmann, 1964). Game ranching was considered to have the greatest comparative advantage in hot arid environments unsuited for conventional livestock where eland (*Taurotragus* sp.) (Skinner, 1967), oryx (*Oryx* spp.), and gazelles (*Gazella* spp., *Antidorcas marsupialis*) (Skinner, von La Chevallerie & van Zyl, 1971) are better adapted physiologically and behaviourally.

Despite the enthusiastic response in some quarters, considerable resistance from those with vested interests in conventional agriculture existed. Much of southern Africa has been relegated to crop monocultures where, particularly in the higher rainfall areas, game ranching is not easily integrated with such farming practices. Also, sheep and cattle form an integral part of the tribal laws and customs of Africans, precluding game from extensive agricultural regions where wild ungulates could well have an ecological advantage. More technical prejudice arose from concerns about harvesting methods and threats of disease.

Early studies indicated losses from wounded animals not recovered after shooting were 8.4%, and bullet damage resulted in a further waste of 13.9% of carcase weight (von la Chevallerie, 1972). Moreover, carcases from animals which were stressed before slaughter developed early rigor mortis. In such carcases, there is a breakdown of glycogen, change in pH, decrease in water binding capacity, and change in colour – such meat usually has poorer keeping qualities. Today, experienced game ranchers take care to reduce these factors to a minimum for cropped game but they are still an important consideration where venison is a by-product of trophy hunting.

Game also was implicated in the spread of diseases such as foot-and-mouth disease, corridor disease, and rinderpest (Neitz, 1965). The foot-and-mouth virus is believed to be endemic in many wild populations of African buffalo (*Syncerus caffer*). The movement of meat products from endemic areas is prohibited, leading to antipathy by farmers and veterinary authorities towards wild ungulates. Game ranching in such areas is frequently regarded as incompatible with conventional agriculture.

Status

Since the initial experiments on Doddieburn Ranch in Zimbabwe over 25 years ago (Dasmann & Mossman, 1960; Dasmann, 1962, 1964), game ranching has increased steadily in areas marginal for dryland cropping (Child, 1985). The area in Zimbabwe devoted to game in the decade since 1974 increased 62%, from 59,700 to 97,000 hectares. A similar trend occurred in South Africa but it is difficult to quantify because game ranching has been integrated with other forms of land use. The number of game ranches in South Africa is estimated at over 8200 with an average of 2531 hectares (Benson, 1986).

The total venison export from South Africa varies between 1446 tonnes (Wilkins, 1982) and 4000 tonnes (Conroy, 1983) with respective values of R 8 million and R 40.5 million. The number of animals removed from private land between 1982 and 1984 is given in Table 15.1 (Benson, 1986). Some 37,000 animals (40%) were hunted rather than cropped in 1984, yielding some R 7.5 million. With expansion of this form of land use, trade in live animals as breeding stock is important.

To place game ranching in perspective, production of springbok and merino sheep in the arid zone can be compared. There are an estimated 275,000 springbok in South Africa (Skinner, 1984a). The potential yield of this population, assuming a 30% crop (as realised in Namibia, Anon., 1981), a mean carcase weight of 12 kg, and a price of R 3/kg, is 990 tonnes worth R 2.97 million. In comparison, over 100,000 tonnes of wool were exported in 1984 of which 72% was produced by 17.7 million merino sheep. Other woolled sheep and mutton breeds make the national flock up to 29 million animals. The national mutton and lamb product for 1984 was 225,000 tonnes valued at R 528 million (Anon., 1985). Similar comparisons could be made between cattle and other wild ungulates in the savanna zone.

Table 15.1. *Numbers of wild ungulates cropped, hunted, and sold live from private lands between 1982 and 1984 in South Africa.[a]*

	1982	1983	1984	Total	%
Number cropped	45,530	37,005	38,934	121,469	47.7
Number hunted	30,374	33,386	37,284	101,044	39.7
Number sold live	10,397	10,818	10,669	31,884	12.5
Total	86,301	81,209	86,887	254,397	

[a]Benson (1986).

Production parameters

Stocking rates

Numerical determination of carrying capacity for mixed-species grazing systems is very complex. Commonly, each species is compared to the standard livestock unit (*LU*) used in conventional agriculture (Mentis & Duke, 1976):

$$LU = \frac{(\text{Mean weight of species})^{0.75}}{450^{0.75}}$$

Scaling according to metabolic weight assumes that differences in energy requirements and forage intake are simply proportional to size. Minimal available data (Rogerson, 1968; Meissner & Hofmeyr, 1972; Meissner, 1982) for African ungulates questions these assumptions, yet we have few other means of determining carrying capacity at this stage.

Meissner (1982) proposed that energy requirements should be based on a knowledge of growth, basal heat production, and efficiency of feed utilisation. One LU is the equivalent of a steer with a weight of 450 kg growing at the rate of 500 g/day on grass pasture with a mean digestible energy concentration of 55%. Values for some mature wild ungulate species are as follows: blesbok ewe or ram, 0.19; springbok ewe or ram, 0.09; impala ewe, 0.14; impala ram, 0.16; kudu cow, 0.39; kudu bull, 0.53; eland cow, 0.96; eland bull, 1.19; giraffe cow, 1.35; giraffe bull, 1.68; warthog sow, 0.18; and warthog boar, 0.22. The LU will vary according to the physiological state of the female, for example increasing during lactation.

The estimated number of a species in an area can then be converted into biomass and the LUs calculated for each species and summed. However, this does not allow for ecological separation. The carrying capacity will also vary with vegetation type (Mentis & Duke, 1976), being lower in areas where the grass sward lignifies rapidly and becomes unpalatable or where trees and shrubs are largely deciduous. Mentis & Duke (1976) estimated the stocking rate in woodland savanna at 6 hectare/LU.

Life history features

Detailed life history features for several species are presented in Table 15.2. It is widely conceded today that, under extensive ranching conditions, wild ungulates reproduce more efficiently than do domestic livestock (Skinner, 1984b). The ideal sex ratio to secure maximum reproductive rates is still a matter of conjecture but some guidelines are presented.

Table 15.2. *Life history features of eight prime species of ungulates used for game farming in southern Africa.*

	Blesbok	Springbok	Impala	Kudu	Eland	Giraffe	Mountain reedbuck	Warthog
Mature weight (kg)								
Female	60	37	40	160	450	828	29	56
Male	70	42	55	230	650	1200	30	80
Age when first breeding (months)								
Female	18	7	18	17	28	56	8	18
Male		12						
Breeding season	autumn	all year	autumn	winter	all year	all year	all year	autumn
Gestation (days)	225	168	196	212	271	457	240	171
Parturition	summer	all year	mid summer	late summer	all year	all year	all year	early summer
Birth weight (kg)								
Female	7	3.4	5	16	25.2	89.0	3.1	0.8
Male		3.2			30.0	95.1		
Annual calving % (mature females)	85	46-100	90	100	83	48	80	96
Calving interval (days)	365		365	365	354.5	645		
Theoretical maximum productivity (offspring)	13	24	13	14	16.2	10.18		
Age of last breeding (yr)	15	13	15	16	18	23		
Adult sex ratio (from Skinner 1984b) M:F	1:25	1:30	1:30	1:30	1:10	1:20		
Source	Skinner (1984b)	Skinner et al. (1971)	Fairall (1983) Skinner (1984a)	Skinner (1984b)	Skinner (1967)	Hall-Martin (1975)	Skinner (1980)	Mason (1982)

Growth and carcase yield

Growth is related to mature size. In smaller species (springbok, mountain reedbuck [*Redunca fulvorufula*], and warthog [*Phacochoerus aethiopcus*]) the point of inflection on the growth curve is reached before one year of age (Skinner, 1980; Mason, 1982), in the blesbok (*Damaliscus dorcas*) and impala (*Aepyceros melampus*) at about 18 months, in the eland at about three years (Skinner, 1967), and in the giraffe (*Giraffa camelopardalis*) at five years of age (Hall-Martin, 1975). Of course, environmental stress slows growth. In the Drakensberg mountains, severe cold and poor nutrition delays attainment of mature weight in eland for a year compared with other regions (Jeffery & Hanks, 1981).

Table 15.3. *Live mass and dressing percentage in eight species of ungulates.*

	Live mass (kg)	Dressing percentage	Reference
Springbok			Skinner et al. (1971)
Male	36	56.0	
Female	30	55.0	
Blesbok			Huntley (1971)
Male	73.4	52.9	
Giraffe			Hall-Martin et al. (1977)
Male	1174.3	61.9	
Female	791.8	56.6	
Eland			von la Chevallerie (1971)
Male	408.5	51.3	
castrate	412.7	63.2[a]	
Kudu			Huntley (1971)
Male	236.3	56.6	
Impala			van Zyl et al. (1969)
Male	77	58.8	
Mountain reedbuck			Skinner (1980)
Male	30.2	55.3	
Female	28.6	51.4	
Warthog			Mason (1982)
Male	79.6	55.1	
Female	56.5	52.9[b]	

[a]Starved before slaughter.
[b]Non pregnant; pregnant females dressed 2.4% lower.

Dressing percentages of several wild ungulate species are higher than those recorded for domestic livestock grazing extensive pastures (Table 15.3). African ungulates are typically lean (Table 15.4). In springbok prior to maturity there is a continuous increase in carcase lean, reaching a peak of 84% in adult males and 82% in adult females (von La Chevallerie, 1971). A similar tendency was found in mountain reedbuck (Skinner, 1980). Meat quality has been estimated for males in four antelope species (Table 15.5).

Castration is acknowledged as a method of improving fat content of domestic livestock (Turton, 1962). Few studies have been carried out with

Table 15.4. *Influence of season on fat reserves in male ungulates.[a]*

	Kidney fat (g)			Percent fat in buttock		
	May	July	Sept	May	July	Sept
Springbok	44.7	23.7	29.9	1.50	1.47	1.13
Impala	359.4	23.2	17.9	2.59	0.55	0.54
Blesbok	91.5	94.3	66.8	2.43	1.58	1.33

[a]von la Chevallerie (1971).

Table 15.5. *Meat quality in males in four antelope species.[a]*

	Springbok	Eland	Impala	Blesbok	Standard deviation
Number	72	6	18	23	
Moisture content (%)	74.7	74.8	75.7	75.5	1.4
Fat content	1.7	2.4	1.4	1.7	1.0
Colour (colorimeter) unit)	7.3	5.9	7.4	7.9	0.6
Fibre diameter (μm)	45.5	66.3	56.7	53.8	3.2
Toughness (g/cm)	1181	3366	2751	2323	3.2
Taste panel scores for flavour					
Number	36	3	9	11	
Intensity (out of 10 points)	4.2	4.4	4.0	4.9	0.8
Acceptability (out of 10 points)	6.1	5.3	5.2	5.8	0.8

[a]von la Chevallerie (1972).

wild ungulates. In eland, average slaughter weight has varied from 412 kg for stall-fed castrated mature animals from the grassland arid zone to 314 kg for veld-raised castrates from the same region (Keep, 1972). Carcases from castrated eland contained almost 10% more fat than carcases from bulls.

Technology and management

Fencing

Fencing is a major development cost. In the arid zone, the prime species (springbok and blesbok) generally can be contained by standard stock fences of 1.5 m in height with an extra strand of wire 50 mm above ground level. For many other species, fencing must be 2.3 m high with the bottom strand 50 mm above ground, followed by four strands 100 mm apart, five strands 125 mm apart, three strands 150 mm apart, three strands 175 mm apart, and the last strand 200 mm.

The spacing of poles for 2.3 m fencing will depend on whether wooden or steel poles are used but 10 m spacing usually is prescribed. Droppers are placed at 2 m intervals to keep the strands taut and correctly spaced. Total costs of fencing materials are in the order of R 3/metre.

Internal fencing need not follow the same strict specifications but there is not much latitude if fencing is to be effective. Paddocks must be very large for effective veld management. Experience is the best guide and a game ranch should only be subdivided when the farmer has clearly defined objectives.

Capture and cropping

Capturing live animals is an expensive undertaking and is best left to professionals, who can be contracted for this purpose. Chemical immobilisation is seldom used. Animals are driven by helicopter into a boma (Hofmeyr & Lenssen, 1975), loaded immediately onto a truck, transported directly to their destination, and off-loaded immediately.

Cropping ungulates for meat production also can be contracted. In such cases, abattoir facilities are invariably provided by the contractor who usually uses a helicopter to shoot antelope. However, the farmer may crop animals himself. Shooting is usually done at night using a powerful spotlight and the animals dressed in an abattoir on the farm (Fig. 15.2).

There is little flexibility in scheduling such activities. If, for example, an opportunity is missed for catching or cropping it may be some time before a contractor is available. This may have serious implications for the profitability that year or for veld conservation if the rainfall has been disappointing.

Veld management

Optimum utilisation of vegetation in the savanna or arid zones is not achieved with domestic livestock. Cattle, for example, do not browse efficiently. Even indigenous breeds (*Bos indicus*) such as the Africander browsed only new leaves of one bush (*Ochna pulchra*) during one month in spring (Skinner, Monro & Zimmermann, 1983). On the other hand, impala (*Aepyceros melampus*) varied their diet, grazing succulent grass in spring and early summer and switching to browse as the grass lignified. Similarly, in the Karoo arid zone, springbok fed on a much wider variety of plant species than merino sheep (Davies, Botha & Skinner, 1986).

Considerable expertise is required to devise appropriate stocking combinations. Since these skills are frequently lacking among farmers, it is more practical to maximise sustained productivity with a mix of domestic and wild ungulates. Essentially, it is a matter of supplementing domestic grazers with wild browsers – the question is how many of each species to include.

Management must encourage ecological complementarity and obviate overgrazing, bush encroachment, and erosion. Seasonal adjustment of stocking rates is easily attained with domestic livestock which can be moved or marketed at assured prices; but is more difficult with wild ungulates where capture invariably entails considerable expense. Moreover, replacements are not always available for subsequent restocking.

Fig. 15.2. Diagram of a farm abattoir (A. M. Conroy, personal communication).

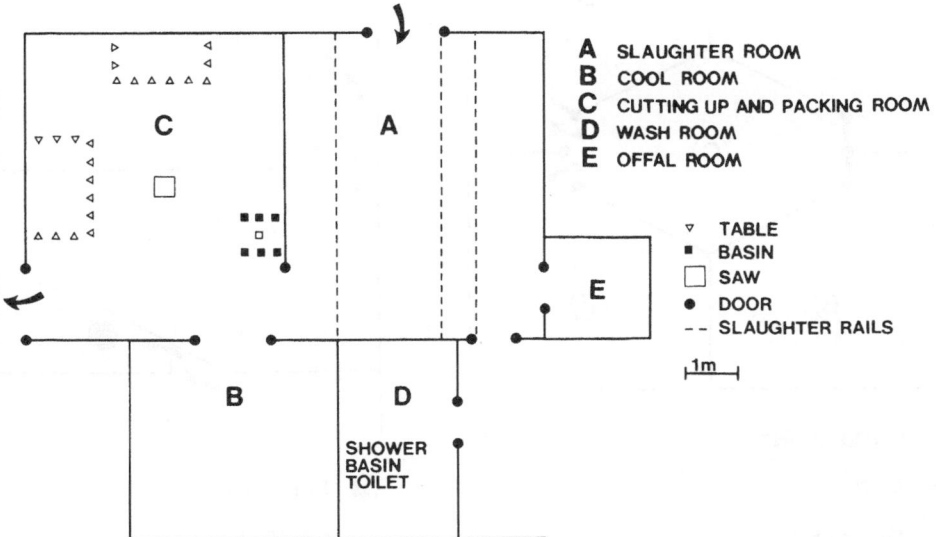

A SLAUGHTER ROOM
B COOL ROOM
C CUTTING UP AND PACKING ROOM
D WASH ROOM
E OFFAL ROOM

▽ TABLE
▪ BASIN
▢ SAW
● DOOR
-- SLAUGHTER RAILS

SHOWER
BASIN
TOILET

Since stocking rates of game are difficult to adjust in the short term, *slight* understocking is advisable to hedge against seasonal pasture shortages. Excess grass cover in good seasons can be removed by grazing cattle for brief periods (this also can be used to reduce ticks through regular cattle dipping). Another buffer against pasture shortage is a crop such as cowpeas, tame pasture, or lucerne hay. Such areas are of necessity only a few hectares in size and require fencing so access by game is allowed only when the farmer deems this necessary.

Rotational grazing of game can be difficult. Young (1984) provides two schemes which can be applied with skill (Fig. 15.3). Both depend on attracting game to specific areas through a gate and then closing the gate to ensure that the animals cannot return. Carefully placed watering points can facilitate control. By opening water points in specific areas and closing them in others according to veld condition the farmer can manipulate movements of water-dependent species.

Fig. 15.3. Two fencing systems to enable rotational grazing with wild ungulates (Young, 1984).

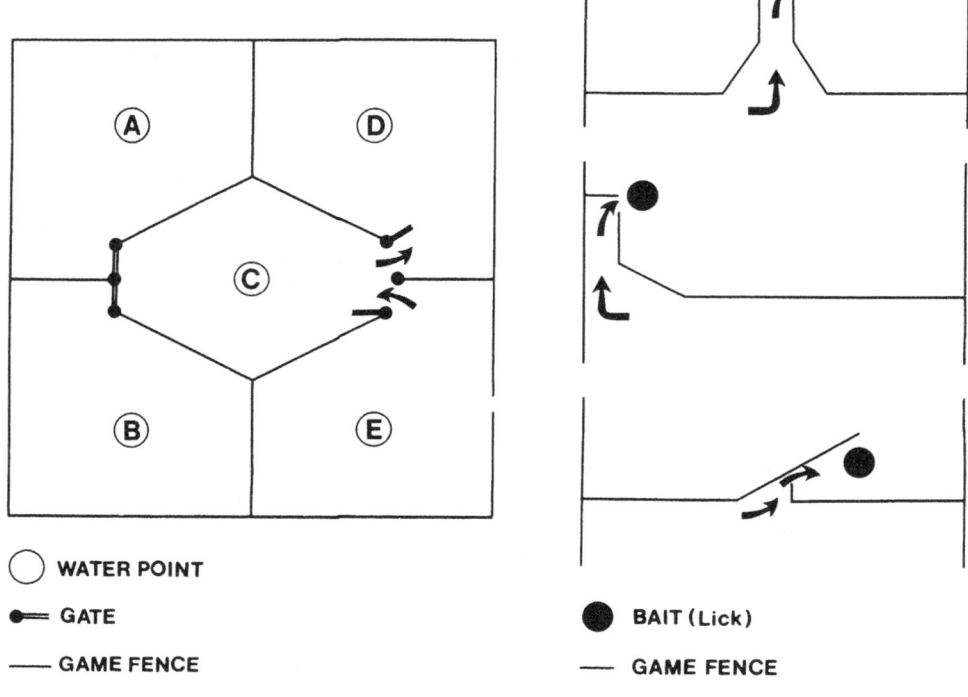

WATER POINT

GATE

GAME FENCE

BAIT (Lick)

GAME FENCE

Fire can be a major hazard. It is essential to maintain firebreaks of 25 m around the perimeter and at critical locations. The veld-type should be taken into account in placing interval firebreaks. For example, areas of palatable grass should be separated from unpalatable species as the latter are likely to be undergrazed causing a greater fire hazard.

Disease control

One of the great advantages of wild ungulates is their resistance to most endemic diseases and parasites. Although, under natural conditions, wild animals can harbour large populations of internal parasites without apparent symptoms, they can succumb when confined (Ortlepp, 1971). Moreover, domestic livestock and ungulates may act as reciprocal sources of infestation (Horak, 1979).

Treatment of disease is not only difficult but generally not recommended. Likewise treatment for parasites is contraindicated because this will eventually lead to the breeding of game which are no longer adapted to withstand parasitic infections under extensive conditions, eliminating one of their main advantages. Heavy parasite loads may indicate that wild ungulates are malnourished or that they are not well suited to the particular area (Underwood, 1975).

Anderson (1983) recommends quarantining game for a short period before loading and treating with non-toxic anthelmintics such as fenbendazole or albendazole. He further warns that high stocking rates not only lower grazing conditions but also increase the parasitic load. In such conditions, non-toxic anthelmintics can be added to watering points particularly in dry winter months (Horak, 1979). A number of medicated mineral licks can also be considered.

Herd management

Sex ratios have been recommended in the past with little attention to the social systems of different species. For example, territory size of springbok is about 0.22 km^2, placing an upper limit on the number of territorial rams that an area can accommodate. Bachelor herds in confined areas appear to disrupt territorial rams resulting in a drastic lowering of the calf crop in springbok and gemsbok (Skinner, unpublished data). All copulations involve territorial springbok rams (Skinner *et al.*, 1987) so surplus rams serve only as territorial replacements. The appropriate number of territorial rams can be calculated by dividing the ranch area by the mean territory size. The stocking rate of ewes is a function of veld carrying capacity.

Maximum sustained yield (Dasmann, 1964) assumes a certain constancy of environmental conditions seldom realised in Africa, particularly in arid regions. For this reason, harvesting or cropping programmes should follow the feedback principles of Stocker & Walters (1984). The implementation of such a strategy requires a knowledge of available food and the changing energy requirements of the populations. With regard to vegetation, a visual estimate would be the simplest appraisal. More scientific monitoring would be difficult for the game ranch as the vegetation would show varying responses to factors such as rainfall and herbivory. To obtain the second parameter, regular censusing is essential. Counting ungulates in arid zone or grassland savanna is feasible but in wooded savanna only estimates can be made.

Monitoring of the herds and classification into age groups should be conducted in early winter to facilitate winter culling. Population estimates and structures are best carried out along standard roads, counting all individuals within a specific distance (Davis & Winstead, 1980), say 50 m in bushveld and 100 m in open country on either side of the road. Another feasible method is to count all animals which drink at watering points, but this is not suitable for those species less dependent on drinking water, such as springbok or gemsbok.

Often, accurate population estimates are unnecessary; trends may be sufficient. Therefore, the census should be carried out at the same time each year and under the same climatic conditions. Culling rates should be determined not only on the estimates of each age-class made but also on veld conditions. In dry years, more animals should be culled if the veld is in poor condition.

Economics

Investments

The investment requirements for game ranching in southern Africa are centred primarily on the ranch, enclosures, watering points, abattoir facilities, and the purchase of suitable ungulates plus a small amount of equipment. With an inflation rate during recent years of almost 20%/annum, these costs can be quite high, and on a 2000 hectare ranch at a stocking rate of 6 hectare/animal unit would amount to R 1,083,000 in the savanna ranching eight species, and R 370,000 in the arid zone ranching springbok and blesbok. In the savanna, of this cost the land would represent 55%, perimeter fencing 7.5%, watering points 3%, abattoir facilities 2.5%, game 28%, and equipment 4%. This amount does not include the price of a homestead and wages for a manager and staff (perhaps R 40,000/annum).

Production of commodities
Relative productivity of game and livestock is available for a 12,832 hectare ranch in the Karoo arid zone (Skinner *et al.*, 1986). Approximately 3735 merino sheep were ranched over the entire area, primarily to produce fine wool but also were sold as lamb and mutton. Approximately 1060 springbok were managed on 35% of the area.

The mean annual weight of springbok harvested during the seven-year period (1978–85) was 7608 kg/annum. This represented 31% of the total biomass. Dressing percentage was calculated at 56% (cf. Skinner *et al.*, 1971; Conroy & Gaigher, 1982), so the annual yield was 4210 kg of venison. The mean annual wool produced over the same period was 23,257 kg. Lamb sales to abattoirs amounted to 39,656 kg/annum to give an annual merino gross product of 62,913 kg representing 41% of the total biomass. At a dressing percentage of 48% (A. M. Conroy, personal communication), the lamb meat yield was 19,035 kg/annum.

Annual variation in venison, wool, and lamb production is illustrated in Fig. 15.4 along with annual rainfall which was below average in all years except 1981/82; this peak in rainfall coincided with the maximum venison produced. The two years preceding the study had very high precipitation, but drought necessitated the provision of supplementary feed to sheep throughout the study period; on average 65,013 kg of maize was provided per annum. Springbok were not observed to utilise these supplements.

Stock values
To avoid the confounding influence of fluctuating market prices, 1984 values and prices are applied to incomes and costs throughout this analysis, venison market prices being particularly subject to fluctuations. At a mean price of R 31/animal, the merino flock is valued at R 119,017. Natural mortality of merino sheep removed an average of R 1547/annum; however, this was replaced by lambing. The mean price of a springbok in 1984 was R 35, placing a value of R 36,995 (24% of the total value) on the springbok herd. Natural mortality removed R 1480/annum and this again was replaced by lambing. However, a population crash in August 1984 (Skinner *et al.*, 1987) represented a capital loss of R 26,320.

Gross income
The price of wool in 1984 was R 3.02/kg. The mean annual wool production of 23,257 kg thus earned R 70,236. Lambs sold at R 1.26/kg liveweight; the mean annual lamb sale of 39,656 kg thus yielded R 49,967. Together with a mean annual sale of skins of R 425, the mean gross annual income from merinos was R 120,628.

Venison sold at R 3.00/kg in 1984. Allowing for wastage and domestic use each kilogram of springbok liveweight removed from the veld yielded R 1.43. An average of 7608 kg liveweight of springbok harvested per annum thus yielded R 10,879. Sales of skins yielded a further R 107 per annum, to give a mean gross income from springbok of R 10,986/annum.

Fig. 15.4. Annual yield of farm products (kg) between 1978 and 1984 (Skinner *et al.*, 1986).

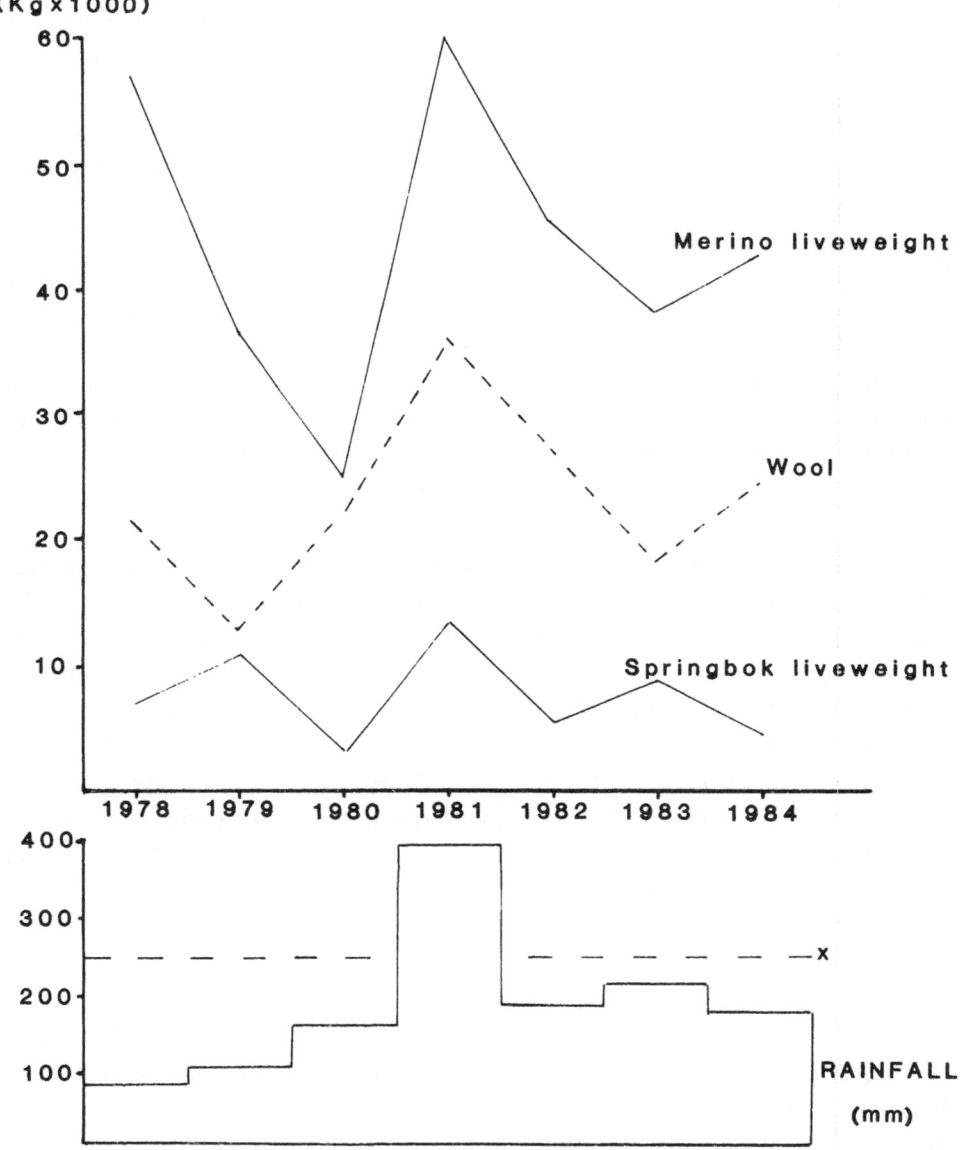

Table 15.6. *Cost Analysis. Figures represent an assessment between 1983 and 1985 and are expressed as the percentage of springbok or merino sheep gross income. 1984 prices were used in the analysis.*[a]

Cost	% Springbok	% Merino sheep
Supplementary feed		14.50
Marketing		9.16
Manager's salary		6.43
Stock purchases		0.12
Maintenance of grounds		0.73
Diesel		0.63
Fuel, electricity, labour	2.42	6.16
Production requirements	12.19	5.47
Packing material	3.25	0.66
Vehicle upkeep	1.76	0.82
Maintenance of veld	1.45	1.35
Maintenance of buildings	1.72	0.53
Labour	2.86	0.18
Petrol	1.24	1.39
Power	1.18	0.33
Telephone	2.12	0.37
Office	0.87	0.16
Diverse	0.29	2.12
Total costs	31.35	51.11

[a]Skinner et al. (1986).

Table 15.7. *Economic comparison of impala and beef cattle in the savanna (Rands).*[a]

	Impala	Beef cattle
Gross income/hectare (a)	5.03	10.00
Allocated costs/hectare	0.68	2.00
Overhead costs/hectare	0.00	5.20
Total costs/hectare (b)	0.68	7.20
Net income/hectare [(a)–(b)]	4.35	2.80
Net income/AU stocked	52.25	15.40
Net income/impala stocked	7.97	—

[a]Collinson (1979).

The annual combined income of springbok and merino sheep totals R 131,614 of which 91.65% is derived from merino sheep.

Net income
Table 15.6 lists the costs of production as a percentage of gross income derived from both species. At 31.35% costs, the gross springbok income (R 10,986) is reduced to a mean net income of R 7547/annum. The gross merino income (R 120,628) is reduced by 51.1% to R 58,987/annum. The combined net income from both species was thus R 66,634 of which the merino income comprised 89%. The annual variation in net income from springbok and merino sheep is plotted in Fig. 15.5 along with feed costs (included in net merino income assessment) and a valuation of the springbok stock loss (not included in net springbok income assessment).

In the savanna zone, Collinson (1979) made a three-year comparison of the economics of ranching impala and cattle in Natal (Table 15.7). He used a stocking rate of 2 animal-units (AU) cattle:1 AU impala (6.1 impala = 1 AU) on a 630 hectare ranch. Although beef production (kg carcase/hectare) was 2.6 times that of venison, impala earned greater profits. On the basis of meat production (kg/AU), beef cattle performed better than impala. Nevertheless, a monoculture of impala would be far less profitable than combined stocking with cattle. First, all overhead costs (roads, watering points, vehicles, etc.) would be directly attributed to the impala enterprise and the profit margin therefore would decrease substantially. Secondly, impala are highly selective in their grazing habits and they are impractical to manage in rotational grazing. Thirdly, cattle utilise the abundant fibrous herbage.

Conclusions
Despite recent acceptance of game ranching in southern Africa as a complementary agricultural land use, this has still to be translated into active support for research. There are still qualifications and reservations, but production systems based on selected wild ungulate species with domestic livestock have been most encouraging.

Acknowledgements
I wish to acknowledge the financial assistance of the University of Pretoria and the Department of Agriculture and Water Affairs who have supported for many years the research on which this review is based.

Fig. 15.5. Net income from springbok and merino sheep on a Karoo ranch between 1978 and 1985 (Skinner *et al.*, 1986).

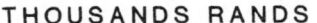

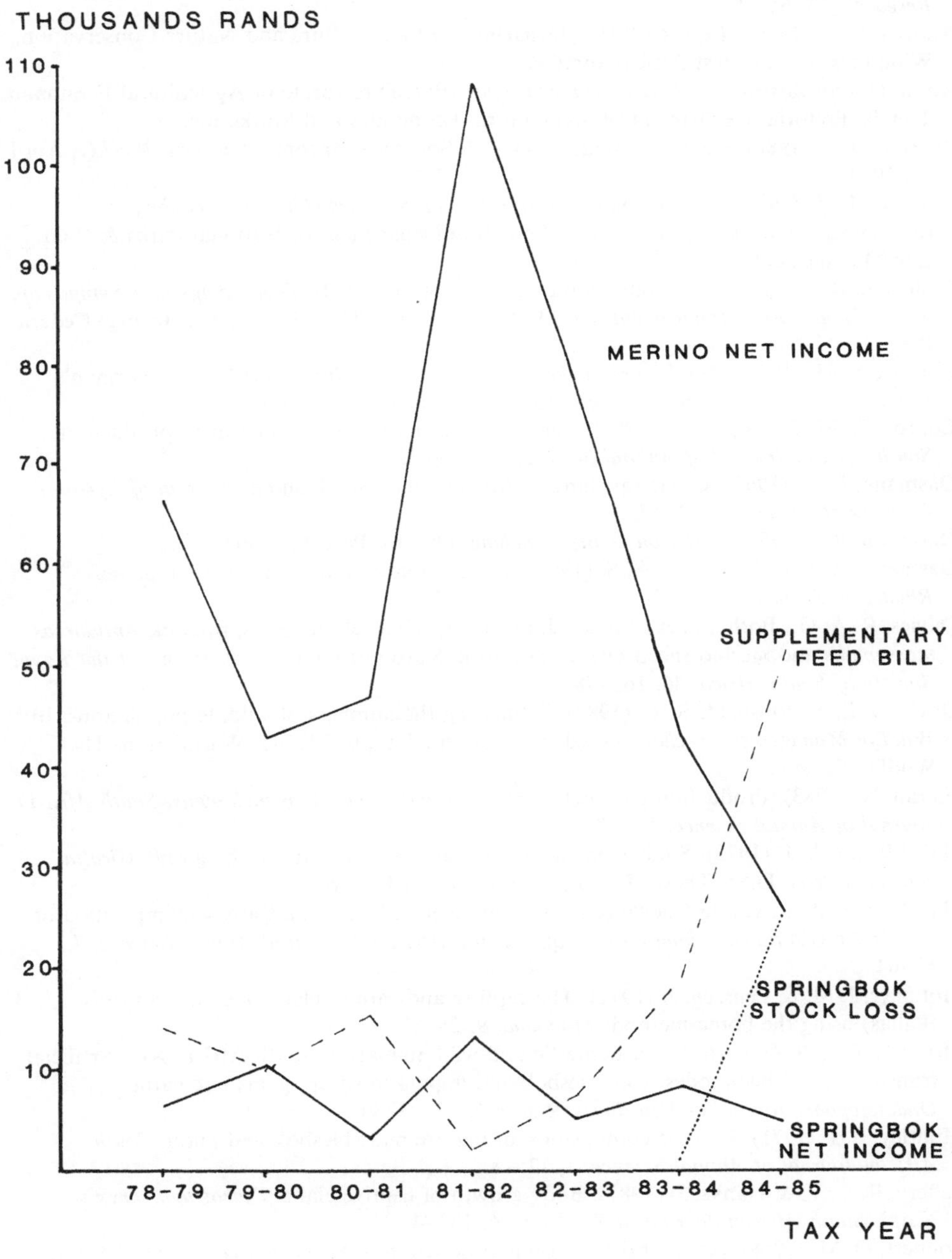
THOUSANDS RANDS
110
100
90
80
70
60
50
40
30
20
10
MERINO NET INCOME
SUPPLEMENTARY FEED BILL
SPRINGBOK STOCK LOSS
SPRINGBOK NET INCOME
78-79
79-80
80-81
81-82
82-83
83-84
84-85
TAX YEAR

References

Anderson, I. G. (1983). The prevalence of helminths in impala *Aepyceros melampus* (Lichtenstein 1812) under game ranching conditions. *South African Journal of Wildlife Research*, 13, 55–70.

Anon. (1981). *Annual Report (1980)*. Department of Agriculture and Nature Conservation, Windhoek, South West Africa/Namibia.

Anon. (1985). *Abstract of Agricultural Statistics 1985*. Directorate of Agricultural Economic Trends. Pretoria: Department of Agricultural Economics and Marketing.

Benson, D. E. (1986). Game farming survey – 2. Sources of income. *Farmers' Weekly*, April 11, 10–1.

Child, G. F. T. (1985). *A preliminary investigation of game ranching in Zimbabwe* (unpublished mimeograph). Harare: Zimbabwe Department of National Parks & Wild Life Management.

Collinson, R. F. H. (1979). Production economics of impala. In *Proceedings of a Symposium on Beef and Game Management*, ed. M. A. Abbott, pp. 90–103. Pietermaritzburg: Cedara Press.

Conroy, A. M. (1983). *The Future of the Game Industry, an Individual View*, Chairman's address. Kimberley: Game Farmers Association, North Cape Branch.

Conroy, A. M. & Gaigher, I. (1982). Venison, aquaculture and ostrich meat production. *South African Journal of Animal Science*, 12, 219–22.

Dasmann, R. F. (1962). Game ranching in African land use planning. *Bulletin of Epizootic Diseases of Africa*, 10, 13–17.

Dasmann, R. F. (1964). *African Game Ranching*. Oxford: Pergamon Press.

Dasmann, R. F. & Mossman, A. S. (1960). The economic value of Rhodesian game. *Rhodesian Farmer*, 30, 17–20.

Davies, R. A. G., Botha, P. & Skinner, J. D. (1986). Diet selected by springbok *Antidorcas marsupialis* and merino sheep *Ovis aries* during Karoo drought. *Transactions of the Royal Society of South Africa*, 46, 165–78.

Davis, D. E. & Winstead, R. L. (1980). Estimating the numbers of wildlife populations. In *Wildlife Management Techniques*, ed. S. D. Schemnitz, pp. 221–45. Washington: The Wildlife Society.

Fairall, N. (1983). Production parameters of the impala *Aepyceros melampus*. *South African Journal of Animal Science*, 13, 176–9.

Hall-Martin, A. J. (1975). Studies on the biology and productivity of the giraffe *Giraffa camelopardalis*. D.Sc. Thesis. Pretoria: University of Pretoria.

Hall-Martin, A. J., von la Chevallerie, M. & Skinner, J. D. (1977). Carcase composition of the giraffe *Giraffa camelopardalis giraffa*. *South African Journal of Animal Science*, 7, 55–64.

Hofmeyr, J. M. & Lenssen, J. (1975). The capture and care of eland *Taurotragus oryx* (Pallas) using the boma method. *Madoqua*, 9, 25–33.

Horak, I. G. (1979). Parasites of domestic and wild animals in South Africa. XII. Artificial transmission of nematodes from blesbok and impala to sheep, goats and cattle. *Onderstepoort Journal of Veterinary Research*, 46, 27–30.

Huntley, B. J. (1971). Carcase composition of mature male blesbok and kudu. *South African Journal of Animal Science*, 1, 125–8.

Jeffery, R. C. V. & Hanks, J. (1981). Body growth of captive eland *Taurotragus oryx* in Natal. *South African Journal of Zoology*, 16, 183–4.

Joubert, D. M. (1968). An appraisal of game production in South Africa. *Tropical Science*, 10, 200–11.

Keep, M. E. (1972). The meat yield, parasites and pathology of eland in Natal. *Lammergeyer*, 17, 1–19.

Mason, D. R. (1982). *Studies on the Biology and Ecology of the Warthog* Phacochoerus aethiopicus sundevalli *Lonnberg, 1908 in Zululand*. D.Sc. Thesis. Pretoria: University of Pretoria.

Meissner, H. H. (1982). Theory and application of a method to calculate forage intake of wild southern African ungulates for purposes of estimating carrying capacity. *South African Journal of Wildlife Research*, 12, 41–7.

Meissner, H. H. & Hofmeyr, H. S. (1972). A preliminary note on the fasting metabolism and feed intake in the blesbok *Damaliscus dorcas phillipsi* and the South African mutton merino under comparable conditions. *South African Journal of Animal Science*, 2, 73–4.

Mentis, M. T. & Duke, R. A. (1976). Carrying capacities of natural veld in Natal for large wild herbivores. *South African Journal of Wildlife Research*, 6, 65–74.

Neitz, W. O. (1965). A checklist and hostlist of the zoonoses occurring in mammals and birds in South and South West Africa. *Onderstepoort Journal of Veterinary Research*, 32, 198–376.

Ortlepp, I. J. (1971). 'n Oorsig van Suid-Afrikaanse helminte veral met verwysing na die wat in ons wildeherkouers voorkom. *Tydskrif vir Naturwetenskap*, 1, 203–12. (in Afrikaans)

Rogerson, A. (1968). Energy utilization by the eland and the wildebeest. *Symposium of the Zoological Society of London*, 21, 153–61.

Skinner, J. D. (1967). An appraisal of the eland as a farm animal in Africa. *Animal Breeding Abstracts*, 35, 177–86.

Skinner, J. D. (1980). Productivity of mountain reedbuck at the Mountain Zebra National Park, Cradock. *Koedoe*, 23, 123–30.

Skinner, J. D. (1984a). Selected species of ungulates for game farming in South Africa. *Acta Zoologica Fennica*, 172, 219–22.

Skinner, J. D. (1984b). Mating and calving seasons, sex ratios and age groups, and monitoring ungulate populations for game farming. In *Proceedings of a Workshop on Conservation and Utilization of Wildlife on Private Land*, ed. P. R. K. Richardson & M. P. S. Berry, pp. 64–7. Pretoria: Southern African Wildlife Management Association.

Skinner, J. D., Davies, R. A. G., Conroy, A. M. & Dott, H. M. (1986). Productivity of springbok *Antidorcas marsupialis* and merino sheep *Ovis aries* during a Karoo drought. *Transactions of the Royal Society of South Africa*, 46, 149–64.

Skinner, J. D., Dott, H. M., Van Aarde, R. J., Davies, R. A. G. & Conroy, A. M. (1987). Population dynamics of springbok in the Karoo. *Transactions of the Royal Society of South Africa*, 46, 179–98.

Skinner, J. D., Monro, R. H. & Zimmermann, I. (1983). Comparative food intake and growth of cattle and impala on mixed tree savanna. *South African Journal of Wildlife Research*, 15, 1–9.

Skinner, J. D., von la Chevallerie, M. & van Zyl, J. H. M. (1971). An appraisal of the springbok as a farm animal in Africa. *Animal Breeding Abstracts*, 39, 215–24.

Smithers, R. H. N. (1983). *The Mammals of the Southern African Subregion*. Pretoria: University of Pretoria.

Stocker, M. & Walters, C. (1984). Dynamics of a vegetation:ungulate system and its optimal exploitation. *Ecological Modelling*, 25, 151–65.

Turton, J. D. (1962). The effect of castration on meat production and quality in cattle, sheep and pigs. *Animal Breeding Abstracts*, 30, 447–56.

Underwood, R. (1975). *Social Behaviour of the Eland (*Taurotragus oryx*) on Loskop Dam Nature Reserve*, M.Sc thesis. Pretoria: University of Pretoria.

van Zyl, J. H. M., von la Chevallerie, M. & Skinner, J. D. (1969). A note on the dressing percentage in the springbok (*Antidorcas marsupialis* Zimmermann) and impala (*Aepyceros melampus*). *Proceedings of the South African Society of Animal Production*, 8, 199–200.

von la Chevallerie, M. (1971). *Carcase Quality in Different Species of Game and the Influence of Season*, unpublished progress report. Potchefstroom: Agricultural Research Institute.

von la Chevallerie, M. (1972). Meat quality in seven wild ungulate species. *South African Journal of Animal Science*, 2, 101–4.

Wilkins, J. (1982). Opening address. First National Game Congress. Sept. 1982. South African Agricultural Union.

Young, E. (1984). *Wildboerdery en Natuurreservaat Bestuur*. Swartklip: Eddie Young Publications.

SECTION F

Intensive containment systems: game farming

K. R. DREW

Game farming is a production mode in which game animals are managed, handled, and slaughtered in similar ways to traditional livestock. Inherent in this definition is the use of supplementary feeding, weighing, individual animal identification, veterinary procedures, modern genetic selection, and organised slaughter at abattoirs where *ante-* and *post-mortem* inspection is possible. Although farmed venison still makes a modest contribution to international production and trade, it is a rapidly growing production mode and promises to dominate emerging international markets.

Although game farming might seem not to capitalise on the unique ecological adaptations of wild animals, many species (particularly deer of the subfamily *Cervinae*) are amenable to intensive husbandry. The practical advantages of intensive husbandry include the ability to ensure high standards of disease control and hygiene which facilitates development of new specialty markets, and the option to use relatively small land bases which allows diversification of existing agricultural enterprises.

Game farming is both very old and very new and is practised in many parts of the world (Table F.1). Deer have been farmed on a commercial scale in China for many decades. The industry had a later start in Korea but interest is strong and populations of farmed deer continue to grow rapidly. On the other hand, deer farming in North America and Japan are embryonic industries. The Soviet Union is among major deer producing countries but the diversity of operation and the enormous size of the country make it difficult to determine the relative contribution of game farming, ranching, or commercial hunting. Deer farming in New Zealand has been spectacularly successful, starting in 1970 and growing to more than 300,000 animals by 1986. The breeding herd has doubled every 3–4 years.

In the Orient, velvet antlers for pharmaceuticals are the main product of the deer industry. A number of Western countries serve this lucrative market although the removal of antlers in the velvet is prohibited in the United Kingdom. Despite current prices, the antler trade cannot be perceived as a growth industry for Western countries. International deer/game farming organisations must establish attractive meat markets to ensure continued growth. With knowledge and skill, game producers will be able to establish a new generation of meat products which are rich in all the best attributes of red meat but lacking in the real or perceived disadvantages of excess fat.

The following chapters review deer farming on a regional basis; Oceania, Europe, and Asia. The firmly established North American bison industry also is described. Other intensive production systems such as those for muskoxen and musk deer which remain at an experimental stage are covered in Section G.

Table F.1. *Some game farming systems.*

Geographical area	Animal species	Approx. numbers farmed	Grazing management	Main products
China	Sika and malu deer	300,000	Intensive	Velvet antler
New Zealand	Red, fallow, and wapiti	300,000	Extensive/ intensive	Meat and velvet antler
USSR	Sika and maral deer	90,000	Extensive	Velvet antler
North America	Bison	88,000	Extensive/	Meat
	Wapiti	6,000	intensive	Meat and velvet antlers
Australia	Fallow, red, and rusa deer	40,000	Extensive/ intensive	Meat, velvet antlers
Korea	Sika deer	52,000	Intensive	Velvet antlers
	Red deer	2,000		
	Wapiti	3,000		
Mauritius	Rusa deer	4,000	Intensive	Meat for local markets and by-products
United Kingdom	Red deer	10,000	Intensive	Meat
Alaska	Muskoxen	100–200	Intensive	Fibre (qiviut)
Japan	Sika deer	Few	Extensive	Meat and velvet antler

16

Deer farming in Oceania

P. F. FENNESSY & P. G. TAYLOR

Abstract

Deer farming in Oceania is confined largely to New Zealand and Australia. Deer were introduced to both countries in the 19th century, so successfully in New Zealand that control operations were required to protect vegetation and watersheds. Marketing venison from control operations led to the farming of deer as feral populations declined. In 1985, there were approximately 374,000 farmed deer on 4000 farms in New Zealand. Deer respond to intensive management and husbandry similar to that applied to cattle and sheep. Strong demands and speciality prices for venison create a favourable economic picture and confidence in the industry remains high. In Australia, about 43,000 deer are supported on 400 farms and the infrastructure is less developed. Growth of the industry is restricted mainly by the availability of breeding stock.

Introduction

Oceania includes Papua New Guinea, Australia, New Zealand, and neighbouring Pacific islands. Deer are not endemic but were introduced by European settlers mainly during the 19th century. Various species are now found in New Zealand (mainly red deer (*Cervus elaphus*) and fallow deer (*Dama dama*) but also North American wapiti (*C. e. nelsoni*), white-tailed deer (*Odocoileus virginianus*), sika (*Cervus nippon*), sambar (*Cervus unicolor*), and rusa (*Cervus timorensis*)). In Australia, the main species farmed are fallow deer and red deer but also sambar, rusa, chital (*Axis axis*), and hog deer (*Axis porcinus*). Papua New Guinea and New Caledonia have rusa. Currently, deer farming is virtually confined to New Zealand and Australia but there is considerable interest in Papua New Guinea (Fraser-Stewart, 1985) and New Caledonia. Consequently, this chapter is concerned only with deer farming in New Zealand and Australia.

New Zealand

Deer were first introduced to New Zealand in 1851 with several more liberations following over the next 60 years. Favoured by plentiful food supplies, mild climate, and absence of predators, populations grew rapidly. By early this century, damage to native forests and grasslands was such that they were regarded as pests (Wodzicki, 1950). Of the several species introduced, only red deer (Fig. 16.1) and to a lesser extent fallow deer prospered although there are small local populations of wapiti, sika, rusa, sambar, and white-tailed deer.

Because of damage to vegetation and concern about erosion, there have been several attempts at population control. In initial attempts, bounties were paid to private hunters. Then, in the 1930s, the government employed their own hunters to shoot deer. A substantial proportion of the skins were recovered for export (Challies, 1985). Although Challies (1974, 1985) considers it unlikely that early control operations in forested areas significantly reduced deer numbers, commercial hunting for skins undoubtedly did have a significant impact in more accessible areas in the 1940s and 1950s. From the late 1950s, the greater hunting pressure applied in priority areas (i.e., those selected on the

Fig. 16.1. Red deer, which comprise about 85% of farmed deer in New Zealand. (Photograph by Audio-Visual Unit, MAFTech, Invermay Agricultural Centre.)

basis of downstream values at risk (Riney, 1956)) appears to have been effective in reducing browsing on preferred plant species (Challies, 1974).

Subsequently, the recovery of deer carcases proved economic and by the early 1960s venison was being exported to Europe and by-products such as antlers, tails, pizzles, and sinews to Asia for traditional medicines and health products. Following the success of the first commercial helicopter venison operation in 1963, it became apparent that pressure on feral populations could become so great that the resource would soon become limiting. Venison exports increased from 480 tonnes in 1962/63 to 4400 tonnes only 10 years later, but by 1979/80 exports had declined to about 1000 tonnes. It was this pressure on wild stocks which led to the development of deer farming and new markets for breeding stock.

The first serious attempts to capture live deer were made in 1967 but it was not until 1969 that deer farming was legalised, with the first licence issued in March 1970. Initially, growth was slow and by 1975 there were only about 25 farms with a total of 5000 deer. However, the high prices received at the first deer auctions in 1977 encouraged helicopter operators to seize the opportunity for live capture. For example, in the 1979/80 season 20,000–25,000 feral deer were captured (Wallis & Hunn, 1982; Fennessy & Drew, 1983). Although live capture is now much less important, a number of deer are still captured each year.

Deer farming initially developed to provide venison for export. However, lucrative prices for velvet antler on the Korean market in the late 1970s temporarily diverted the industry from this main course. With the promise of a lucrative velvet antler market, prices for stags and breeding hinds increased substantially. Other factors stimulating development of deer farming include the ready availability of feral deer, adaptability of red deer to farm conditions, development of efficient capture techniques, and relatively low prices for other livestock products, especially sheep meat and wool. The last factor was particularly important in the 1960s and 1970s when New Zealand farmers were looking to diversify their operations, a situation which has persisted. Another important stimulus was the infusion of capital by urban investors formerly encouraged by favourable taxation policies. However, even without tax advantages, there is considerable interest in a variety of arrangements. For example, deer farmers often manage animals owned by investors, charging a grazing fee or by taking a share of the progeny and products.

Structure and size of the industry

In New Zealand, farm animals are pastured throughout the year. In times of pasture shortage, such as during winter or summer drought, they are fed supplements such as hay, silage, or grain. Thus, the same husbandry systems are used for deer as for sheep and cattle and in fact most deer are run on mixed farms where the predominant animals are likely to be sheep or dairy cows.

Currently, the New Zealand deer industry is expanding maximally and virtually all young females are retained for breeding. According to a survey by the New Zealand Deer Farmers' Association, there were about 374,000 farmed deer in mid 1985, of which 188,000 females were mated. About 8% were fallow and 84% were red deer; most of the remainder were wapiti or red x wapiti hybrids. Also, about 47,000 older males were farmed for velvet antler production (mainly red deer as fallow deer are not used for velvet production to any extent) and breeding. Slaughter animals are mainly young stags, with the balance being older stags which are usually killed following the rut when they are very low in carcase fat. In the absence of female culling, the number of breeding females can be expected to double about every 3 years. The deer are run on about 4000 farms with about 64% of all deer in the North Island and the remainder in the South Island.

The industry is well organised nationally. The New Zealand Deer Farmers' Association (NZDFA), through its national body (established in 1975) and local branches, represents a majority of those farming deer. The NZDFA keeps its members well informed through a newsletter and an annual conference, while it also has the major interest in the publication of the monthly magazine, *The Deer Farmer*. The NZDFA's role has been primarily political in representing the industry to government. The Game Industry Board, established by government statute in 1985, has equal representation from producers and from the processing industry together with a government representative. It therefore draws together all sectors of the deer farming industry with the stated function of promoting the orderly development of this new industry, particularly its overseas markets.

Technology

Most New Zealand deer farmers have considerable experience with pastoral production systems and have developed appropriate husbandry methods for deer. Although there is a wide range in climate and hence the annual pattern and total amount of pasture production, much of New Zealand is well suited to animal production from pasture (see Nicol, 1987) and deer are raised on all types of country from the intensive dairying areas of the North Island to the extensive high country of the South Island. Compared with farming sheep, deer farming has a relatively low labour requirement, with the most intensive periods being weaning, mating, and during supplementary feeding.

Regulations require 2-m perimeter fencing which is usually a purpose-built deer netting. Internal fencing may be the same or may be an extended sheep fence, whereas temporary electric fencing is often used to ration pasture. Many farms have a laneway system so that the animals can be shifted simply. Good yarding facilities are essential for effective management (e.g., for anthelmintic treatment, ear-tagging, testing for tuberculosis, velvet antler removal, etc.). Frequently, these yards will include a body crush, weigh scales, and often a small handling pen which will hold five to eight hinds. Deer became easy to handle and are apparently very amenable to the systems developed in New Zealand. For example, deer are safely trucked all over New Zealand by specialist deer transport operators (Trask, 1986), and many farms also have their own deer transport trailers.

Strong research and extension programmes coupled with the enthusiasm of deer farmers and infusions of off-farm capital have had a major collective impact on the development of deer farming. Major research contributions include the following:

1. Guidelines have been established on management procedures (including animal health) to achieve high levels of production (Fennessy, 1982; Mason, 1984; Macintosh & Beatson, 1985; Moore, Cowie & Bray, 1985; Macintosh, 1986; Pearse, 1987a,b).
2. Research on the control of tuberculosis has been vital (Beatson & Hutton, 1981; Carter *et al.*, 1986; Griffin & Cross, 1986).
3. An understanding of some of the factors which influence antler growth and development has assisted farmers involved in velvet antler production (Fennessy & Suttie, 1985).
4. Comparative data on different species of deer and on hybridisation (e.g., wapiti and red deer; Moore, 1985) is expected to increase interest in wapiti hybrids greatly over the next few years.
5. Guidelines for the selection of genetically superior deer are being developed along with a recording and genetic improvement system (Cowie, 1985; Fennessy, 1986). Bloodtyping also may have a place in improving assessments of genetic value (Dratch, 1986).
6. Reproductive technologies, such as advancement of the breeding season and artificial insemination, have been developed (Moore & Cowie, 1985; Fennessy *et al.*, 1986; van Reenen, 1986).
7. New meat processing technologies are being adopted (Drew & Fennessy, 1986).

Management of farmed deer is a skilled operation. This is well illustrated on intensive farms where it is essential to regularly treat young deer through their first autumn and winter with anthelmintics to ensure that performance does not suffer from lungworm (*Dictyocaulus viviparus*) infestation (Mason, 1985). With young deer in their first winter, adequate nutrition minimises the risk of the stress-related diseases such as yersiniosis (Macintosh, 1986) and malignant catarrhal fever (Oliver *et al.*, 1986). Similarly, an appreciation of seasonal feed requirements is essential (Fennessy, 1982). For example, adult stags have very low fat reserves following the rut. Consequently, high quality diets are required to ensure that they do not succumb under difficult climatic conditions and to improve velvet antler production the following spring (Fennessy & Suttie, 1985). Adequate quantities of high quality feed should be supplied to lactating hinds so their calves thrive (Fennessy & Milligan, 1987). Although with proper management there are few problems with animal health, malignant catarrhal fever, yersiniosis, and bovine tuberculosis do cause problems on some farms. A scheme for tuberculosis control run by the NZDFA and with cooperation of the Deer Branch of the NZ Veterinary Association is in operation (Hunter, 1986).

Farmed deer are slaughtered at specialised deer slaughter premises (DSP) of which there are four in the South Island and five in the North Island. At the DSPs, deer are subject to both *ante-* and *post-mortem* examination by a veterinarian. Since deer are not classified as domestic stock, they cannot be slaughtered at abattoirs along with sheep, cattle, goats, and pigs. The venison is processed and packaged for export at Game Packing Houses, a number of which are attached to the DSPs. Feral (wild-shot) venison is also processed through the Game Packing Houses, with the separation of farmed from feral venison being rigidly controlled by government inspectors.

Productivity

Deer are efficient converters of grass to meat on good grazing land; the use of such high-class land contrasts with the traditional habitat of the deer in New Zealand. Body growth is strongly seasonal. Basic information on feed requirements has enabled more efficient use of feed and improved production from pasture (Fennessy, Moore & Corson, 1981; Harbord, 1982; Fennessy & Milligan, 1987). The meat is very lean compared with traditional livestock (Drew, 1985; Drew & Fennessy, 1986) and the eating qualities of farmed and feral deer are similar (Forss *et al.*, 1979).

The principal factors determining herd productivity are reproductive and growth rates (and hence slaughter weight) of the progeny and the amount of feed consumed by the herd. As well as being farmed for meat production, some red deer (and wapiti) are kept specifically for velvet antler production and to a lesser, but increasing extent for trophy hunting.

Red deer and fallow deer normally breed for the first time at 16 months of age, thus calving at two years. For red deer, there is a threshold bodyweight of about 70 kg (about 70% of mature weight) below which hinds are less likely to become pregnant (Fennessy *et al.*, 1986). A corresponding threshold weight for the onset of puberty has not been defined for fallow does (Asher, 1986). As both red and fallow deer have low natural twinning rates (probably less than 0.5%), high conception rates are essential (>90%). Therefore, it is important to have adult hinds in good condition for mating. Consequently, pre-rut weaning may help in ensuring high herd fertility (Fennessy & Milligan, 1987). Although pre-rut weaning is general on intensive farms, weaning after the rut is not uncommon on the more extensive properties (Yerex & Spiers, 1987). Single sire mating in red deer is commonly practised with the recommendations being one stag per 50 hinds (Moore *et al.*, 1985).

Good growth rates in female deer are vital in terms of yearling fertility. They also are very important for slaughter stags as the NZ venison price system is based on carcase weight and leanness. The premium carcase weighs 50–70 kg with less than 10% fat (Drew & Fennessy, 1986). Most red deer stags fulfil these criteria at 12–18 months of age although this is currently too late to enjoy the seasonal premiums which prevail in the spring, when these animals are about 9–10 months old. Consequently, many older stags are slaughtered in the early spring. From a productivity point of view, this is not the optimum time for the farmer as the stags do not grow over winter but must be fed well to maintain weight (Fennessy *et al.*, 1981). Overall, the objective is to maximise growth rates at times of the year when high quality grazing feed is available and the animals have the propensity to grow. Spring pasture in most parts of New Zealand is adequate but summer pastures are limiting in some regions so irrigation or costly supplements may be necessary.

The use of such expensive procedures ultimately depends on the market return for venison and by-products such as skins, testes, pizzles, and tails (Kong & But, 1985). The international market for velvet antler is relatively small and well-supplied (Hughes, 1986). Since it is a buyer's market, many New Zealand deer farmers have little if any interest in velvet production. However, for some it is very profitable, with highly selected groups of red x wapiti hybrids and wapiti being retained for this purpose.

There is considerable interest among deer farmers in widening the genetic base of farmed deer. Consequently, since 1981 there have been several importations of wapiti from Canada and red deer from Europe (particularly Germany and the United Kingdom). Small numbers of fallow deer and Pere David's deer (*Elaphurus davidianus*) also have been imported. Much of the initial interest was stimulated by the possibilities for both wapiti-type velvet in the velvet antler trade and larger carcases in the venison trade.

Economics

The economics of deer farming in New Zealand are difficult to assess in the long term without several assumptions, mainly relating to the longterm relativity between the price for weaner hinds destined for breeding and weaner stags for slaughter (currently for red deer weaners of 3–6 months of age, the relativity is about 6:1 in favour of hinds). Most commentators recognise that the price relativity will eventually settle to about 1:1, but the question for investors is when.

With the high price for animals and facilities (fencing at about NZ$ 8000/km and a set of yards for about NZ$ 20,000), the cost of establishment is high. However, currently returns on investment are good, especially when compared with traditional farming enterprises such as sheep and cattle (Harbord, 1986; Pearse, 1986).

Prospects

The New Zealand deer industry is generally optimistic about the future for venison internationally. Currently deer farming is expanding at the maximum possible rate (no females are being slaughtered) while deer, particularly red deer from Europe, are being imported to broaden the genetic base of the local herd. However, the long-term future depends absolutely on the ability to develop new markets for venison. The future for velvet antler must be seen as limited unless there is a major breakthrough in sales to Western consumers.

Australia

Australia lacks indigenous deer and early settlers imported many species for aesthetic and sporting purposes, attempting to establish wild populations of about 14 species (Bentley, 1978) of which only six remain: fallow, red, rusa, sambar, hog deer, and chital.

Recreational hunting of deer is popular and, despite its illegality, wild-shot venison has long supplied the restaurant trade. Recognition of potential markets for farmed venison, and of similar interest in New Zealand, led to the establishment of two commercial deer farms in Victoria in 1971/72. By 1980, commercial herds were established in all eastern states and the industry received much publicity. Rapid escalation of stock prices accompanied the expansion (Anderson, 1982).

Whereas export and some local markets for antler were developed, high prices received in New Zealand were rarely attained in Australia. Difficulties associated with marketing small amounts of venison, an influx of lower priced venison from New Zealand, and the failure of velvet antler to provide anticipated returns, led to reduced investment in the early 1980s and a marked decline in livestock values.

However, since about 1984, there has been increasing acceptance of venison by the local market as a low-fat, high quality meat although its use is largely confined to the better restaurants. Renewal investment has resulted from farmers wishing to diversify away from traditional enterprises which are experiencing a period of low profitability.

Currently, production is oriented to local markets although it is likely that as the industry grows, increasing attention will be given to higher-priced export opportunities.

Structure and size of industry

Accurate data regarding size and composition of the Australian deer industry are unavailable. However, Woodford (1986) estimates that there are about 43,000 head on 400 farms (Table 16.1), with a capacity for growth of some 25% per annum. In contrast to New Zealand, about 65% are fallow and 18% are red deer. A significant proportion of the total herd is made up of a small number of large units, there being many extremely small herds. Hybridisation between rusa and sambar has been practised at least since 1976, and more recently between red and wapiti. Some importations of wapiti have been made from New Zealand.

Owners of most of the large and many of the small herds belong to state industry organisations established to represent deer farmers in discussions with government and to provide information to members. These state bodies

Table 16.1. *Estimates of farmed deer in Australia, 1986.*[a]

	Queensland	New South Wales	Victoria	South Australia	Western Australia	Tasmania	Total
Fallow	300	11,000	7,000	2,500	2,500	4,400	27,600
Red	5,000	800	1,000	500	500	—	7,800
Rusa	1,000	1,500	2,000	200	100	—	4,800
Chital	300	1,500	300	20	50	—	2,170
Other	—	200	110	—	—	—	310
Total	6,600	15,000	10,410	3,220	3,150	3,400	42,680

[a]Woodford (1986).

are in turn represented nationally by the Australian Deer Breeders Federation formed in 1979.

Technology

Many Australian deer farmers are newcomers and some even have little experience with conventional livestock. There is wide variation in standards of management and use of available knowledge and technology, but active dissemination of such information by industry leaders is rapidly raising standards.

The industry is based on several species, each with particular requirements. It also is practised in a wide range of environments in which rainfall and other climatic factors influence pasture growth. Traditional livestock production in Australia characteristically involves sale or purchase of breeding stock in keeping with feed availability, an option difficult to exploit with deer. In only a few areas does seasonal variation in pasture production coincide with nutritional requirements of any given deer species and supplementary feeding and/or irrigation of pastures are essential on most fully stocked enterprises.

Most deer farms are still being developed and labour input is thus generally high but will decrease substantially relative to herd size as farms become fully fenced, equipped, and stocked. Labour requirements vary substantially with needs to conserve fodder or manage irrigation programmes.

Deer on most farms require infrequent yarding and handling; routine administration of anthelmintics is not widely practised since internal parsitism generally has remained low (Presidente, 1984). However, deer in much of Queensland need regular treatment for infestation of cattle tick (*Boophilus microplus*) (MacKenzie, 1984). Although a variety of viral and bacterial diseases have been reported (English, 1984), their incidence is sporadic and of little significance on most farms.

Legislation in most Australian states requires minimum fence standards; usually about 2.1 m in height and constructed of mesh or netting specifically designed for deer (Taylor, 1984). Farm layout typically involves small paddocks connected by laneways to handling yards, the design of which often permits no more than basic confinement of deer. However, the use of scales and crushes for tagging and veterinary attention is becoming more widespread.

While many deer are still transported in makeshift or poor quality containers, standards are improving rapidly and there are several deer transport contractors in operation (see Stevens, 1986). Losses and injuries during transport are infrequent.

Only one abattoir, at Muswellbrook in New South Wales, has been constructed specifically for deer with other venison being processed in conventional abattoirs (Scarf, 1986). Sale of farm-killed venison is generally illegal. Deer brought to the Muswellbrook works are held before slaughter but elsewhere are killed upon arrival, usually in the transport crate. With industry expansion, it is likely that some abattoirs will construct permanent facilities for deer although relevant authorities generally regard present techniques as adequate for meeting humane and quality standards.

Productivity

Reproductive performance and growth rates vary widely among farms and between successive years. Few quantitative data are available but nutrition is a major factor determining reproductive success. Shortage of breeding stock and retention of aged and poor quality females depresses overall herd performance and rate of genetic improvement. Only recently have breeding males been subjected to much selection and proven sires are generally not available. Artificial breeding techniques are used on a limited scale; the long-term interest arises mainly with importation of semen or embryos from outside Australia.

Before 1984, high proportions of non-breeding males were retained for antler production but most are now slaughtered. Some farms with established venison markets buy young or unfinished stock for slaughter after further growth, which may involve lot-feeding. Seasonal fluctuations in male body-weight and aggressive behaviour in the breeding season interrupt continuity of supply of venison, alleviated to a small degree by use of species whose seasonal patterns differ from fallow and red deer. Castration is not widely practised because growth is slowed (Drew, Fennessy & Greer, 1978; Mulley & English, 1985) although there is interest in the strategic use of castration so that fallow deer are available for slaughter during the rut and winter. Woodford (1986) estimated about 2300 deer were slaughtered in 1985/86, representing only 5% of the estimated total herd, suggesting a capacity to markedly increase production in the short term.

Antler production has declined markedly since 1984. Commercial drying facilities do not exist and the previous major outlet for frozen antler, Taiwan, does not now permit imports in unprocessed form. With increased slaughtering of males for venison it is unlikely that any processing venture could obtain sufficient quantities for viability at present values. A market still exists in the local Chinese·community supplied by some individual farmers, but many farms no longer harvest antler for sale, removing it, if at all, only for stock management reasons.

Economics

Deer prices currently represent several times a female's worth as venison. Consequently, acquisition of a herd requires substantial investment and confidence that females produced can be sold for comparable returns. On such a basis, profitability can be superior to many other livestock but there is general recognition (Woodford, 1986) that eventually returns will more closely reflect venison prices.

Prospects

Growth of the Australian industry and its markets is severely restricted by lack of breeding stock, but its long-term future depends largely upon achievement and maintenance of high standards of venison quality whether for local or export consumption. Antler production will probably continue to decline, although hides and other slaughter by-products may be marketed more profitably as quantities increase. However, deer are most unlikely ever to usurp the position of any of the more traditional livestock species in Australia.

References

Anderson, R. (1982). Australia. In *The Farming of Deer: World Trends and Modern Techniques*, ed. D. Yerex, pp. 23–30. Wellington: Agricultural Promotion Associates.

Asher, G. W. (1986). *Studies on the reproduction of farmed fallow deer (*Dama dama*)*. Ph.D. thesis. Canterbury: Lincoln College, University of Canterbury.

Beatson, N. S. & Hutton, J. B. (1981). Tuberculosis in farmed deer in New Zealand. *Proceedings of a Deer Seminar for Veterinarians, Queenstown*, Deer Advisory Panel of New Zealand Veterinary Association. pp. 143–51.

Bentley, A. (1978). *An Introduction to the Deer of Australia*. Victoria: Koetong Trust Service Fund Forests Commission.

Carter, C. E., Corrin, K. C., de Lisle, G. W. & Kissling, R. C. (1986). An evaluation of the comparative cervical test in deer. *Proceedings of a Deer Course for Veterinarians*, 3, 65–70. Deer Branch of the New Zealand Veterinary Association.

Challies, C. N. (1974). Trends in red deer (*Cervus elaphus*) populations in Westland forests. *Proceedings New Zealand Ecological Society*, 21, 41–50.

Challies, C. N. (1985). Establishment, control, and commercial exploitation of wild deer in New Zealand. In *Biology of Deer Production*, ed. P. F. Fennessy & K. R. Drew, Bulletin 22, pp. 23–36. Wellington: Royal Society of New Zealand.

Cowie, J. (1985). Practical deer recording and its benefits. *The Deer Farmer*, 26, 38–9.

Dratch, P. A. (1986). A marker for red deer-Wapiti hybrids. *Proceedings New Zealand Society of Animal Production*, 46, 179–82.

Drew, K. R. (1985). Meat production from farmed deer. In *Biology of Deer Production*, ed. P. F. Fennessy & K. R. Drew, Bulletin 22, pp. 285–90. Wellington: Royal Society of New Zealand.

Drew, K. R. & Fennessy, P. F. (1986). Venison research – carcass features, proceedings and packaging. *Proceedings of a Deer Course for Veterinarians*, 3, 17–34. Deer Branch of the New Zealand Veterinary Association.

Drew, K. R., Fennessy, P. F. & Greer, G. J. (1978). The growth and carcass characteristics of entire and castrate red stags. *Proceedings New Zealand Society Animal Production*, 38, 142–4.

English, A. W. (1984). Veterinary aspects of deer farming in New South Wales – an update. *Proceedings Post Graduate Committee in Veterinary Science, The University of Sydney*, 72, 427–59.

Fennessy, P. F. (1982). Nutrition and growth. In *The Farming of Deer: World Trends and Modern Techniques*, ed. D. Yerex, pp 105–14. Wellington: Agricultural Promotion Associates.

Fennessy, P. F. (1986). Genetic selection. In *Deer Farming into the Nineties*, ed. P. Owen, pp. 141–7. Brisbane: Owen Art & Publishing.

Fennessy, P. F. & Drew, K. R. (1983). The development of deer farming in New Zealand. *Philippine Journal of Veterinary and Animal Sciences*, 9, 197–202.

Fennessy, P. F., Fisher, M. W., Webster, J. R., Macintosh, C. G., Suttie, J. M., Pearse, A. J. & Corson, I. D. (1986). Manipulation of reproduction in red deer. *Proceedings of a Deer Course for Veterinarians*, 3, 121–31. Deer Branch of the New Zealand Veterinary Association.

Fennessy, P. F. & Milligan, K. E. (1987). Grazing management of deer. In *Livestock Feeding on Pasture*, ed. A. M. Nicol, Occasional Publication No. 10, pp. 111–18. Hamilton: New Zealand Society of Animal Production.

Fennessy, P. F., Moore, G. H. & Corson, I. D. (1981). Energy requirements of red deer. *Proceedings New Zealand Society Animal Production*, 41, 167–73.

Fennessy, P. F. & Suttie, J. M. (1985). Antler growth: nutritional and endocrine factors. In *Biology of Deer Production*, ed. P. F. Fennessy & K. R. Drew, Bulletin 22, pp. 239–50. Wellington: Royal Society of New Zealand.

Forss, D. A., Manley, T. R., Platt, M. P. & Moore, V. J. (1979). Palatability of venison from farmed and feral red deer. *Journal of the Science of Food and Agriculture*, 30, 932–7.

Fraser-Stewart, J. W. (1985). Deer and development in southwest Papua New Guinea. In *Biology of Deer Production*, ed. P. F. Fennessy & K. R. Drew, Bulletin 22, pp. 381–5. Wellington: Royal Society of New Zealand.

Griffin, J. F. T. & Cross, J. P. (1986). *In vitro* tests for tuberculosis in farmed deer. *Proceedings of a Deer Course for Veterinarians*, 3, 71–7. Deer Branch of the New Zealand Veterinary Association.

Harbord, M. (1982). Southland winter feed trials break new ground. *The Deer Farmer*, 13, 27–30.

Harbord, M. (1986). Investment analysis: your return on capital. In *Deer Farming – A Profitable Alternative*, pp. 10–11. New Zealand Deer Farmers Association Inc.

Hughes, R. (1986). Velvet marketing: Long term Korean market prospects poor. *The Deer Farmer*, 31, 33–5.

Hunter, J. W. (1986). The accreditation programme for tuberculosis control in farmed deer. *Proceedings of a Deer Course for Veterinarians*, 3, 43–8. Deer Branch of the New Zealand Veterinary Association.

Kong, Y. C. & But, P. P. H. (1985). Deer – the ultimate medicinal animal (antler and deer parts in medicine). In *Biology of Deer Production*, ed. P. F. Fennessy & K. R. Drew, Bulletin 22, pp. 311–24. Wellington: Royal Society of New Zealand.

Macintosh, C. G. (1986). Yersiniosis: a bloody waste. *The Deer Farmer*, 33, 45–7.

Macintosh, C. G. & Beatson, N. S. (1985). Relationships between diseases of deer and those of other animals. In *Biology of Deer Production*, ed. P. F. Fennessy & K. R. Drew, Bulletin 22, pp. 77–82. Wellington: Royal Society of New Zealand.

MacKenzie, A. R. (1984). The diseases of deer in Queensland. *Proceedings Post Graduate Committee in Veterinary Science, The University of Sydney*, 72, 595–602.

Mason, P. C. (1984). Lungworm in red deer: biology and control. *AgLink* FPP248. Wellington: Information Services, MAF.

Mason, P. C. (1985). Biology and control of the lungworm *Dictyocaulus viviparus* in farmed red deer in New Zealand. In *Biology of Deer Production*, ed P. F. Fennessy & K. R. Drew, Bulletin 22, pp. 119–21. Wellington: Royal Society of New Zealand.

Moore, G. H. (1985). Management – mating, calving, lactation, weaning. *Proceedings of a Deer Course for Veterinarians*, 2, 155–70. Deer Branch of the New Zealand Veterinary Association.

Moore, G. H. & Cowie, G. M. (1985). Advancement of breeding in non-lactating adult red deer hinds. *Proceedings New Zealand Society Animal Production*, 46, 175–8.

Moore, G. H., Cowie, G. M. & Bray, A. R. (1985). Herd management of farmed red deer. In *Biology of Deer Production*, ed. P. F. Fennessy & K. R. Drew, Bulletin 22, pp. 343–55. Wellington: Royal Society of New Zealand.

Mulley, R. C. & English, A. W. (1985). The effects of castration of fallow deer (*Dama dama*) on body growth and venison production. *Animal Production*, 41, 359–61.

Nicol, A. M. (ed.) (1987). *Livestock Feeding on Pasture*. Occasional Publication No. 10. Hamilton: New Zealand Society of Animal Production.

Oliver, R. E., Sutherland, R. J., Saunders, B. W. & Poole, W. S. (1986). Observations on the pathogenesis of malignant catarrhal fever of deer. *Proceedings of a Deer Course for Veterinarians*, 3, 146–55. Deer Branch of the New Zealand Veterinary Association.

Pearse, A. J. T. (1987a). Fundamental deer management, parts 1 and 2. *The Deer Farmer*, 35, 45–8.

Pearse, A. J. T. (1987b). Fundamental deer management, parts 3 and 4. *The Deer Farmer*, 36, 45–8.

Pearse, J. (1986). Investment report: deer farming offers a very competitive return. In *Deer Farming – A Profitable Alternative*, pp. 28–32. New Zealand Deer Farmers Association Inc.

Presidente, P. (1984). Parasites of farmed and free-ranging deer in Australia. *Proceedings Post Graduate Committee in Veterinary Science, The University of Sydney*, 72, 623–43.

Riney, T. (1956). Comparison of occurrence of introduced animals with critical conservation areas to determine priorities for control. *New Zealand Journal Science & Technology*, 36B, 1–18.

Scarf, M. (1986). Production and presentation of venison. In *Deer Farming into the Nineties*, ed. P. Owen, pp. 130–3. Brisbane: Owen Art & Publishing.

Stevens, M. (1986). Transportation of deer. In *Deer Farming into the Nineties*, ed. P. Owen, pp. 113–14. Brisbane: Owen Art & Publishing.

Taylor, P. G. (1984). Twelve years with deer farmers. *Proceedings Post Graduate Committee in Veterinary Science, The University of Sydney*, 75, 575–93.

Trask, R. (1986). Preparation and transport of deer for slaughter. Technical Papers, 11th Annual Conference, pp 60–1. Palmerston North: New Zealand Deer Farmers Association.

van Reenen, G. (1986). On the cervus side: AI – why no takers. *The Deer Farmer*, 31, 27–31.

Wallis, T. & Hunn, R. (1982). Helicopter live capture. In *The Farming of Deer: World Trends and Modern Techniques*, ed. D. Yerex, pp. 84–9. Wellington: Agricultural Promotion Associates.

Wodzicki, K. A. (1950). *The Introduced Mammals of New Zealand*, Bulletin No. 98. Wellington: DSIR.

Woodford, K. (1986). Economic analysis. In *Deer Farming into the Nineties*, ed. P. Owen, pp. 13–19. Brisbane: Owen Art & Publishing.

Yerex, D. & Spiers, I. (1987). *Modern Deer Farm Management*. Carterton (New Zealand): Ampersand Publishing Associates.

17

Deer farming in Europe

T. JOHN FLETCHER

Abstract

The farming of deer can be traced to Ancient Greek and Roman civilisations. Enclosed deer parks numbered over 2000 in mediaeval Britain but declined to several hundred by this century. Modern deer farming emerged in the 1970s with the establishment of research programmes at Glensaugh in 1970 and the founding of the British Deer Farmers' Association in 1978. In Britain, currently 15,000 breeding hinds, almost exclusively red deer (*Cervus elaphus*), are held on 250 farms. Larger numbers of fallow deer (*Dama dama*) are farmed in continental Europe. Management systems are designed to minimise costs of winter feeding. About 85% of farmed red deer hinds successfully wean calves. In Britain, red deer calves are normally weaned in late September before the rut and housed for the winter. Breeding adults usually are out-wintered with supplemental feeding. Despite the small size of wild Scottish red deer, their potential for growth is high and yearling stags on fully stocked lowland pastures yield liveweight gains of up to 780 kg/hectare between April and November. On deer farms, slaughter by rifle of 15–18 or 26–29 month-old stags is the traditional method but increasing numbers are slaughtered in abattoirs.

Historical context

Deer enclosures are described by Aristotle referring to fallow deer and later in great detail by Columella writing about AD 65. The latter indicates that deer enclosures, especially in parts of Gaul (probably modern France), could be of great size and were constructed both for pleasure and financial gain. Stone or brick walls and wooden fences were described, supplementary feeding detailed, and hand-rearing of young as a tame nucleus advocated.

Such farming of deer in parks was widespread, gaining impetus throughout the Roman empire by *res nullius*, the principle laid down by Justinian, whereby game belonged to whoever killed it regardless of land ownership.

Clearly a demarcated reserve in which *res nullius* could be suspended was a great advantage and within the Frankish empire, as later in the Norman, the ruler's permission to establish a hunting park or reserve was highly valued. Once deer were confined within a park, management was required to prevent destruction of trees and to maximise production of venison. Within Sweden and Norway, at least, the Roman principle of *res nullius* did not apply and a landowner had sole legal right to game killed on his property; he could use law to establish a game reserve on his land without the expense of containing it. Perhaps because of this, the tradition of maintaining deer parks is almost absent from Sweden and Norway, but present to a varying extent throughout Denmark and West Germany, southwards towards Italy, and westwards through France and Spain. The system had disintegrated in much of Europe by the 10th century but was re-established by the Normans and carried under their influence into Britain.

Mediaeval parks were usually surrounded by a ditch and bank topped with a wooden palisade, although increasingly walls of brick and stone were used. Leaps that allowed deer to jump in but prevented their escape usually required royal licence but were often incorporated. In northern Europe, the enclosed deer were red deer but fallow deer gradually infiltrated, reaching Scotland by 1288. Management entailed maintenance of the park pale, control of wolves (*Canis lupus*), feeding hay, oats, etc., hand-rearing occasional calves, restocking by capture and transport of wild deer, as well as killing for meat (Gilbert, 1979). Within Britain over 2000 such parks appear to have existed in the medieval period but the number declined as the human population was devastated during the Black Death of 1348/49, and as the feudal system disintegrated, removing the labour required to maintain the park perimeter. Many new parks were formed in both England and Europe during the 17–18th centuries as the country house became a vogue, but even so by 1867 only 334 parks remained in England and by 1949 only 143 (Shirley, 1867; Whitehead, 1950).

The vogue for deer stalking in Victorian Britain, combined with the vagaries of the home sheep trade, led to large treeless tracts of Scotland being cleared of sheep and reserved as deer 'forests' (Orr, 1982). Since the close of the 19th century, populations of wild red deer grew steadily in the Highlands. Competition was reduced as cattle and sheep declined and human depopulation left the deer almost undisturbed except by the seasonal impact of the sporting lairds. Further south, deer parks fell into disrepair as the resources of the landed gentry dwindled and in particular as the great country houses were assigned to the War Department during the two world wars. Military exercises destroyed the walls and fences allowing the deer to escape to the surrounding countryside.

By the early 1970s, 150 years of protection for sport in the Scottish Highlands had increased the wild red deer herd to about 200,000. Due to demand in West Germany, the value of red deer venison suddenly increased to twice that of lamb. Consequently, the Rowett Research Institute, in conjunction with the Hill Farming Research Organisation, commenced a deer farming project at Glensaugh, Kincardineshire, in the Grampian mountains of Scotland in 1970. This enterprise is still thriving and has been the source of much published research (Blaxter *et al.*, 1974, 1987). Using week-old hind calves collected from the wild and reared on artificial milk, a nucleus of 100 tame hinds was established by 1974. Private deer farms soon followed using deer caught from the wild as adults or obtained from the few remaining long-established English deer parks.

While Glensaugh, by using hand-reared stock, was able to collect invaluable data on the physiology of red deer, private farms soon showed that deer taken from the wild could be rendered sufficiently tractable to permit regular worming, weaning of calves, splitting into different groups for rutting, calving, etc., and eventually winter housing and passage through abattoirs.

It may be asked why no attempt has been made to engage in commercial free-range deer ranching in the Scottish Highlands. In practice, the problems of such an enterprise would be considerable: land ownership is not heeded by animals which, unlike in the days of *res nullius*, may be shot by the landowner. Consequently, such systems would demand remarkable cooperation among (often absentee) landowners with different priorities but for whom deer stalking is generally the main reason for possessing the land. Secondly, the impoverished and robust terrain and difficult climate of the Highlands require 2–3 hectares to support a single hind. The practicalities of slaughter and carcase extraction are much the same as they are in the wild population.

Structure and size of the industry

In 1978, the British Deer Farmers' Association (BDFA) was founded and by 1986 had some 400 members of whom nearly half were active deer farmers. The BDFA publishes a journal, arranges conferences and farm visits, and also acts as a political spokesman where changes of legislation affect deer farming. Similar bodies have been formed in other European nations, including Denmark, France, Ireland, Sweden, West Germany, and Switzerland.

At the time of writing, some 250 farms carrying approximately 15,000 hind are in existence in Britain with the hinds all retained for breeding and the stags slaughtered for domestic venison consumption or sold on the hoof. In Britain, industry stratification has developed with many hill farms in the north selling their calves to lowland farms in the south where the winter is shorter, feed and housing more readily available, and the eventual market closer. In addition to being readily available, early experience indicated that red deer are more tractable than the fallow deer and consequently red deer became the predominant species on British farms.

In continental Europe, particularly West Germany, interest centres around fallow deer which appear more resistant to helminths, withstand hot dry summers, produce preferred venison, and are more available than red deer (Reinken, 1980). Research institutes maintain herds of fallow deer in West Germany and Denmark, and in Switzerland and Austria state support is provided for fallow venison production enterprises.

Technology

The management of northern European deer farms is dictated by the long winters. Few farms have yet been established in areas where summer drought is commonly a serious problem and most strategies have been evolved to reduce the expenses of winter feeding. Traditionally, deer parks have maintained very low stocking densities of around 2.5 adults/hectare and the deer have wintered on the accumulated rank grass growth of summer, while hay, roots, and occasionally corn, are usually provided in late winter. For profitability, moderately high stocking densities of between 5–13 hinds and followers/hectare are required and this necessitates handling to medicate against parasites, wean calves, assign females to mating groups, etc.

On British red deer farms, calves are normally weaned in late September before the rut though some farmers leave the calves with their mothers for another month or so. Calves can then be grazed after weaning for a few weeks supplemented by a concentrate ration before being housed for the duration of the winter. Hind calves attract more than twice the price of stags, as all hinds are required for breeding. In 1987, weaned stag calves sold for £ 2/kg liveweight and hinds around £ 4/kg. Consequently, to ensure fertility at 15 months, hind calves are likely to be fed *ad libitum* high protein concentrate until turned out in April or May. Stag calves may have their ration reduced during the photoperiodically induced mid-winter inappetence period and also to capitalise on compensatory growth when calves are turned out on pasture (Kay, 1979; Adam, 1986). Hay or silage usually make up the bulk of the ration and some farmers also use roots.

On Scottish hill farms, adult stock are wintered on rough hill grazing dominated by heather, in housing, in 'sacrifice' paddocks, or in mature woodlands. The first system has proved most profitable at Glensaugh (Blaxter *et al.*, 1987).

With the commencement of spring growth, deer are turned out on fertilised pasture from the buildings or wintering areas. They are then rotationally grazed, usually on conventional rye grass swards. The herd may be split into small groups of stud stags, adult calving hinds, yearling hinds, and yearling stags. Generally, electric fencing is not used for rotational grazing or subdivisional fencing in Britain although it has proven effective.

Anthelmintic treatment administered by injection or orally is given to at least the yearling stock, usually at or shortly after spring turn out, one or more times during the summer, and again at the autumn gathering. At this last gathering, calves are weaned and hinds split into groups of around 40 to be run with a single stag. During the rut, this stag is often replaced by another after two or three weeks to ensure fertilisation. Stags are removed by late November to prevent late conceptions.

Currently, stag selection is based on rather unscientific principles of body size with little consideration being paid to food conversion or conformation which might favour high priced cuts. However, the BDFA is now establishing a national herd register which should encourage better selection.

In Britain, the removal of velvet antlers from live deer was made illegal in 1978 on welfare grounds and this has forced farmers to risk running antlered stags with their deer through the winter or to remove hard antlers just prior to the rut. The dart gun is still the most widely used means of handling adult stags for this purpose with Immobilon being the drug of choice. For transport to abattoirs, antlers must be removed from yearling stags and the drop-floor deer crush is the most suitable technique for restraint.

High fixed costs are among the major disadvantages of a deer farming enterprise. In fact, the purchase of breeding red hinds is likely to exceed the cost of deer fencing a new farm on good low ground pasture, by about 4:1. Paddocks of around four hectares are probably best suited to low ground management and it is helpful if these paddocks open into a central raceway of 10 m width, enabling it to be grazed. This raceway leads to a handling system in which the deer can be dosed, ear-tagged, weighed, drafted into groups, and loaded into vehicles. Dosing by injection or by mouth is easily carried out by confining up to ten individuals in a small pen. Red deer can often be blood sampled in the same way, but for the more elaborate procedures, such as tuberculin testing, the animal is walked into a narrow padded crush, the floor dropped, and the deer is suspended by its thorax and abdomen. For fallow deer, a similar principle is often used but the race should be smaller, roofed, and have a small exit hole to accommodate the animal's head. Encouraging deer to move from darkness into light is helpful, especially in handling fallow deer.

The transport of deer has posed few problems. Conventional cattle trucks are suitable if well designed with partitions high enough to prevent deer jumping them and sufficiently solid to prevent deer being injured. Care must be taken to avoid packing too tightly and, as for conventional livestock haulage, a sensible driver is a vital consideration. In general, it seems wise to avoid unloading deer during long journeys but rather to give them a little more room and make provision for feeding and watering within the vehicle.

Productivity

Reproductive performance

Wild Scottish red deer produce 40–46 calves/100 hinds of which perhaps 30 survive to become yearlings (Mitchell, Staines & Welch, 1977). Once introduced to a farm and fed adequately, the calving percentage rapidly increases to 70–80%. After culling unproductive hinds and once the hinds have become fully accustomed to farm techniques, then 90% can reliably be achieved from hinds originally taken from the wild. Well-managed deer parks usually achieve around 70% and wild red deer in good wooded habitat in northern England have been recorded as producing 65–70 calves/100 hinds (Mitchell, Staines & Welch, 1977). For budgetary purposes, 85% of farmed hinds can be expected to rear calves to weaning and over 90% should be achieved in well-managed herds.

Growth rates

Records of growth rates in Europe vary regionally due in large part to climatic and nutritional rather than genetic factors. Thus, for example, wild Scottish red deer whose calves average only 6.7 kg for stags and 6.4 kg for hinds may weigh 120 kg as adult stags of 7–9 years or 78 kg as 6–8-year-old adult hinds weighed during the autumn (Anon., 1971; Mitchell & Brown, 1974; Clutton-Brock, Guinness & Albon, 1982). However, if adequately fed and sheltered, young Scottish stags may weigh up to 185 kg at 26 months, with adult hinds reaching 115 kg (Mitchell, Staines & Welch, 1977). In Hungary, hand-reared red deer calves reached 100-day liveweights of 64 kg for stags and 52 kg for hinds and 120-day weights of 75 and 60 kg, compared to Scottish stag calves' 100-day weights of only 45 kg (Blaxter *et al.*, 1974; Horn, Laszlo & Horn, 1986). However, Scottish hinds introduced to Denmark and mated to Danish stags yielded 100-day weights of 51 kg for hind calves and 57 kg for stag calves (F. Vigh-Larsen, personal communication). In well-wooded and sheltered parts of northern England, wild adult stags were between 67 and 73% heavier and adult hinds 34–38% heavier than those on Scottish hill ground and their antlers were about twice as heavy (Mitchell, Grant & Cubby, 1981).

Meat production

For much of the year, European farmers cannot rely on sufficient pasture and must hand-feed their deer. Much work on feeding levels has been carried out by Dr Clare Adam at the Rowett Research Institute in Aberdeen, Scotland. From mid summer, protein levels fall in the grass and hinds may benefit from a concentrate supplement during lactation. At the Rowett, hinds are fed from late pregnancy 0.5 kg/head/day of a barley based concentrate of 87% dry matter (DM) yielding about 13 MJ of metabolisable energy/kg DM. Replacing 13% of the barley with fishmeal increased average birth weight by 0.3 kg and liveweight gain by 17 g/day. On farms further south with earlier spring grass growth, great care must be taken to avoid allowing hinds to become over-fat before parturition as calving problems will certainly result. Dr Adam has explored the practicality of advancing calving dates by oral administration of melatonin as a way of matching the needs of the hind to seasonal pasture growth.

Feeding lactating hinds facilitates weaning as calves which have experienced concentrates will much more readily adjust to a milk-free diet. Left at pasture after weaning, Rowett calves grew initially at 140 g/day when receiving 0.25 kg and 200 g/day when being fed 0.5 kg/head. Later in the year, growth at pasture remained at 140 kg/day whether receiving 0.9 kg or 0.5 kg concentrate/day, suggesting that at the higher rate pasture intake was being reduced. If calves were left on their hinds until after the rut, growth on good lowland pasture was 230–280 g/day but on hill pasture only 50–120 g/day.

On many farms, calves are housed at weaning and fed *ad libitum* to give maximum growth before winter inappetence commences. They are then likely to take daily around 1 kg concentrates and 0.6 kg hay. Between December and February, calves voluntarily restrict their daily intake to 1–1.3 kg DM/head, but thereafter with increasing daylength, appetite increases to 1.5–2.2 kg/head. For meat production, compensatory growth on spring pasture is so pronounced that it may be uneconomic to feed to appetite during late winter but hind calves may warrant *ad libitum* feeding throughout winter housing to maximise their breeding performance. On summer pasture, calves can be expected to achieve liveweight gains of 150–200 g/day, attaining 16-month weights of 80–90 kg.

The Glensaugh deer farm has a stocking rate of 0.66 hind with calf/hectare throughout the year, selling calves shortly after weaning at 4–5 months to farms on better grassland for slaughter at around 15 months. Glensaugh is between 200–450 m above sea level and at first the grazing was predominantly of heather (*Calluna vulgaris*) though upland grass has now been added. Hay is purchased for feeding in storm conditions and some concentrate is fed during late gestation and lactation at around 0.5 kg/head daily.

On improved upland hill farms where liming, fertilising, and occasional re-seeding is carried out, stocking rates of 10–12 hinds and calves or 15–18 yearling stags/hectare can be achieved from mid May to late November (Hamilton, 1986).

On the best arable lowland pasture where stocking is possible from April to November, 12–16 hinds and calves or 18–22 yearling stags/hectare can be carried. In the last two categories, some hay or silage can be cropped in a normal growing season but around 100 kg of nitrogen/ha should be applied (Hamilton, 1986). Yearling stags stocked at over 20/hectare should yield up to 780 kg of liveweight increment between April and November on arable lowland pasture (Table 17.1).

Marketing

In Europe, deer farmers are usually close to markets, and often establish their own outlets. Regulations for handling farmed deer meat are either being formulated or, in most cases, have not yet been considered, so many deer farmers continue to slaughter their deer as has been practised for millenia in the deer parks. The deer are normally shot at 15–18 months or 26–29 months using a high-powered rifle. This on-farm slaughter usually requires investment in butchering facilities, refrigeration equipment, etc., and in many cases, farm shop enterprises have been established or the meat is sold

Table 17.1. *Stocking rates and liveweight gains of red deer on British pastures.*[a]

Land type	Stocking rate (deer/hectare)		Liveweight yield (kg/hectare)	
	Hinds & calves	Yearling stags	Yearling Calves	stags
Rough hill	0.66	—	16.83	—
Improved grassland	10-12	15-18	280.5	660
Arable lowland	12-16	18-22	344	780

[a]Assuming: 85% calving; 30 kg calf; turnout at 50 kg; off grass at 90 kg.

locally to catering outlets. Domestic regulations in France and Belgium prohibit the sale of venison during many months of the year. This legislation, designed to protect wild deer, is likely to be changed as deer farming develops.

Many farmers who have joined the British Deer Producers' Society Ltd (BDPS) are now sending their deer for slaughter to abattoirs. These are conventional red meat abattoirs with lairage, raceways, and stunning crates adapted to red deer. In England, the law does not exclude deer from conventional red meat abattoirs and three abattoirs have adapted their lairage, fitted a purpose-built stunning pen, and use the cattle lines for the slaughter of deer. In Scotland, a legal anomaly precludes the use of abattoirs for deer except under specific licence. Elsewhere in Europe, deer are killed on farm by rifle although in Denmark plans are being made to construct an abattoir for fallow and red deer. There is little doubt that abattoir slaughter will be required by the larger retail outlets which demand veterinary meat inspection and the highest standards of hygiene.

Meat yield

Throughout Europe, all farmed red deer hinds are recruited to the breeding herd so all red deer slaughtered are stags. These are being slaughtered in Britain at 15 months in most cases, but also occasionally at 27 months. Dressed carcases of red deer average 50 and 65 kg at 15 and 27 months, respectively. This 15 kg difference does not justify the additional winter feeding and summer grazing required.

Research projects to induce earlier calving have not yet reached commercial trials. The advantage would be that slaughter at the end of the first autumn would become realistic. However, there is concern that use of hormones even in the breeding herd could alienate consumers, that the cost of treatment might be too high, and that disruption of seasonality could impair the ability of deer to survive the winter.

Economics

The capital costs of establishing a deer farm are large but no more than for most other agricultural enterprises. On a 40 hectare deer farm grazing 400 hinds, fencing costs are likely to exceed £ 30,000 but this figure is dwarfed by expenditure on the purchase of hinds which are likely to cost around £ 400/head. Even with the development cost of handling facilities at £ 4000, the purchase of breeding stock is likely to exceed all fixed costs by a factor of at least 4:1.

Most modern deer farms use high-tensile deer net of 1.8 or 2 m in height at a cost of £ 3/m erected. Assistance for fencing and other capital costs still exists in Britain and most other European nations but it seems likely in the current economic and political climate that this will not be increased nor perhaps even maintained. Rates of grant within Britain are between 15 and 50% depending on the classification of the farm – those in 'less favoured areas', predominantly hill farms, receiving the higher rate.

For a commercial red deer farm, handling facilities should be simple and functional. They require a race of perhaps 10 m in width and at least 30 m in length leading preferably round a corner into a boarded yard of only perhaps 6 m in diameter, from which doors open into a number of other pens. Roofed complexes have a calming effect on deer and are more convenient for handlers. Weighing scales are essential for good management and a crush may be required for removal of antlers from yearlings prior to transport to abattoirs. Such a complex is likely to cost £ 3000–5000 including a crush and weighing scales.

Labour costs are difficult to calculate. If the farm is a breeding unit then one man should be able to manage 750 hinds assuming seasonal labour is available for silage production at weaning, and provided he has adequate facilities and equipment for feeding and bedding in the winter. A finishing system requires labour in feeding housed weaned calves, but during the summer, demands on labour are minimal. Given adequate housing and tractors, etc., it would not be unreasonable for one man to manage 1000 or more calves from weaning to slaughter.

The future

The future of European deer farming depends on economics. Even in the present European agricultural climate where EEC dairy, beef and sheep meat regimes together with headage subsidies render conventional livestock production artificially profitable, venison production is still just able to compete. If the level of support for conventional enterprises is reduced, deer farming would become more profitable.

In cattle and sheep markets there is a specific, and quite constant, relationship between the price of breeding stock and the value of the meat, usually around 1.5:1. It has been argued that red deer hinds are over-valued in this context as they exceed the value of the carcase by 3:1. However, this argument ignores the fact that red deer are in short supply. Should deer farming be more profitable than cattle or sheep even by only a very small margin, then demand is likely to force prices much higher as it has in New Zealand.

The market for venison is at least as important as the cost of production.

Only very small amounts of farmed venison are yet available and the market is, therefore, dominated throughout Europe by venison produced by hunters. Within Britain, a traditional demand for venison has not existed, but attempts by the deer farming cooperative (BDPS) and a number of farm shops have been successful in creating a premium for the farmed product. In the rest of Europe, farmed venison is likely to be absorbed initially into the game market but efforts by New Zealand exporters and growing farmed production may eventually create a separate sales route which is likely to carry a premium due to its superior quality. The value of 'wild' meat as a selling point is unlikely to stand up where quality of the farmed products is superior, as it inevitably is.

References

Adam, C. (1986). Feeding. In *Management and Diseases of Deer, a Handbook for the Veterinary Surgeon*, ed. T. L. Alexander, pp. 25–37. London: Veterinary Deer Society.

Anon. (1971). The weights of new-born to one-day-old red deer calves in Scottish moorland habitats. *Journal of Zoology (London)*, 164, 250.

Blaxter, K. L., Kay, R. N. B., Sharman, G. A. M., Cunningham, J. M. M. & Hamilton, W. J. (1974). *Farming the Red Deer*. Edinburgh: Her Majesty's Stationery Office.

Blaxter, K. L., Kay, R. N. B., Sharman, G. A. M., Cunningham, J. M. M., Eadie, J. & Hamilton, W. J. (1987). *Farming the Red Deer*. Edinburgh: Her Majesty's Stationery Office.

Clutton-Brock, T. H., Guinness, F. E. & Albon, S. D. (1982). *Red Deer Behavior & Ecology of Two Sexes*. Chicago: University of Chicago Press.

Gilbert, J. M. (1979). *Hunting and Hunting Reserves in Mediaeval Scotland*. Edinburgh: John Donald.

Hamilton, W. (1986). Structure of the industry. In *Management and Diseases of Deer, a Handbook for the Veterinary Surgeon*, ed. T. L. Alexander. London: Veterinary Deer Society.

Horn, P., Laszlo, S. & Horn, A. (1986). Deer breeding. *Kongresszus Elott*. (Hungarian Agriculture) III, 26–7.

Kay, R. N. B. (1979). Seasonal changes of appetite in deer and sheep. *Agricultural Research Council Research Review*, 5, 13–15.

Mitchell, B. & Brown, D. (1974). The effects of age and body size on fertility in female red deer (*Cervus elaphus* L.). *Transactions XI International Congress Game Biology*, 89–98.

Mitchell, B., Grant, W., & Cubby, J. (1981). Notes on the performance of red deer, *Cervus elaphus*, in a woodland habitat. *Journal of Zoology (London)* 194, 279–84.

Mitchell, B., Staines, B. W. & Welch, D. (1977). *Ecology of Red Deer*. Cambridge: Institute of Terrestrial Ecology.

Orr, W. (1982). *Deer Forests, Landlords and Crofters*. Edinburgh: John Donald.

Reinken, G. (1980). *Damtierhaltung auf Grun- und Brachland*. Stuttgart: Verlag Eugen Ulmer.

Shirley, E. P. (1867). *Some Account of English Deer Parks*. London: John Murray.

Whitehead, G. K. (1950). *Deer and their Management*. London: Country Life.

18

Deer farming in Asia

K. R. DREW, Q. BAI & E. V. FADEEV

Abstract

Deer are farmed to varying degrees throughout Asia. In China, over 260,000 deer are raised mainly for velvet antler production. Farm management is quite sophisticated with a mixture of feedlot and controlled grazing, and deer farming appears to be a profitable form of land use. Velvet antler production is believed to be about 115 tonnes (dry weight)/annum. Primitive but apparently effective cooking/drying procedures are used to prepare pharmaceuticals. In the Soviet Union, deer also are farmed mainly for velvet antler production with annual yields of 28 tonnes (dry weight) some of which is made into the medicinal liquid *pantocrin*. Because of the long cold Soviet winter, deer are fed for extended periods on roughage and some grain. Korea farms small numbers of deer for the local velvet antler market. Holdings are very small and intensively managed with 8–10 animals/farm. Income is derived from the sale of velvet antler and a small number of live animals. In Mauritius, venison production is the main objective of deer farming and there is much opportunity for expansion to meet local demands. Most deer operations are on marginal land, integrated with sugar cane operations. Deer farming in Japan is limited to a small number managed experimentally since 1983 and may provide an economic return from the presently under-utilised hill country, supplying lean meat and velvet antlers.

Introduction

Examples of both the oldest and youngest forms of deer farming are found in Asia. Containment of deer in China, where the animals are regarded as a symbol of longevity and bliss, can be traced to 200 BC (Kao, 1973) although deer have only been farmed systematically for the production of medicines since the 17th century. Since the 1950s, farmed deer numbers in China have risen sharply by a factor of 30. Other deer farming countries are the Soviet Union, Korea, and Mauritius. Japan, a major market for deer products, is currently investigating the farming of deer on hill country abandoned by traditional farmers in the 1960s.

Industry structure

China

Two main types of deer are farmed in China; sika deer (*Cervus nippon*) with six subspecies, and red deer (*Cervus elaphus*) with four subspecies. Most important is the Northeastern sika (*C. n. hortulorum*) commonly called *meihualu*, which was found mainly in the forests of the Changbai and Xinganling mountains. In recent times, the meihualu has been introduced to most parts of China and is particularly important on the borders with Korea and far eastern parts of the Soviet Union. The male has a mature body weight of 120–150 kg and produces a high yield of good quality velvet antler.

The most important red deer type is the Asiatic wapiti (*C.e. xanthopygus*) or *malu* which is distributed in the region of the big and small Xinganling mountains. Mature males weigh 250–350 kg. Xinjiang red deer (*C. e. sogarius*), numerous in the Xinjiang region, are smaller than their northeastern counterparts but grow very large antlers.

In recent years, controlled hybridisation between sika and red deer has been practised using semen collection, freezer storage of semen, synchronisation of females, and artificial insemination. Hybrid vigour is usually found and complete pedigree and production records are kept.

The population of farmed deer in China was estimated to be 260,000 (75% sika) in 1981 (Table 18.1) and increasing at 10–20%/annum (Pinney, 1981). Deer farming falls within the sector of Agriculture, Forestry, Health and Foreign Trade. Deer farms are state operated and wild deer are managed by the Forestry Department. Farm production of antlers and tails is marketed for either home consumption or export through the Pharmaceutical Adminis-

Table 18.1. *Numbers of farmed deer in China.*[a]

Province	Meihualu	Malu	Total
Jilin	140,000	10,000	150,000
Heilongjiang	50,000	25,000	75,000
Xinjiang		30,000	30,000
Hepei, Beijing	3,000		3,000
Sichuan, Yunnan and Qinghai	1,000		1,000
TOTAL	195,000	65,000	260,000

[a]Pinney (1981).

tration (Pinney, 1981). Income is also derived from the sale of live animals and some meat for local consumption.

Soviet Union

In the late 19th and early 20th centuries, significant numbers of red deer were transplanted from Poland and Germany to the Baltic region and central part of the Soviet Union, but after World War I and the Civil War numbers in these regions had been reduced to a few hundred. After World War II, many deer were relocated from the Voronezhsky Reserve to the European part of the Soviet Union where they thrived under careful management including supplemental feeding and track formation during heavy snow (Pavlov *et al.*, 1974).

Because of the immense size of the country, diverse nature of the republics, and difficulty separating deer managed for hunting from those that are farmed, it is almost impossible to establish accurate statistics of farmed deer. Table 18.2 summarises regional populations of mid-European and Caucasian

Table 18.2. *European red deer populations in the Soviet Union.*

Regions	Subspecies	
	Mid-European	Caucasian
Azerbaijan SSR		1,300
Byelorussian SSR	5,000	
Georgian SSR		1,400
Kaliningrad	1,100	
Krasnodar territory		8,200
Latvian SSR	8,750	
Lipetzk	550	
Lithuanian SSR	2,150	
Moscow	450	
North Ossetian ASSR		750
Rostov	1,300	
Saratov	550	
Stravropol territory		500
Ukranian SSR	5,400	
Volgograd	600	
Voronezh	2,000	
Other areas	2,800	750
Total	30,650	12,900

(*C.e. hippelaphus*) subspecies of red deer. Many of the 54,000 animals are ranched and hunted rather than farmed.

In addition to the mid European red deer, the Ukranian SSR has 8700 Carpathian and 1900 Crimean red deer. Asiatic wapiti (*C. e. xanthopygus* and *C. e. maral*) are reported to number 130,000–160,000 (Siviridov, 1978). Unspecified numbers of sika deer are farmed in the Soviet Far East and at least one group showing poor reproductive performance was shifted in 1968 to the Caucasus region (a journey of some 10,000 km) where production has greatly improved. It is believed that some 60,000–70,000 are currently held in the Caucasus region where many are farmed.

The primary purpose for farming deer appears to be production of velvet antler which is used in local medicines as well as exported. The Ministry of Agriculture is responsible for running the production units as well as cutting and drying antlers. There are convoluted and confusing links between the deer producers and agencies responsible for the processing of antlers, the quality control of the medicines, and the sale of final products.

Korea

Deer farming in Korea is quite recent although deer products have been used in medicine for many years. The principal species is the Taiwanese sika (*C. nippon taiouanus*). Projections based on Table 18.3 suggest that 80,000 deer will be farmed by 1989 (Kim Yong Kook, 1981). Production units are small (8–9 deer) and are generally owned by wealthy merchants and staffed by salaried employees. Velvet antler and the sale of part of the calf crop directly to consumers provides the farm income (Kim Yong Kook, 1981).

Table 18.3. *Deer (all species) and deer farm numbers in Korea.*[a]

Year	Number of deer	Number of deer farms	Deer per farm
1970	680 (est.)	unknown	unknown
1972	1,400	unknown	unknown
1974	2,800	294	9.5
1976	4,450	447	10.0
1978	6,400	780	8.2
1979	7,100	887	8.0

[a]Kim Yong Kook (1981).

Mauritius

The tropical island of Mauritius in the Indian Ocean is very small (185,000 hectares) and about half is under cultivation. In 1639, the Dutch introduced rusa deer (*C. timorensis rusa*) for meat hunting, and the people have now combined intensive deer farming with the main economic activity of sugar cane production. Some of the sugar by-products are fed to the deer which are used mainly for local meat production. The island supports a population of over 30,000 feral rusa deer providing an annual harvest of 5000 and a current farmed deer population of about 10,000 (Lalouette, 1985). The island population of about 1 million people are mainly Moslems or Hindus. Some do not eat beef, and others do not eat pork, but they all eat venison. Consequently, venison production probably could be expanded ten-fold to meet local demand.

Operational methods and productivity

China

In the high mountains, hills, grasslands, and islands, deer are managed in grazing units, whereas on the plains where crops are grown and in forest regions, feedlot systems are most common. In both systems, a remarkable degree of domestication has been achieved by close contact with man and much hand feeding from an early age. One group of 200 deer apparently were transferred by two herdsmen more than 100 km along a highway without a single loss. The animals were not frightened by trucks, streams of people, or noise and had an overnight stopover in a railroad station. Where land is available, herdsmen take the deer out for grazing once or twice a day, controlling their movements with dogs or more recently by solar powered electric fences. The deer respond to a variety of stimuli such as poles, stock whips, calls, and musical instruments (Pinney, 1981).

On all farms, deer are separated by age and sex for feeding and management. Concentrates containing high protein soyabean meal (SBM) as a major portion are widely fed especially to improve antler production. Experience rather than documented experimentation seems to determine selection of diets for maintenance, growth, lactation, and antler production (Pinney, 1981). Table 18.4 outlines the concentrates used to supplement roughage such as chopped or ground maize and oak leaves. Very high levels of protein are frequently fed to stags in late winter to accelerate hard antler casting and initiation of antler growth.

Sika and red deer seldom have twins, and Chinese deer farms manage a calving rate in sika of 85–90% (surviving calves/hinds mated) and about 5% less for red deer. One farm reported a 5–8% twinning rate in sika deer

(Pinney, 1981). Age at first calving in sika is usually two years, and three years for red deer.

Velvet antler production for local consumption and export is a substantial and highly developed industry. Meat production can be considered a by-product and is used by farm workers or sold in small quantities to local communities. Other by-products like sinews, tails, and pizzles are used domestically.

Estimated velvet antler production and yield is given in Table 18.5 (Pinney, 1981). Home consumption might account for about 80 tonnes of dried antler and the remaining 35 tonnes is exported through Hong Kong, Singapore, South Korea, and Thailand. The Pharmaceutical Administration claims that annual production is only 30 tonnes, half of which is exported. The differences will not be resolved until deer and production statistics are better documented.

Velvet antler is harvested and processed according to the specifications of

Table 18.4. *Concentrates fed to Chinese stags during antler growth.*[a]

	Concentrates kg/head/day	Soyabean meal kg/head/day	Estimated % of feed requirements	% as soyabean meal
Sika deer	0.7–2.0	0.4–1.6	20–55	up to 45
Red deer	1.5–2.5	0.7–2.5	25–45	up to 45

[a]Pinney (1981).

Table 18.5. *Estimated production of Chinese farmed antler velvet, 1981.*[a]

	Head	Average kg/head (unprocessed)	Dry yield (%)	Dry weight (kg)	Total dry weight (kg)
Sika deer					
2-point velvet	98,000	1.0	30	0.3	29,400
3-point velvet	46,000	2.4	37	0.9	41,400
Total	144,000				70,800
Red deer	36,000	3.5	38	1.23	44,280
Grand total					115,080

[a]Pinney (1981).

traditional Chinese medicine. Antler from sika deer can be cut at the two-point stage (blood removed for local consumption) or three-point stage (blood remaining for export). Red deer antlers are usually cut at a four or five-point stage for the export market as long as calcification is minimal. Stag handling has improved from manual restraint to the use of a whole body crush with the feet off the ground. Once the stag is immobilised, the antlers are quickly sawn off, bleeding minimised, and the animal set free. All procedures are aimed at minimising stress with stags immobilised for only 3–4 minutes. Chinese medicinal herbs or Western drugs can be used to stop bleeding, diminish inflammation, and facilitate healing. After removal, the antlers are processed to either retain or remove the blood. A vacuum is applied to the cut end of the antler shortly after removal to produce the bloodless *snow-white* product. Export specification into the Korean market requires *blood in*, and this is achieved by heat-sealing the cut. Both antler products are dipped into boiling water (tip first) and left for 30–50 seconds. After 2–4 minutes, the procedure is repeated for about an hour. Larger antlers are left longer in boiling water than smaller ones. Top quality processed antler (export) is pink-red in colour, aromatic, and without decay or cracks. Final drying is done in either a well-ventilated room for 8–12 weeks or in a forced draught oven. Some Chinese believe that the medicinal and nutritional value of velvet antler is damaged by heat drying.

In traditional Chinese medicine, velvet antler is the most important animal product (Kong & But, 1985). It is used as a general tonic and specific treatment for a large number of human ailments such lumbago, impotence, amenorrhoea, and anaemia.

Soviet Union

Most deer are managed in unfenced areas or, at best, large enclosures. Winters are long and harsh and a key feature of production is winter feeding. Deer do not appear to be held in feedlots but extensive trough feeding systems are built in the field to which the deer are signalled by a variety of high pitched calls. Winter rations comprise 1 kg of roughage and 1 kg of concentrates (mainly grain and acorn)/head/day. In some areas, grazing capacity is increased by cultivating and fertilising local areas and sowing alfalfa, meadow fescue, timothy, meadowsweet, or lupine. With adequate rainfall, kale and Jerusalem artichokes can be grown for extra feed. Digestive disorders are common when animals come off winter rations onto lush spring growth, particularly in northern regions. Apart from supplementary winter feeding, deer are managed rather extensively and only the stags are yarded and handled during the velvet harvest.

Velvet antlers are removed in late spring for export (mainly red deer type – maral) or for the production of pantocrin from sika deer. Stags are herded into yards, individually restrained in a body crush and antlers are quickly removed with a saw. No drugs are used and bleeding is minimal. Antlers are processed by immersing in boiling water for 40 seconds at intervals of several minutes. The procedures are repeated for about an hour and then the antlers are laid flat in a forced draught oven (70–80°C) for 6–8 hours. Oven drying is repeated periodically over 12 days (2–3 hours at a time) until judged to be dry and the antler tip still soft. Production of dried antler is thought to be about 28 tonnes from maral and 12 tonnes from sika deer.

Pantocrin is a dilute medicinal liquid made from a mixture of dry antlers, rectified spirit, and water. After mixing and stirring for 21 days, the liquid is placed in 50 ml bottles and either consumed domestically as a health tonic or sold in the Far East by Medexport.

Korea

Farming deer is very much a cottage industry in Korea with a large number of very small holdings. About half the daily ration is a mixed concentrate (dairy ration) while a third of the ration is barley bran. Thus, 90% of the total feed supply is in the form of concentrate. Many people think more roughage should be used and experiments are in progress to establish a better balance of ration components. Typically, sika deer are fed about 1.8 kg of feed/day and red deer about 3.1 kg, both on an air dry basis (Kim Yong Kook, 1981).

The deer are managed almost entirely in feedlot systems with space allowances of only 33–400 m^2/deer. There is very little opportunity for improvement through breeding because of the small herds and the difficulty in exchanging animals. Labour requirements are high. Farm production from these small units consists of velvet antler cut from three or four stags and the sale of two or three pairs of calves.

Mauritius

Rusa deer are managed intensively at high stocking rates on subtropical pasture. Rotational grazing using electric fencing is now common, and during the drier months of September and October, molasses and urea together with sugar cane tops are offered as supplements. During the last five years, deer handling yards have been built on some properties and the animals are now regularly handled like conventional livestock with regular weighing, ear-tagging, weaning, and treatment for parasites.

In this subtropical environment, rusa deer are not strongly seasonal in their calving pattern. Calving rates can be as high as 80% after entrapment of feral deer on farms but calf mortality approaches 50% in some years. After adaptation to farming conditions, reproductive rates improve greatly and the longer established operations now achieve 80–100% calving, and birth weights have improved from 3.5 kg to 5.5 kg over 4–5 years.

The objective is to provide meat for local consumption. Some venison is obtained from organised hunting which takes place each year and surplus yearlings from farms are slaughtered by shooting. Carcase weights for yearlings range from 40–60 kg. Velvet antler from rusa stags is saleable for the medicine trade and there have been recent moves to develop this potential source of income.

Production economics
China
Farm productivity can only be an 'educated guess' as reported by Pinney (1981). No estimates of the capital costs for buildings, yards, water supply, and fencing are available. Income is earned from the sale of surplus young stock, velvet antler, and a few slaughtered stags. Feed and wages costs/deer were reported in 1981 to be about US\$ 65 on a property farming nearly 700 sika deer and about US\$ 153 on another farming 900 deer most of which were the large red deer type (Pinney, 1981). The value of production from both farms was about US\$ 310/head. These two farms both gave gross contributions of about US\$ 154,000 and thus appear to be very profitable even allowing substantial capital and maintenance cost for facilities. Since the northeastern part of China is now largely self-supporting in food, deer in that part of the country will benefit from surplus grain, maize, and by-products from food processing industries. Low prices paid for feed and labour seems to ensure high profitability for Chinese deer farms.

Soviet Union
One state farm enterprise comprises 280 hectares of land and cost about US\$ 30,000 to fence to a height of 2.7 m. The fence was made of massively strong wire to hold deer and prevent access by predators. Farm profit was about US\$ 112,000/annum or about US\$ 160/animal. Six people managed about 700 deer and it was the least profitable of the deer, rabbit, mink, and nutria units on the same farm. Since antler from Soviet maral deer is judged to be the finest in the world by the influential Korean market, it commands premium prices and should ensure profitability in south central regions where these animals are raised.

Korea

An economic appraisal of a typical Korean deer farm carrying five pairs of sika deer is given by Kim Yong Kook (1981) with translation and editing arranged by the late J. R. Luick. Expenses associated with the deer herd and fencing are much greater than those for land and labour.

Production costs are low when compared with the high prices paid for deer and their products. Table 18.6 summarises the economic returns from a typical enterprise (Kim Yong Kook, 1981). Depreciation of the deer is based on 50% for culled deer value and a longevity of 10 years. No allowance appears to have been made for labour.

Mauritius

Table 18.7 is derived from 1978 unpublished data and relates to the establishment of a deer farm of 360 animals on 14 hectares of pasture land. After allowing for all costs, the deer farm is likely to return about 11% on invested capital.

Prospects

China is clearly the leading producer of velvet antlers but further development of deer farming in that country depends on improved

Table 18.6. *Economics (US$) of a typical Korean deer farm carrying five pairs of sika deer.*[a]

Expenditure		
Depreciation of deer	1,100	
Feed Costs:		
Concentrates	1,610	
Hay	1,210	
Depreciation on shed and facilities	147	
Cost of cutting velvet antler	294	
Other costs (e.g. veterinarian, drugs)	294	
		4,950
Income		
Sale of calves two pairs @ 4400	8,800	
Sale of products from four stags	2,350	
		11,150
Net farm income		6,200

[a]Kim Yong Kook (1981).

marketing. Increased production will need to be absorbed by increased local consumption, a decrease in international export price, or significant penetration into Western markets which are sceptical of the medical claims made for velvet antler. With improved supplies of stock feed, the Chinese will have the capacity to substitute velvet antler production from the heavy red deer type (malu) for the lighter sika (meihualu). This should earn farmers greater income from export production but put more pressure on the major international market of South Korea. Venison is likely to remain very much a by-product of the antler industry.

Very little is known about recent changes in Soviet deer farming and although there is enormous capacity to extend existing herds it is impossible to predict what changes will occur in this state-owned industry.

Table 18.7. *Expenditure and income estimates (in US$) from a deer farm in Mauritius.*[a]

Capital expenditure		
360 deer × 130	46,800	
Sowing of pasture	2,200	
Fencing 2750 m × 3.30	9,075	
Feeding troughs, water, molasses		
tank, store, sundries	6,100	
		64,175
Recurrent expenditure		
Labour (1 stockman)	1,560	
Feed supplement (Molasses, urea,		
copra cake, fish meal and minerals)	2,496	
Pasture maintenance and fertiliser	1,365	
Repairs and maintenance	182	
Interest on capital (5%)	3,220	
Depreciation and sundries	1,638	
		10,461
Income		
42 carcases × 45 kg × US$ 1.82/kg		
(wholesale)	3,440	
100 carcases × 45 kg × US$ 3/kg (retail)	13,500	
140 kg offals and pieces × US$ 3.25/kg	455	
		17,395
Farm profit		6,934
Return on capital		10.8%

[a]Meat Producers Association of Mauritius (personal communication, 1978).

Korean deer farming will continue to grow with assistance from the Korean Deer Farming Research Council established in 1980 (Kim & Han, 1985). There is interest in importing wapiti or elk because of the superior velvet antler production, but there has been no progress in changing regulations that prohibit importations. One of the major constraints on expansion is poor coordination in marketing products from Korean farms made more difficult because of the large number of small holdings.

Because the small island of Mauritius with a large population must import most of its meat requirements at high cost, the deer farming industry is likely to further expand in size and profitability. With good management, the subtropical climate will produce high yields of pasture and the sugar cane industry is a good source of feed supplements. Deer farms are relatively few in number and operated by the European minority. This situation tends to provoke poaching and may somewhat inhibit the industry's growth.

Japan is a highly industrialised country representing a significant market for Oriental medicines. The 5 million hectares of hill country (Satoyama), once the main site of Japanese hill farming, is now largely abandoned. Sika deer are naturally part of the Satoyama's ecosystem and some people believe that deer farming on this hill country will be an attractive and economic land use (Tamate, personal communication). The Japanese Deer Farmers' Association was established in 1985 and is supported by Professor Tamate and the staff of the Department of Animal Science at Tohopu University. Japanese people appreciate lean meat, many use velvet antler products, the population is large, and has a very strong economic base.

References

Kao, Y. F. (1973). The animal carcasses among the funeral food found in Han tomb No. 1 at the Ma-wang-tui, changshee. *Wen Wu*, 1973(a), 76–8.

Kim Yong Kook (1981). The farming of White Spotted Deer (*Cervus nippon taiouanus*) in Korea. *Korea Deer Farming*, 4, 8–15.

Kim, D. & Han, K. H. (1985). Deer farming and the velvet antler industry in Korea. In *Biology of Deer Production*, ed. P. F. Fennessy & K. R. Drew, Bulletin 22, p. 390 (abstract only). Wellington: Royal Society of New Zealand.

Kong, Y. C. & But, P. P. (1985). Deer – the ultimate medicinal animal (antler and deer parts in Medicine). In *Biology of Deer Production*, ed. P. F. Fennessy & K. R. Drew, Bulletin 22, 311–324. Wellington: Royal Society of New Zealand.

Lalouette, J. A. (1985). Development of deer farming in Mauritius. *Biology of Deer Production*, ed. P. F. Fennessy & K. R. Drew, Bulletin 22, pp. 379–80. Wellington: Royal Society of New Zealand.

Pavlov, M., Korsakova, I. & Lavrov, H. (1974). *Transplantation of Game and Birds of the USSR*. Part 2. Kirov (in Russian).

Pinney, B. (1981). Delegation to China. *The Deer Farmer*, Spring 1981, 22–35.

Sviridov, N. (1978). *Maral*. Krupnye Khischniki i Kopytnye zveri, Moscow (in Russian).

19

Bison farming in North America

ALEX W. L. HAWLEY

Abstract

In 1985, there were about 90,000 bison (*Bison bison*) in Canada and the United States, most of which were privately owned. About 11,000 are slaughtered annually to serve specialty meat markets. Bison are generally less productive than cattle when feed quality is high but can be more productive when feed quality is low, probably because of their greater capacity to digest low quality feeds. Greater cold hardiness and less selective grazing contribute to their utility on native range. Crossbreeding bison with cattle has failed to produce a hybrid with outstanding production characteristics. Calving rates are often over 80% in intensively managed bison herds. Cows usually become sexually mature at 2 years of age and remain reproductive until 12 years of age, after which pregnancy rate gradually declines. Growth rates of calves or yearlings under experimental conditions are up to 0.4–0.6 kg/day. Bison achieve a slaughter weight similar to that of cattle (about 450 kg) at 2.5 years of age. Bison meat attracts a price 1.3–1.5 times that of beef. There are markets for a number of by-products and for hunting. The primary constraints on growth of the industry are the availability of breeding stock and the high initial capital outlay. However, the industry is growing and demand for bison and their products will probably exceed supply for many years.

Introduction

Two subspecies of bison are recognised in North America, the plains bison (*B. b. bison*) and the wood bison (*B. b. athabascae*) (McDonald, 1981; van Zyll de Jong, 1986). Most bison are plains bison, the only subspecies presently produced commercially (Fig. 19.1). Wood bison are protected as an endangered animal but may be produced commercially in some areas of Canada in the near future.

There are large genetic differences among herds of plains bison (Peden & Kraay, 1979). Many herds have developed from a few animals (Dary, 1974), creating genetic bottlenecks and increasing the potential for founder effect and genetic drift. Therefore, there may be wide variation in productivity and other characteristics among herds of plains bison.

Historical perspective

Estimates of the number and distribution of bison at the time of European exploration of North America vary greatly. Seton (1909) estimated that there were 75 million bison on almost 8 million km² of range, while McHugh (1972) made a perhaps more reasonable estimate of 30 million animals based on his estimate of available range.

Bison were the primary source of food and shelter for several Indian tribes who were utterly dependent upon bison for survival (Roe, 1970; Dary, 1974). Europeans extirpated bison on the Great Plains of the United States during the 19th century to expand range for cattle and to weaken Indian resistance. The major commercial use of bison was as a source of hides for fur and leather industries. This was followed by collection of bones for refining sugar (Dary, 1974). Bison were reduced by the late 19th century to less than 100 plains bison in private herds in the United States and a remnant wild herd of about 300 wood bison near Great Slave Lake, in northern Canada (Banfield & Novakowski, 1960; Dary, 1974).

Fig. 19.1 Adult bull and cow plains bison on a commercial operation in western Canada.

An attitude of preservation was adopted towards bison and the few remaining captive animals prospered, providing parent stock for several government herds established on federal reserves and in national parks. Dary (1974) traced the increase in bison numbers from 800 animals in 1895 to over 30,000 in 1972. The near extermination and high profile of bison could have led to a permanent orientation of preservation. However, except during the brief period at the turn of the century, bison were perceived as an economic animal. A market developed in the early 1900s which motivated many people to raise bison for economic gain (Rorabacher, 1970).

Plains bison are unique among American native ungulates in that they are not generally considered to be wildlife. This may be partly because they were extirpated from virtually their entire range and partly because settlement left few places where wild bison could live. Also, the preservation of bison was initially achieved through predominantly private efforts and repatriation of original ranges has been in the form of confined animals under private ownership. These circumstances led to the perception of bison as a private rather than a public resource. Public ownership of wildlife is an underlying principle of wildlife management in North America (Peek, 1986), and commercialisation of wildlife is restricted. Commercial production of bison was facilitated in many areas because the species was not classified as wildlife.

The legal classification of bison is not consistent. For example, at this writing bison are classified as wildlife (big game) in British Columbia, domestic animals in Alberta, wildlife (big game) in Saskatchewan unless raised domestically in which case they are neither domestic nor subject to wildlife regulations, neither domestic nor wildlife in Manitoba, and wildlife (furbearer) in Ontario. Similar jurisdictional variation in classification exists in the United States (J. Hebbring, personal communication).

Although bison on federal lands (e.g. reserves and national parks) are a public resource, surplus animals have been sold routinely to producers and this remains a major source of commercial breeding stock. Wood bison can be obtained from the Government of Canada and commercial use of this subspecies may be possible when wood bison are no longer considered endangered (H. W. Reynolds, personal communication).

The variation in legal classification is important to the industry because regulations governing the sale, slaughter, and transport of animals, and the processing and sale of meat differ between wildlife and domestic species. Legal status will likely become more uniform as the industry develops. It appears that provinces or states in which commercial bison farming is of significant interest are able to make the legislative changes necessary to avoid impeding development of the industry. For example, Manitoba and Saskatchewan are in the process of changing provincial regulations so captive bison can be recognised officially as domestic animals.

Structure and size of the industry

The total number of bison in Canada and the United States in 1985 was nearly 90,000 animals, of which about 85% were privately owned (Table 19.1). Farm herds range in size from hobby herds of a few animals to commercial enterprises of up to 3000 head (Jennings & Hebbring, 1983). Several organisations have been established to support the industry including the National Buffalo Association and the American Buffalo Association in the United States, and the Western Canada Buffalo Association and the Canadian Bison Association in Canada.

Bison meat produced in North America is destined primarily for domestic markets. Infrastructures for processing and marketing bison meat exist in some areas and are more common in the United States than in Canada. Marketing is generally *ad hoc* with a high proportion of product going to service trades such as restaurants, hotels, and caterers. The number of bison slaughtered annually in the United States is presently estimated at 10,000 animals (J. Hebbring, personal communication). A similar estimate for Canada is likely no more than 1000, based on the proportion of privately owned bison that are in Canada (Table 19.1).

Existing regulations for *ante* and *post mortem* inspection of domestic animals apply to bison where they are legally defined as domestic. This permits bison to be processed through cattle abattoirs, a distinct advantage over other game species which require special facilities and regulations. Animal and carcase inspections are especially important for export markets. For example, the European Economic Community requires federal approval of abattoirs and *ante mortem* inspection of animals. *Ante mortem* inspection is potentially expensive and difficult if bison are not shipped to abattoirs.

Table 19.1. *Estimates of bison populations in Canada and the United States, 1985.*[a]

	Farms/ranches	Free-ranging	Total
Canada	4,800-5,000	6,800-8,800	11,600-13,800
United States	70,600	4,400	75,000
Total	75,400-75,600	11,200-13,200	86,600-88,800

[a]Hawley & Bunnage (1985).

Provincial and state regulations permit on-site slaughter and inspection as long as the meat is not exported out of state or province.

Management and technology

Although bison can be considered legally as a domestic animal, they are biologically a wild one. There are major differences between bison and domestic cattle in tractability, hardiness, and commercial products. Bison farming or ranching is compatible with other forms of agriculture if bison are confined by fencing. Bison could be produced on open range but they wander widely and often destroy fences, crops, and hay stacks (Jennings & Hebbring, 1983; Hawley, 1987b).

There are few experimental data upon which to base comprehensive recommendations for bison management. Most management is by best guess and experience. Ideas are often spread by word of mouth or are formulated into publications from producer associations (e.g., Jennings & Hebbring, 1983).

Feeding

Feed requirements are generally similar to those for cattle, but there are some important differences in relation to feed quality and feeding schedules. Bison routinely eat less than cattle, especially in winter, and usually have lower rates of growth and lower production efficiencies (Peters, 1958; Young, Schaefer & Chimwano, 1977; Christopherson, Hudson & Richmond, 1978; Christopherson, Hudson & Christophersen, 1979; Hawley, Peden & Stricklin, 1981b). Lower feed consumption by bison relative to cattle in winter may be an adaptive strategy to reduce growth and limit activity and intake at a time of year when forage quality is low and energetically expensive to obtain (Hawley *et al.*, 1981b; Hawley, 1987a). This seasonal adjustment in intake is conducive to reducing feed costs during winter, such as in a minimum-input production system. Yearling and older bison could be fed to sustain a 10–15% winter weight loss. High quality and quantity of range forage in the spring would result in rapid compensatory weight gain.

Supplemental feeding is necessary if winter weight loss is going to be excessive. Feeding high quality supplements in winter can enhance production under intensive production systems. For example, bison calves continue to grow through their first winter (Peters, 1958; Christopherson *et al.*, 1978) and Hawley (unpublished data) observed an average growth rate of about 0.5 kg/day from September to April for six bison steer calves under feedlot conditions. Feeding high quality supplements to weaned calves should, therefore, improve growth and perhaps enhance reproductive performance when the animals are yearlings.

There is a trend towards 'finishing' bison on grain for 60–90 days to standardise and increase carcase fat (R. J. Hudson, personal communication). This may enhance product attractiveness in the domestic market but may be counterproductive and unnecessary in the longterm because the low fat content of bison meat is considered to be a primary marketing feature.

Bison digest high-fibre, low-protein rations to a greater extent than do cattle (Hawley *et al.*, 1981a,b). It has been suggested that this difference is due to the greater capacity of bison to recycle nitrogen (Peden *et al.*, 1974). This difference can impart a production advantage to bison when feed quality is low (Hawley *et al.*, 1981b) but the production advantage may be reversed when feed quality is high (Reynolds, Glaholt & Hawley, 1982). The digestibilities of protein and energy in low quality feed (less than 7% crude protein) can be 5% higher in bison than in cattle, and this could be taken into account when evaluating feeds for bison (Reynolds *et al.*, 1982).

Range management

One of the bison's greatest assets is the ability to graze native range throughout the year. They are able to forage through deep snow and are very resistant to cold weather (Hawley & Reynolds, 1987). In contrast, winter feeding and provision of shelter are primary requirements of cattle production in cold climates (Pringle & Tsukamoto, 1974). Harassment from biting flies can also be a problem on some northern ranges and bison may be more tolerant than cattle (Hawley & Reynolds, 1987). There is also some subjective evidence that bison may be better adapted than European cattle to hot arid conditions on southern ranges (Jennings & Hebbring, 1983).

The superior digestion of low quality forages allows bison to use range forage that would be considered marginal for cattle. Bison are also less selective grazers than cattle (Peden *et al.*, 1974; Rice, Dean & Ellis, 1974). Consequently, it is common to stock bison at a slightly higher rate (approximately 3:2) than would be recommended for cattle (Jennings & Hebbring, 1983). Another view is that carrying capacity is the same for bison and cattle under most conditions, but that stocking bison at the same rate as would be recommended for non-winter grazing of cattle allows the producer to leave bison on the range all year with little or no requirement for supplemental feeding.

Equipment and labour

Bison have unruly dispositions and are impressively agile. There are many anecdotes of their quickness, strength, and intractability (Dary, 1974). Consequently, they require special facilities and techniques for handling. Designs of several handling facilities are described by Jennings & Hebbring (1983). Fencing and partitions in holding areas should be made of wooden planks, steel, or material of similar strength and should be at least 2 m high. Chutes, squeezes, and other restraint areas should be covered or darkened to prevent the animals from attempting to escape. Young animals have been known to climb in the corners of plank fences (Hawley & Peden, 1977). It is possible, but not reliable, to control animals using only visual barriers such as tarpaulins (Fig. 19.2; Hudson & Tennessen, 1978). Handling areas must have an avenue of escape for workers.

Various commercial crushes are available. It is very difficult to catch and hold bison in a head gate because the animals accelerate quickly and their head and horns are wide compared to their shoulders. The problem can be solved by mounting a stop gate several feet beyond the head gate.

Herding bison can be dangerous when people are on foot or on small motorised vehicles. There is disagreement about the advisability of using horses to herd bison on open range (Jennings & Hebbring, 1983). Large motorised vehicles are effective but require suitable terrain. Bison can suffer extensive injuries and stress when driven from the range into holding

Fig. 19.2 Bison herded along an alley using only a tarpaulin as a visual barrier at Wood Buffalo National Park, Alberta.

facilities, even when these facilities are especially designed for bison (Hudson & Tennessen, 1978). During periods of adverse range conditions, good forage can be used to lure animals into handling areas, reducing the stress and expense of roundups.

A variety of fencing products are used. Since bison do not often challenge fences unless driven to do so from lack of food or water or by harassment, standard cattle fences of 3 or 4 strands of barbed wire have been used with some success. However, they will not withstand a challenge from bison and are usually not recommended for this reason. Only woven wire fences 2 m high or higher, or fences of equivalent strength and resistance, will contain bison reliably. Producers should ensure adequate food, water, and shelter. Corners and other potential stress points should be reinforced or bevelled.

Bison that are not habituated to handling can be extremely difficult to transport and pose a high risk of injury. These animals should be held for several days after roundup to calm them before further processing. Heat stress is a particular problem when transporting wild bison (Reynolds *et al.*, 1982).

Bison usually are transported and slaughtered in the autumn or winter when they have highly insulative winter coats and are susceptible to heat and stress. Spring is probably the worst time of year because of hazards to fetuses or newborn calves and because the animals may not have shed their winter coats and ambient temperature may be high (H. W. Reynolds, personal communication). Transport should be during cool weather, preferably at night, and water should be provided *ad libitum* before and after transport (Reynolds *et al.*, 1982). Double-decked cattle trucks are not advisable (Jennings & Hebbring, 1983) but straight-decked cattle liners are adequate (Reynolds *et al.*, 1982).

Locating abattoirs and animal inspection facilities on farms eliminates the need to transport live animals with its attendant risks. However, this is a costly alternative warranted only when large numbers of animals are slaughtered, such as in the past at Elk Island and Wood Buffalo National Parks, and on some large bison ranches in the United States.

Labour requirements can be lower for bison than for cattle because construction of shelters may not be necessary, handling can be reduced by leaving the animals on range all year, and supplemental feeding during winter or drought may be minimal or absent, depending on the quality of the range. On the other hand, greater effort is required to process animals and to construct and maintain handling facilities. Requirements for fence construction and repair can also be considerable.

Many practices in cattle husbandry cannot be applied directly to bison farming because of differences in tractability and behaviour. Habituation of bison to handling procedures is very important. However, this conflicts with the desire to reduce labour input and minimise the frequency at which animals and workers are exposed to potentially injurious situations. The balance struck will depend on the characteristics of the production unit. Bison kept under feedlot conditions adjust to daily routines of operation. Range bison should be handled at least once a year for calf identification, disease control, and animal habituation.

Crossbreeding

Attempts at crossbreeding bison with domestic cattle date to the mid 18th century (Dary, 1974). Since then, considerable effort has been made to exploit the production advantages of both bison and cattle by crossbreeding to produce 'cattalo'. Probably the greatest sustained effort was made by the Government of Canada between 1916–64, in a series of experiments in which bison were crossed with Hereford, Angus, Shorthorn, and Holstein cattle (Peters, 1975). Hydromacy in domestic cows bred with bison bulls, mating indifference, and infertility of hybrid bulls were major impediments. Breeding bison cows with domestic bulls alleviated hydromacy, but the problems of sterility of F1 and F2 bulls and breeding indifference remained (Peters, 1975).

Other disadvantages of cattalo compared to domestic cattle included lower calving frequencies, lower birth weights, and lower rates of post-weaning growth of hybrid calves under feedlot conditions (Peters, 1958; Peters & Slen, 1966; Lawson & Keller, 1976; Lawson & Peters, 1976). Pre-weaning growth rates and weaning weights were observed by Peters & Slen (1966) to be higher for cattalo than for purebred domestic cattle, but Lawson & Keller (1976) found no significant differences. Keller (1980) found that milk yield decreased as the percentage of bison in the cattalo dam increased. He suggested that the greater preweaning growth rates and weaning weights of cattalo could be partially explained by more efficient use of range forage. Different percentages of bison in the cattalo being tested may have also contributed to these contrasting results (Lawson & Peters, 1976).

The major production advantage of hybrids was their winter hardiness (Smoliak & Peters, 1955), but the disadvantages outweighed the advantages and cattalo did not develop as an independent breed. The more recent appearance of 'beefalo' (theoretically 3/8 bison, 3/8 Charolais, and 1/4 Hereford; Burnett, 1975) renewed interest in crossbreeding, but the genetic makeup and production advantages of beefalo have not yet been demonstrated.

It may be that the full potential of crossbreeding has not been realised. However, repeated efforts have failed to produce a hybrid with outstanding production characteristics. The genetic incompatability of bison and cattle makes it difficult to create a useful crossbreed and crossbreeding cannot be considered a viable production alternative at this time.

Productivity

Reproduction

Bison are seasonally polyestrous and most animals breed synchronously from July to October, depending on the location (Reynolds *et al.*, 1982; Calef & van Camp, 1987). Calving usually occurs from April to June with peak calving from May to early June. Single calves are the norm and twins are extremely rare. Pregnancy rates range widely from 50%–90% (Reynolds *et al.*, 1982). Intensively managed herds often exhibit a calving rate greater than 80% (Jennings & Hebbring, 1983) and some wild herds have calving rates this high (van Camp & Calef, 1987).

The age of peak reproductive performance of bison cows varies among herds. Wild bison cows are usually sexually mature at 2 years of age (McHugh, 1958). Pregnancy rates may be somewhat lower for animals two and three years old than for older animals in some herds, but McHugh (1958) observed little change in pregnancy rates in free-ranging bison cows 2–12 years of age, a gradual decline in pregnancy rate after 12 years of age, and a marked decline after 24 years. Few cows in wild herds breed as yearlings (Reynolds *et al.*, 1982).

One factor contributing to the lower calf ratio of free-ranging herds may be the propensity of bison cows to suckle young for longer than one year, thereby reducing a newborn calf's chances of survival. This makes weaning in late autumn or early winter essential for maximum productivity (Jennings & Hebbring, 1983).

Body weight and growth rate

Males are heavier than females (Reynolds *et al.*, 1982) with weights varying markedly among herds (Table 19.2). Maximum growth rates under experimental conditions have ranged from about 0.4–0.6 kg/day (Peters, 1958; Christopherson *et al.*, 1978). These rates were observed in calves or yearlings over periods that usually included spring or early summer when a seasonal maximum in growth rate might occur. Richmond, Hudson & Christopherson (1977) recorded growth rates in yearlings of 0.8 kg/day in May after a six-week period of weight loss averaging 0.3 kg/day.

The production advantages of feeding high quality rations may not be realised in bison to the same degree as in cattle. Both Peters (1958) and Hawley (in Reynolds *et al.*, 1982) obtained higher growth rates of cattle than of bison offered finishing rations. The greater growth rate of cattle probably reflects the different selective pressures that have moulded the species. Beef cattle have been selected for continuous growth whereas bison appear adapted to strongly seasonal environments in which periods of climatic adversity and extreme food deprivation are followed by high feed quality and availability.

Behavioural differences between bison and cattle could also affect comparisons of performance. Many comparative studies have been conducted under intensive husbandry. Bison obtained from essentially wild conditions and confined might experience greater stress than cattle and this could adversely affect production (Stott, 1981). In addition, bison seem more active when confined and more aggressive during handling. This could increase maintenance requirements and lower feed conversion.

Carcase and meat characteristics

Slaughtering bison in the autumn will maximise feed conversion, carcase weight, and fat cover at slaughter. Bison are still growing considerably during their second year and some producers recommend slaughtering

Table 19.2. *Approximate ranges of total body weights reported for plains bison from different herds.*

Age	Weight (kg)	Source
birth	14–18	McHugh (1972)
	18–32	Jennings & Hebbring (1983)
4–5 months	113–162	Hawley, unpublished results
7–8 months	150–171	Peters (1958)
8–9 months	136–181	Meagher (1973)
12 months	200–300[a,b]	Hawley, unpublished results
	136–182	McHugh (1972)
14–15 months	247–281[a]	Peters (1958)
yearling	227–318	Meagher (1973)
	309	Nelson (1965)
18 months	275–365[a,b]	Hawley, unpublished results
30 months	444 ± 58[a,b,c]	Hawley (1986)
adults	360–1350	Reynolds *et al.* (1982)

[a]Animals fed high quality rations in confinement.
[b]Steers.
[c]Mean $\pm$ SD.

animals in their third autumn (Jennings & Hebbring, 1983). At about 2.5 years of age, bison achieve a slaughter weight that is similar to that of cattle (about 450 kg, Table 19.2).

The dressing percentage and percentage of marketable meat from bison compare favourably with cattle. Peters (1958) obtained dressing percentages of 60% and 61% for bison bull and cow calves, respectively, that were fed finishing rations for 196 days prior to slaughter at 14–15 months of age. Average carcase weight was 276 kg and 248 kg for bulls and cows, respectively. Dressing percentages and carcase weights of Hereford cattle in the same study were 59% and 57% and 368 kg and 346 kg for bull and cow calves, respectively.

Hawley (1986) described carcase characteristics of six bison steers slaughtered at approximately 2.5 years of age after receiving a finishing ration for 78 days prior to slaughter. The average slaughter weight was 444 kg with a dressing percentage of 60%, providing 205 kg of marketable meat (about 77% of cooled carcase weight). Both Peters (1958) and Hawley (1986) observed that about 46% of the carcase weight of bulls and steers, respectively, was in the hindquarter, compared to about 50% in cattle steers (Berg & Butterfield, 1976).

Carcase grading systems for bison have been developed but not used extensively. The bison carcases described above received a variety of grades using criteria for cattle. The chief detractions of the bison steer carcases examined by Hawley (1986) were poor conformation, dark colour, and excessive animal age. Peters (1958) obtained lower grades for carcases of bison and cattalo heifers compared to Hereford heifers because of a low degree of finish and a high proportion of carcase weight in the forequarters. Carcases of bison, cattalo, and Hereford bulls in the same study were discounted for being 'bully'.

Species differences in carcase conformation and colour need not adversely affect quality. Bison meat may be darker than beef when retailed but the colour difference is not detectable after cooking (Cox, 1978) and the darker colour seems to have little effect on consumer acceptance. Excessive age by cattle standards is a consequence of slaughtering animals at 2.5 years of age in order to achieve a slaughter weight of 450 kg. This size need not be a requirement but could be an advantage if bison are processed through standard cattle facilities.

Bison meat is typically leaner than beef (Dickinson, 1976; Cox, 1978; Morris *et al.*, 1981). However, Lawson & Peters (1976) found no difference between cattalo and Hereford in the depth of fat over the ribeye (*longissimus dorsi* muscle). In fact, certain fat measurements, such as the depth of fat over the ribeye, can be very high in carcases of feedlot bison, but this fat is localised around the kidneys and subdermally over the shoulder and loin (Hawley, 1986). Large amounts of localised fat can also be deposited in the fall and early winter by range bison receiving no grain supplement.

Samples of lean bison and cattle meat examined by Cox (1978) had moisture, protein, and fat contents that were highly variable but similar between species. Taste panelists were unable to consistently distinguish between bison and cattle meats, ranking them as similar in palatability.

Economics
Bison farming is still developing as an industry and it is difficult to make a meaningful economic evaluation. However, a few generalizations are possible. Operating costs may be similar for bison and cattle under intensive or feedlot production systems, but lower for bison under extensive ranching systems. Capital investment (excluding land) for proper facilities will likely be higher for bison in both systems.

At present, prices of stock and meat vary regionally and temporally. Mature bulls, mature cows, and young cows were sold from public herds in 1986 for about US$ 930, US$ 670, and US$ 590, respectively, in the United States (National Buffalo Association, 1986) and US$ 860, US$ 1000, and US$ 800, respectively, in Canada (Elk Island National Park, personal communication).

The price of bison meat has usually been 1.3–1.5 times that of the equivalent cuts of beef. Presently, meat supplies are generally erratic and retail meat sales are often *ad hoc*. Both conditions tend to weaken the retail market. The market for bison meat will likely increase if production increases and meat supply is stabilised. Demand will likely exceed supply for many years to come.

The commercial value of by-products contributes significantly to economic prospects. Bison are valuable attractions in private and public zoos and parks. There are markets in some areas of the United States for hides (US$ 60), skulls (US$ 30), horns (US$ 10), tails, and hooves (Jennings & Hebbring, 1983). However, these markets are likely to be highly regional and limited. Bison heads have a limited market, are often processed for display and, depending on size and quality, valued in hundreds of dollars in the unprocessed state and thousands of dollars for completed mounts.

Hunting is potentially one of the most important sources of income requiring little additional input by the producer. There are both domestic and foreign markets for the opportunity to shoot bison. These opportunities value in the thousands of dollars and are not necessarily constrained by the need to provide an 'outdoor experience'. Recreational shooting of captive bison is contentious and may be opposed by the public in both Canada and the United States.

Future prospects

The greatest potential for bison farming may be as a complement to, rather than as a replacement for, cattle production. Although premium prices can make intensive bison production economically viable, the greatest complement comes from exploiting the environmental hardiness of bison. Thus, bison may be a suitable alternative on extensive range that is marginal for cattle production because of severe winters, insect harassment, predation, low forage quality, or low availability of water (e.g., Hawley & Reynolds, 1987).

The primary constraints to the growth of bison farming are the availability of breeding stock and the high initial capital outlay. Market opportunities and product value will probably both improve with increased production. Bison meat is presently a specialty product and will remain so for many years because of the limited supply. Improvements in the quantity and reliability of supply and quality of product will strengthen the market. Thus, there is good potential for expansion and it is likely that the industry will continue to grow.

Acknowledgements

I thank J. Bunnage, Alberta Department of Agriculture, for his input and H. Reynolds, Canadian Wildlife Service, for his helpful review of this chapter. I also thank J. Hebbring, National Buffalo Association, for her general helpfulness. The legal classifications of bison were provided by the following individuals: H. Jenkins, Ontario Ministry of Natural Resources; H. Payne, Manitoba Department of Natural Resources; F. Baker, Manitoba Department of Agriculture; J. Kinnear, Saskatchewan Department of Tourism and Renewable Resources; R. Lind, Saskatchewan Department of Agriculture; B. Saunders, British Columbia Ministry of Environment and Parks; and J. Hebbring, National Buffalo Association, South Dakota.

References

Banfield, A. W. F. & Novakowski, N. S. (1960). *The Survival of the Wood Bison (Bison bison athabascae Rhoads) in the Northwest Territories.* Natural History Papers 8. Ottawa: National Museum of Canada.

Berg, R. T. & Butterfield, R. M. (1976). *New Concepts of Cattle Growth.* Letchworth: Sydney University Press.

Burnett, J. (1975). How I got fertile buffalo/Hereford bulls. *Buffalo!*, 3, 3–6.

Calef, G. W. & van Camp, J. (1987). Seasonal distribution, group size and structure, and movements of bison herds. In *Bison Ecology in Relation to Agricultural Development in the Slave River Lowlands, NWT*, Occasional Paper 63, ed. H. W. Reynolds & A. W. L. Hawley, pp. 15–20. Ottawa: Canadian Wildlife Service.

Christopherson, R. J., Hudson, R. J. & Christophersen, M. K. (1979). Seasonal energy expenditures and thermoregulatory responses of bison and cattle. *Canadian Journal of Animal Science*, 59, 611–17.

Christopherson, R. J., Hudson, R. J. & Richmond, R. J. (1978). Comparative winter bioenergetics of American bison, yak, Scottish Highland and Hereford calves. *Acta Theriologica*, 23, 49–54.

Cox, B. L. (1978). *Comparison of Meat Quality from Bison and Beef Cattle*. Undergraduate Thesis. Saskatoon: Department of Home Economics, University of Saskatchewan.

Dary, D. A. (1974). *The Buffalo Book*. Chicago: Avon Books/Swallow Press.

Dickinson, C. E. (ed.) (1976). *Carcass Characteristics of a Bison Steer* (Bison bison). Natural Resource Ecology Laboratory Technical Report 302. Fort Collins: Colorado State University.

Hawley, A. W. L. (1986). Carcass characteristics of bison (*Bison bison*) steers. *Canadian Journal of Animal Science*, 66, 293–5.

Hawley, A. W. L. (1987a). Bison and cattle use of forages. In *Bison Ecology in Relation to Agricultural Development in the Slave River Lowlands, NWT*, Occasional Paper 63, ed. H. W. Reynolds & A. W. L. Hawley, pp. 49–52. Ottawa: Canadian Wildlife Service.

Hawley, A. W. L. (1987b). Cattle farming by the Oblate missionaries in the Slave-Mackenzie River corridor. In *Bison Ecology in the Slave River Lowlands, NWT*, Occasional Paper 63, ed. H. W. Reynolds & A. W. L. Hawley, p. 72. Ottawa: Canadian Wildlife Service.

Hawley, A. W. L. & Bunnage, R. J. (1985). Bison farming in North America. *IV International Theriological Congress, August, 1985, Edmonton*.

Hawley, A. W. L. & Peden, D. G. (1977). Canada's buffalo renaissance. *Canadian Geographical Journal*, 95, 32–7.

Hawley, A. W. L., Peden, D. G., Reynolds, H. W. & Stricklin, W. R. (1981a). Bison and cattle digestion of forages from the Slave River Lowlands, Northwest Territories, Canada. *Journal of Range Management*, 34, 126–30.

Hawley, A. W. L., Peden, D. G. & Stricklin, W. R. (1981b). Bison and Hereford steer digestion of sedge hay. *Canadian Journal of Animal Science*, 61, 165–74.

Hawley, A. W. L. & Reynolds, H. W. (1987). Management alternatives for ungulate production in the Slave River lowlands. In *Bison Ecology in the Slave River Lowlands, NWT*, Occasional Paper 63, ed. H. W. Reynolds & A. W. L. Hawley, pp. 63–66. Ottawa: Canadian Wildlife Service.

Hudson, R. J. & Tennessen, T. (1978). Observations on the behaviour and injuries incurred by bison during capture and handling. *Animal Regulation Studies*, 1, 45–53.

Jennings, D. C. & Hebbring, J. (1983). *Buffalo Management and Marketing*. Custer: National Buffalo Association.

Keller, D. G. (1980). Milk production in cattalo cows and its influence on calf gains. *Canadian Journal of Animal Science*, 60, 1–9.

Lawson, J. E. & Keller, D. G. (1976). Pre- and post-weaning growth of cattalo, Hereford, and 1/4 Brahman–3/4 Hereford calves. *Canadian Journal of Animal Science*, 56, 489–96.

Lawson, J. E. & Peters, H. F. (1976). Growth and carcass traits of cattalo, Hereford and 1/4–Brahman bull calves. *Canadian Journal of Animal Science*, 56, 193–9.

McDonald, J. N. (1981). *North American Bison: Their Classification and Evolution*. Berkeley: University of California Press.

McHugh, T. (1958). Social Behaviour of the American Buffalo (*Bison bison bison*). *Zoologica*, 43, 1–40.

McHugh, T. (1972). *The Time of the Buffalo*. New York: Alfred A. Knopf.

Meaghcr, M. M. (1973). *The Bison of Yellowstone National Park*. National Park Service, Scientific Monograph Series 1. Washington: US Department of the Interior.

Morris, E. A., Witkind, W. M., Dix, R. L. & Jacobson, J. (1981). Nutritional content of selected aboriginal foods in northeastern Colorado: buffalo (*Bison bison*) and wild onions (*Allium* spp.). *Journal of Ethnobiology*, 1, 213–20.

National Buffalo Association. (1986). 1986 sales results. *Buffalo!*, 14, 4.

Nelson, K. L. (1965). Status and habits of the American buffalo (*Bison bison*) in the Henry Mountain area of Utah. *Utah State Department of Fish and Game Publication* 65-2.

Peden, D. G. & Kraay, G. J. (1979). Comparison of blood characteristics in plains bison, wood bison, and their hybrids. *Canadian Journal of Zoology*, 57, 1778–84.

Peden, D. G., Van Dyne, G. M., Rice, R. W. & Hansen, R. M. (1974). The trophic ecology of *Bison bison* L. on shortgrass plains. *Journal of Applied Ecology*, 11, 489–98.

Peek, J. M. (1986). *A Review of Wildlife Management*. Englewood Cliffs: Prentice-Hall.

Peters, H. F. (1958). A feedlot study of bison, cattalo and Hereford calves. *Canadian Journal of Animal Science*, 38, 87–90.

Peters, H. F. (1975). Inter-species hybridization: the Canadian cattalo experiment. Iowa State University cow–calf producer's day program, Ames, July 1975.

Peters, H. F. & Slen, S. B. (1966). Range calf production of cattle x bison, cattalo and Hereford cows. *Canadian Journal of Animal Science*, 46, 157–64.

Pringle, W. L. & Tsukamoto, J. Y. (1974). Wintering beef cows in the far north. *Canadian Journal of Animal Science*, 54, 709–11.

Reynolds, H. W., Glaholt, R. D. & Hawley, A. W. L. (1982). Bison. In *Wild Mammals of North America: Biology, Management and Economics*, ed. J. A. Chapman & G. A. Feldhammer, pp. 972–1007. Baltimore: Johns Hopkins University Press.

Rice, R. W., Dean, R. E. & Ellis, J. E. (1974). Bison, cattle, and sheep dietary quality and food intake. *Journal of Animal Science*, 38, 1332.

Richmond, R. J., Hudson, R. J. & Christopherson, R. J. (1977). Comparison of forage intake and digestibility by American bison, yak and cattle. *Acta Theriologica*, 22, 225–30.

Roe, F. G. (1970). *The North American Buffalo: A Critical Study of the Species in its Wild State*. 2nd edn. Toronto: University of Toronto Press.

Rorabacher, J. A. (1970). *The American Buffalo in Transition: A Historical and Economic Survey of the Bison in America*. St Cloud: North Star Press.

Seton, E. T. (1909). *Life-Histories of Northern Animals*, Vol. 1. New York: Charles Scribners Sons.

Smoliak, S. & Peters, H. F. (1955). Climatic effects on foraging performance of beef cows on winter range. *Canadian Journal of Agricultural Science*, 35, 213–16.

Stott, G. H. (1981). What is animal stress and how is it measured? *Journal of Animal Science*, 52, 150–3.

van Camp, J. & Calef, G. W. (1987). Population dynamics of bison. In *Bison Ecology in Relation to Agricultural Development in the Slave River Lowlands, NWT*, Occasional Paper 63, ed. H. W. Reynolds & A. W. L. Hawley, pp. 21–24. Ottawa: Canadian Wildlife Service.

van Zyll de Jong, C. G. (1986). *A systematic study of recent bison, with particular consideration of the wood bison (*Bison bison athabascae *Rhoads 1898)*. Publications in Natural Sciences, 6. Ottawa: National Museums of Canada.

Young, B. A., Schaefer, A. & Chimwano, A. (1977). Digestive capacities of cattle, bison, and yak. *University of Alberta Feeder's Day Report*, 56, 31–4.

SECTION G

Experimental systems

J. BRAD STELFOX

Experimental systems are, by definition, recent wildlife production schemes that have not attained extensive acceptance and implementation. Constraints affecting the emergence and viability of these pilot projects are diverse and include such considerations as animal tractability, available technology, economic performance, political barriers, marketing protocol, and public acceptance. A major goal is to discover the optimal intensity of management. Husbandry options include free-ranging, semi-habituated, or domesticated animals; management may be either intensive or relatively passive. Ultimately, the production features that are adopted will represent a compromise of the needs of animals and various human interests.

There appears to be considerable scope for discovering additional ways of rationally exploiting wild ungulates. Of the approximate 200 extant species, only a few dozen have been examined seriously as candidates for commercial production. Many species that are presently utilised in an extensive way, such as in sport hunting, may prove useful under more intensive husbandry. However, both livestock and wildlife interests have impeded the diversification of wildlife production systems. Existing infrastructure for animal production, including handling, processing, and marketing, have been created by, and ensure a monopoly for, conventional agriculture. But, recent economic woes experienced by the livestock industry have stimulated the evaluation of new agricultural livelihoods. Environmentalists have argued that production systems based on wildlife are unethical, damage wilderness aesthetics, and create a legalised market for poached animals and their products.

The emergence of new production systems generally follows a predictable pattern, and initial attempts at exploring novel technologies often encounter a host of biological, socioeconomic, political, religious, ethical and/or logistical impediments. Any fledgling industries may flounder repeatedly until constraints are identified and corrected, be they public acceptance, economic viability, or political expedience. Although many ventures may be inherently unsound, and their failure inevitable, even ultimately viable systems will experience growing pains. It is clear that the merits of experimental production systems must await decades of scrutiny.

The following chapters describe recent experimental systems utilising wild ungulates for the commercial production of natural fibre (qiviut from muskoxen), cosmetic bases (musk from musk deer), and meat, milk and draught (from moose). These case histories, which detail the rationale, technology, and logistics, lend insight into how biological and socioeconomic variables determine the emergence and viability of fledgling enterprises. Whereas the gradient of management intensity is adequately represented, the taxonomic and geographical scope is not. The following additional experimental programmes also deserve mention.

East African ungulates

Two experimental projects in East Africa provide a useful illustration of contrasting production approaches. Located in Kenya's semi-arid rangeland, Galana Game and Ranching Ltd domesticated eland (*Taurotra-*

Fig. G.1. Domesticated fringe-eared oryx in Kenya.

gus oryx) and fringe-eared oryx (*Oryx beisa*, Fig. G.1) to evaluate their tractability, productivity, and water requirements. Interest in these indigenous species as commercial meat producers arose from their physiological and behavioural adaptations to arid environments (Taylor, 1968, 1969), disease resistance (see Karstad, Grootenhuis & Mushi, 1978), and the marginal economic returns from beef cattle. The aim was to introduce indigenous ungulates into the herds of local pastoralists (King, Heath & Hill, 1977). Captured neonates provided the foundation stock for herds which were then habituated to humans, sheltered in corrals at night, and pastured in areas selected by herders during daylight hours. Other management practices included dehorning, ear tagging, and castration. Comparative research (Lewis, 1977, 1978) has shown that many of the environmental adaptations were compromised by altered activity patterns and by removal of forage selectivity, leading to increased water consumption and lower productivity. Specifically, the ability of animals to select optimal foraging areas, and to use moisture condensed on hygroscopic plants at night, was removed. In addition to these reasons, the experiment terminated because of cautious acceptance by pastoralists and a nation-wide ban on the sale of wildlife products. Similar disappointing findings from domesticated eland were recorded in Zimbabwe (Posselt, 1963).

A contrasting approach, which exploits the advantages of a free-ranging ungulate community, has proven feasible and economically attractive (Hopcraft, 1980). Wildlife Ranching and Research, located 40 km southeast of Nairobi, maintains low operating costs by limiting management to night-cropping, prescribed burns, and selective placement of water boreholes (Stelfox, 1985). Cumulative and sustained annual harvest rates of 25–35% for Thomson's gazelle (*Gazella thomsonii*), Grant's gazelle (*Gazella granti*), Coke's hartebeest (*Alcelaphus buselaphus*), wildebeest (*Connochaetes taurinus*), impala (*Aepyceros melampus*), and eland have produced 4 kg of meat/hectare, selling for US$ 4.5/kg. Favourable economic returns from native herbivores have led to a 60% reduction of the cattle herd, creating additional wildlife habitat.

South American camelids

In the Andean highlands of South America, new production systems for camelids are being sought. Although the domestication of llama (*Lama glama*) as beasts of burden, and alpaca (*Lama pacos*) for wool date back several thousand years, new uses may emerge for the related vicuna (*Vicugna vicugna*, Fig. G.2). The greatest potential for camelids appears to be on the altiplano rangelands above 3900 m, where introduced cattle and sheep encounter poor forage and succumb to altitude-related illnesses. Despite unsophisticated management, 17,000 tonnes of meat were produced annually in Peru from a national herd of 3.3 million alpaca and 1.5 million llama (Kyle, 1987). At La Raya station in the Peruvian altiplano, a research programme is identifying aspects of the reproductive and social biology of camelids that, once incorporated into local management, should increase productivity of wool and meat.

Vicuna wool, considered by many to be the finest, attracts a strong international market. In the 1960s, lucrative wholesale prices caused the near decimation of vicuna through unregulated harvest. From a residual herd protected in Peru's Pampa Galeras, the vicuna have staged an impressive recovery, increasing from 1200 in 1967 to 48,000 by the early 1980s (Otte & Hofmann, 1981). Given the existing healthy population, many advocate a sustained-yield harvest of vicuna for wool and/or meat (see Rabinovich, Hernandez & Cajal, 1985). A cropping programme would generate critical revenue for landowners who presently realise no benefit from vicuna, and reduce the devastating effects of a burgeoning vicuna population on their rangeland resources. Obstacles to the development of such programmes include the listing of vicuna on CITES' Appendix I and a vocal and disapproving conservation community (see Eltringham & Jordan, 1981).

Fig. G.2. Vicuna at the Pampa Galeras Reserve in Peru.

Researcher Julio Sumar at La Raya presents the more moderate proposal that excess breeding vicuna males from the Pampa Galeras be sent to alpaca breeding centres, where hybridisation would yield animals which grow wool of both quantity (from alpaca) and vicuna-like quality (Kyle, 1987). It is hoped that the economic rewards of improved camelid husbandry will provide incentives to conserve existing herds, and discourage the expansion of cattle and sheep ranges.

Asian bovines and suids

A fascinating exploratory review by the National Research Council (1983) has speculated on the draught and meat potential of the little-known bovines native to tropical Asia. Although some have been domesticated, such as the banteng (*Bos javanicus*), mithan (*Bos frontalis*), and yak (*Bos grunniens*), there are several other species that are apparently well-adapted, disease-resistant, and possess a tractable disposition. These wild bovines, many whose populations are currently threatened by warfare, overhunting, and habitat destruction, include the kouprey (*Bos sauveli*), gaur (*Bos gaurus*), tamaraw (*Bubalus mindorensis*), and the anoas (*Bubalus depressicornis* and *B. quarlesi*).

Similar potential is suggested by the National Research Council for such indigenous suids as the bearded pig (*Sus barbatus*), Sulawesi warty pig (*S. celebensis*), Javan warty pig (*S. verrucosus*), pigmy hog (*S. salvanius*), and the babirusa (*Babyrousa babyrussa*). If adaptive native suids were domesticated, higher productivity and lower maintenance input might be realised.

Rodents

Although this volume focuses on ungulates, other mammals are currently entering the arena of commercial production, or exhibit considerable potential. Important candidates include the capybara (*Hydrochoerus hydrochaeris*) of South America, and the cane rat (*Thryonomys swinderianus*) of west Africa. The capybara, the world's largest rodent, may bolster the economics of ranching in parts of South America (Kyle, 1987). If the limiting effects of poaching and prowling dogs were controlled, meat productivity from capybara could be 50% higher than cattle (Gonzalez-Jimenez, 1977). Attributes of this species include high fecundity, early sexual maturity, and higher digestive efficiencies than sheep or rabbits. Since preferred forage is aquatic or riparian, competition with conventional livestock is minimised.

In protein-deficient west Africa, where most wildlife have been displaced by traditional agricultural practices and a burgeoning human population, the cane rat often thrives and contributes considerable meat to rural economies.

Meat of the cane rat is preferred by local residents and its market value is highest (Baptist & Mensah, 1986). Harvest of wild populations is generally prohibited where excessive hunting has reduced numbers. Previous investigations and experimental projects have not been encouraging; however new production technologies may allow for economic rearing of animals, and their subsequent retail at affordable prices. New management directions include artificial selection for production and tractability characteristics. According to Baptist & Mensah (1986), an unlimited market for cane rat meat would exist if prices were comparable to poultry.

References

Baptist, R. & Mensah, G. A. (1986). Benin and West Africa: the cane rat – farm animal for the future. *World Animal Review*, 60, 2–6.

Eltringham, S. K. & Jordan, W. J. (1981). The vicuna of the Pampa Galeras National Reserve. The conservation issue. In *Problems in Management of Locally Abundant Wild Animals*, ed. P. A. Jewell & S. Holt, pp. 277–89. New York: Academic Press.

Gonzalez-Jimenez, E. (1977). The capybara: an indigenous source of meat in tropical America. *World Animal Review*, 21, 24–30.

Hopcraft, D. (1980). Nature's Technology. *Technological Forecasting and Social Change*, 18, 5–14.

Karstad, L., Grootenhuis, J. G. & Mushi, E. Z. (1978). Research on wildlife diseases in Kenya, 1967–1978. *The Kenya Veterinarian*, 2, 29–32.

King, J. M., Heath, B. R. & Hill, R. E. (1977). Game domestication for animal production in Kenya: theory and practice. *Journal of Agricultural Science (Cambridge)*, 90, 445–57.

Kyle, R. (1987). *A Feast in the Wild*. Oxford: Kudu Publishing.

Lewis, J. G. (1977). Game domestication for animal production in Kenya: activity patterns of eland, oryx, buffalo and zebu cattle. *Journal of Agricultural Science (Cambridge)*, 89, 551–63.

Lewis, J. G. (1978). Game domestication for animal production in Kenya: shade behaviour and factors affecting the herding of eland, oryx, buffalo and zebu cattle. *Journal of Agricultural Science (Cambridge)*, 90, 587–95.

National Research Council. (1983). *Little-Known Asian Animals with a Promising Economic Future*. Washington: National Academy Press.

Otte, K. C. & Hofmann, R. K. (1981). The debate about the vicuna population in Pampa Galeras Reserve, In *Problems in Management of Locally Abundant Wild Animals*, ed. P. A. Jewell & S. Holt, pp. 259–75. New York: Academic Press.

Posselt, J. (1963). The domestication of the eland. *Rhodesian Journal of Agricultural Research*, 1, 81–9.

Rabinovich, J. E., Hernandez, M. J. & Cajal, J. L. (1985). A simulation model for the management of vicuna populations. *Ecological Modelling*, 30, 275–97.

Stelfox, J. B. (1985). *Mixed-Species Game Ranching in Kenya*. Ph.D. Thesis. Edmonton: University of Alberta.

Taylor, C. R. (1968). The minimum water requirements of some East African bovids. In *Comparative Nutrition of Wild Animals*, ed. M. A. Crawford. Symposium of the Zoological Society of London, Vol. 21, pp. 195–206.

Taylor, C. R. (1969). The eland and the oryx. *Scientific American*, 220, 88–95.

20

Moose husbandry

E. E. SYROECHKOVSKY, E. V. ROGACHEVA &
LYLE A. RENECKER

Abstract

Moose (*Alces alces*) are common to the boreal forest of the Soviet Union, Scandinavia, and North America. From a world population of approximately 2.5 million animals, at least 500,000 are harvested annually through subsistence, recreational, and commercial hunting. For centuries, moose have been trained for riding or hauling loads and their apparent suitability for domestication has not escaped notice. In the Soviet Union, moose have been reared using a free-grazing system since the 1950s. Their agricultural potential in western Canada has been evaluated more recently under both extensive and intensive containment systems. Although moose are easy to tame, they are susceptible to disease when confined at high stocking rates, stray in free-grazing systems, and are difficult and expensive to maintain during the winter feeding period. However, research and practical experience may lead to improved husbandry methods that allow their agricultural potential to be more fully developed.

Introduction

Moose are the most common wild ruminant of the northern boreal forest of North America, Scandinavia, and the Soviet Union. They winter successfully in some of the coldest regions on earth, tolerating ambient temperatures below -30 °C (Renecker & Hudson, 1986a), and snow depths of over 60 cm (Kelsall & Prescott, 1971). Through subsistence, recreational, and commercial hunting, moose have made important contributions to meat supplies and local economies. The species is very responsive to habitat management and recovers quickly from over-harvest.

The total number of moose in the Soviet Union has been estimated at 1 million with about 10–15% of the population harvested annually (Syroechkovsky & Rogacheva, 1974). This represents a total carcase yield of approximately 12,000 tonnes. In Poland, two world wars drastically reduced

moose populations which subsequently have grown from 10 in the Rajgrod forest district in 1945 to 3264 in 1975 (Bobek & Morow, 1987). Currently, a population of about 6000 animals (Krzywinski, Niedbalska & Krzywinska, 1987) provides an annual harvest of 1600 (Bobek & Morow, 1987).

Populations in Scandinavia have grown steadily since the turn of the century and now approach 600,000 animals. New hunting regulations and forest management systems are the primary factors responsible for the rapid population growth (Markgren, 1974). In 1979, the total carcase yield in Sweden was approximately 12,000 tonnes representing 2–3% of the total meat consumption of the country (Wilhelmson & Sylvén, 1979), but increased to 22,000 tonnes (170,000 moose) by 1982 (Cederlund & Markgren, 1988). In Finland, the sustained annual harvest is about 54,000–56,000 (Nygren, 1988).

There are approximately 900,000 moose in North America (Kelsall, 1987) of which approximately 70,000 are harvested annually by sport hunters (Bisset, 1987). Moose also represent an important source of meat for aboriginal people in North America. For the native community of Fort Resolution in northern Canada, moose meat represents 22% of subsistence foods (Bodden, 1981).

Wild populations of moose are productive and moose meat is considered by many to be unrivalled in flavour and texture. Furthermore, moose are often found as calves in the wild and are easy to tame. These factors naturally raise questions about prospects for domestication.

Historical perspective

Attempts to domesticate moose have been made repeatedly over the last 300 years in various parts of Europe, Asia, and North America (Fig. 20.1). Moose were frequently tamed and used for riding until the practice was prohibited in parts of Europe, since police mounted on horses were handicapped in the pursuit of villains riding moose. Although most of these animals were obtained as calves from the wild, self-replacing domestic populations existed. In 1861, the journal *Acclimatization* reported that a pair of tame moose raised ten calves during their productive life and were used to draw corn from the fields. Another pair on a Kurish estate yielded 14 calves

Fig. 20.1. Throughout their circumpolar distribution, moose have been periodically tamed and used as beasts of burden. Above, a moose pulling an Indian travois in northern Alberta (*c.* 1899). (Photograph by C. W. Mathers; permission granted by Saskatchewan Archives.) Below, moose pulling sled at the Popielno Research Station in Poland. (Photograph by A. Stachurski and A. Krzywinski.)

during their lifetime (Turkin & Satunin, 1902). These early experiments indicated that moose could be reared, bred, and used as beasts of burden.

Soviet Union

Middendorff (1853) was first to recommend that the Russian government promote the domestication of moose. Manteufel (1935) experimented with eight moose at the Moscow zoo and concluded that taming and subsequent domestication was feasible, but such experiments should be conducted in natural environments. The Committee for Reserves of the All-Union Central Executive Committee's Presidium made a decision which led to the establishment of moose nurseries at the Demyanka River in 1934 and in Yakutia in 1935. In 1937, moose domestication experiments were conducted in the Serpukhov game grounds near Moscow and in the Buzulook Forest Reserve (Knorre, 1939, 1959, 1961, 1973).

The first experimental moose farm was established in the Pechora-Illychsky State Reserve during the post-war years by E. P. Knorre and later run by M. V. Kozhukhov. The aim was to evaluate the potential of moose for meat and milk production, but attention also was given to their use for draught and riding. By 1969, the Pechora-Illychsky State Forest Reserve had established a breeding herd of 20–30 tame moose which included 18-year-old animals of the fourth generation. During this period, 174 moose were raised, including 60 adults captured as calves and 113 calves reared by tame cows. Presently, the Pechora-Illychsky project is working with individuals of the fifth generation.

Renewed impetus was provided by the establishment of a moose project at the Kostroma Agricultural Station in 1963, and a station affiliated with the Yaroslavl Research Institute of Livestock Husbandry and Forage Production in 1980. Moose farms also exist in the Bashkir ASSR and in the Gorky region. Frequent requests for offspring from Pechora-Illychsky and Kostroma projects are received from taiga regions of the country. In 1977, the National Committee for Science and Technology established the Coordinating Council for Moose Domestication which was affiliated with the Russian Soviet Federated Socialist Republic (RSFSR) Section of the National Agricultural Academy.

North America

In North America, the tractability of habituated moose was known by native people and quickly attracted attention of European explorers, fur traders, and homesteaders. About 1770, Samuel Hearne noted that moose were 'the easiest of the deer kind to tame' (Moodie & Kaye, 1976) and suggested domestication as a means to augment meat supplies. Many homesteaders used moose for farm chores, attesting to their favourable disposition.

Largely in response to successes in the Soviet Union, the Government of Manitoba conducted a preliminary investment analysis and proposed a pilot project, but it was never initiated. A commission in New York State also investigated the profitability of a private moose venture in the Tug Hill region but this proposal similarly never generated the necessary funds.

The only systematic evaluation of the agricultural potential of moose has been conducted at the Ministik Wildlife Research Station, operated by the University of Alberta. A great deal of research has also been conducted with tame moose at the Moose Research Center at Kenai, Alaska, which, although relevant, has not focused on commercial production.

Management systems
Establishing foundation herds

Research at the Pechora-Illychsky Reserve demonstrated that calves up to three days old are most easily habituated. Calves born to tame cows can spend the first 7–14 days with their dam without becoming less

Fig. 20.2. Moose cow and calf of the Kostroma Agricultural Station in the Soviet Union. (Photograph by L. Baskin.)

tractable (Fig. 20.2). Older calves are more difficult to tame, and adults usually die from stress when captured alive and confined.

Orphan moose calves are relatively easy to raise on ewe milk replacers or 50:50 mixtures of whole and evaporated domestic cow milk (with 5% bovine colostrum, if available). These mixtures are suitable when fed at a maximum daily volume of 2.5 litres (Dodds, 1959; Knorre, 1961; Markgren, 1966; Landowski, 1969; Regelin, Schwartz & Franzmann, 1979; Addison, Mac-Laughlin & Fraser, 1983; Lautenschlager & Crawford, 1983; Welch, Drew & Samuel, 1985). If fed to satiation, moose calves will often suffer digestive disturbances. Thus, smaller volumes are fed at regular intervals 5–6 times/day for the first month; the frequency decreases until the animal is weaned at 12 weeks (Welch *et al.*, 1985). Regardless of the formula or protocol, hygiene is paramount in averting health problems.

Calves sample vegetation during the first days of life which promotes development of rumen microbes. However, free-choice mineral supplement and soil should be provided. Green feeds fully dominate the diet by 1.5 months and replace milk by three months.

Moose calves are easily weaned on pelleted rations, but often do not survive past several years without access to natural forage. In the Soviet Union, the winter diet of tame calves is normally supplemented with potatoes, browse, and salt. In an attempt to circumvent this problem, Schwartz, Regelin & Franzmann (1985) formulated a pelleted ration containing aspen sawdust to replace the lignified fibre component of natural diets (Table 20.1). In both Alberta and Alaska, this ration has successfully maintained tame moose up to 9 years old. Sunflower hulls have also been substituted successfully. Regardless, diet diversity appears as important as nutrient balance and physical form, since voluntary intake of a single food declines with time and changing rations only momentarily stimulates appetite.

Free-grazing sytems

At the Pechora-Illychsky Reserve, moose range freely in the taiga during summer (after calving) and return for supplemental feed in winter. Although hand-reared calves remain tame and even develop an affection towards their handlers (Yazan & Knorre, 1964), yearlings tend to pioneer new areas. Consequently, straying losses on unfenced ranges can be high. From 1949–69, of 175 moose raised at this station, 63 disappeared by straying, poaching, and predation; 29 died of diseases, nutritional problems, or injuries; and 57 were slaughtered or distributed to other projects (Kozhu-khov, 1973). The main problems with a free-grazing system were controlling

Table 20.1. *Composition and analysis of two pelleted moose rations.*

Ingredient	Percentages of composition	
	Aspen-concentrate[a]	Sunflower hull-concentrate[b]
Composition		
Corn, ground yellow	30.0	
Wheat		28.2
Sunflower hulls, whole		26.6
Sawdust 'Fiberlite'	25.0	
Oats, ground	15.0	
Oats, whole		17.2
Barley, ground	12.5	5.6
Beet pulp		5.6
Cane molasses, dry	7.5	
Cane molasses, wet		5.6
Soybean meal (7.4% N)	6.3	7.2
'Pelaid'	1.3	
Pellet binder		0.8
Dicalcium phosphate	1.1	
Biophos		1.4
Sodium chloride	0.5	
Salt (cobalt-iodised)		0.8
Vitamin A, D, E	0.3	1.0
'Mycoban'	trace	
Trace minerals and flavour	trace	
Analysis		
Dry matter	84.9	
Crude protein	12.7	
Cell wall constituents	57.0	
Acid detergent fibre	24.0	
Lignin	5.2	
Ash	1.1	
Gross energy (kcal/g)	4.26	
Calcium	1.1	
Phosphorus	0.58	
Magnesium	0.17	
Zinc (ppm)	114.9	
Copper (ppm)	13.5	
Manganese (ppm)	106.8	
Selenium (ppm)	0.2	

[a]Schwartz et al. (1985).
[b]R. Stewart (personal communication).

distributions, providing adequate rations during winter, and high labour requirements.

Containment systems

In North America, free-grazing systems are impractical in most jurisdictions because regulations require perimeter fences. At Elk Island National Park, the productivity of moose in a 11,000 hectare enclosure is high but more intensive production systems have encountered varying success.

At the Ministik Wildlife Research Station, moose have been maintained in paddocks of 2–200 hectares. Several commercial producers have successfully raised moose calves in pens of only 10–30 hectares. However, self-replacing populations are difficult to establish. The most serious problem in such settings is debilitating scours resulting from unsuitable feeds or gastrointestinal parasites. Another serious problem has been the winter tick (*Dermacentor albipictus*). Although these disorders can be corrected with treatment, the long-term solution is light stocking and pasture rotation to reduce parasite loads.

Biology and productivity

Populations and harvests

Population densities are typically 1–2 animals/km^2 in most northern ranges, but can be as high as 5–6 animals/km^2 under favourable conditions particularly where dispersal is limited by fencing as in Elk Island National Park (Table 20.2). However, the most important determinant of carrying capacity is browse production.

Juvenile dispersal is well developed enabling moose to colonise habitats created by fire or other disturbances. Most dispersal is of yearlings which explore new areas once they gain independence from the cow. Movement tends to be from high to low density and mortality rates of dispersing juveniles are high. Thus, it is difficult to establish high densities by local habitat manipulation.

Females commonly breed as yearlings (16–18 months) and maintain high reproductive rates until they are at least 12 years old. Twins are produced commonly and triplets occasionally providing pregnancy rates of over 110% (Table 20.3), an unexpectedly high reproductive rate for an animal of this size (Geist, 1974).

This high reproductive rate is balanced by relatively heavy mortality (Blood, 1973). About 33% of calves and 16% of yearlings and adults die each year. In the absence of heavy sport hunting, the sex ratio is about 1:1

Table 20.2. *Winter densities of moose.*

Location	Density (moose/km²)	Source
Canada		
Alberta	0.39–1.39	Lynch (1975)
British Columbia	0.7	Goulet (1985)
Newfoundland	0.44–1.9	Mercer & Strapp (1978)
Ontario	0.2–0.9	Feit (1987)
Quebec	0.06–0.3	Crête & Joly (1981)
Saskatchewan	0.38	Stewart (1983)
Yukon	0.01–1.51	Larsen (1983)
Elk Island NP[a]	2.3–7.5	Telfer (1984)
Riding Mountain NP[a]	1.0–1.05	Telfer (1984)
Alaska	0.2–3.0	Bailey (1978)
Minnesota	1.04–1.56	Karns (1982)
Poland	0.45	Jezierski & Lewandowski (1987)
Soviet Union	0.01–0.62	Kistchinski (1974)
Sweden	5.6	Wilhelmson & Sylvén (1979)

[a]National Park.

Table 20.3. *Fecundity rates of moose.*

Location	Fecundity (calves/100 cows)	Source
Canada		
Alberta	0.25–1.22	Hauge & Keith (1981)
British Columbia	0.6	K. Child (personal communication)
Newfoundland	0.5–1.0	Pimlott (1959)
Ontario	0.2–1.42	Simkin (1965)
Quebec	0.36–0.52	Crête & Messier (1984)
Saskatchewan	0.37–0.77	Stewart (1983)
Yukon	0.22	Larsen (1982)
Elk Island NP[a]	0.79–1.29	Blood (1973)
United States		
Isle Royale NP, Michigan,	0.37	Peterson (1977)
Alaska	0.29–1.19	Ballard, Miller & Whitman (1986)
Finland	0.5	Nygren (1987)
Norway	1.11–1.63	Saether & Haagenrud (1985)
Soviet Union	0.96–1.47	Filonov & Zykov (1974)
Sweden	0.7	Cederlund & Markgren (1987)

[a]National Park.

indicating that mortality rates for the two sexes are similar. Predation by wolves (Bergerud, Wyett & Snider, 1983) and bears (Franzmann & Schwartz, 1986) is an important source of mortality throughout much of their range. The winter moose tick also may cause heavy losses each spring. In eastern North America, moose distributions are limited by the meningeal worm (*Parelaphostronglyus tenuis*) which completes its life cycle in white-tailed deer (*Odocoileus virginianus*).

Natural rates of population increase and thus sustained harvests may be as high as 30–36%/year. Potential harvest rates vary with harvest patterns. However, populations can be unstable particularly during severe winters when faced with competition from wapiti (*Cervus elaphus nelsoni*).

Habitat and forage requirements

The most important characteristic of moose range is an abundance of browse, the main winter food. Since the shrub layer is poorly developed under closed canopies, the best ranges are those where the forest has been opened by fires or logging (Krefting, 1974; Peek, 1974). They use relatively large openings in areas where they are not disturbed, but not when human access is high (Tomm, Beck & Hudson, 1981).

Seasonal home ranges of moose are often quite small, usually 2–4/km. In areas with heavy snow accumulations, they often congregate in yards much like white-tailed deer (Telfer, 1970). Seasonal movements between winter and summer ranges are quite short, between 2–10 km (Coady, 1982) in the southern boreal forest, but longer movements occur in mountain habitats where summer range is at high elevations while the winter period is spent along river valleys (Edwards & Ritcey, 1956). In winter, movements are influenced by depth and density of snow (Formozov, 1946). Although moose are extremely cold tolerant, hot weather appears oppressive in both winter and summer (Renecker & Hudson, 1986a). Even during warm spells (above − 5°C) in winter, heat stress leads to higher energy expenditures, suppressed activity, and retreat to cooler microenvironments. In summer, heat stress occurs at 14–20 °C, forcing animals to feed in wetlands to take advantage of wind and water to dissipate heat and mitigate insect harassment (Renecker & Hudson, 1986a). Heat loads may limit the southern distribution of moose (Kelsall & Telfer, 1974; Renecker & Hudson, 1986a).

Moose require relatively large quantities of high quality food. Like other browsing herbivores, they have relatively small digestive tracts which permit rapid passage of food particles (Renecker, 1987a). Their digestive strategy is to lightly digest large quantities of food rather than more complete digestion of smaller amounts. Daily voluntary intake of moose varies seasonally from

38 g/kg body weight$^{0.75}$ during winter to 129 g/kg body weight$^{0.75}$ during summer (Renecker & Hudson, 1985).

Nutritional quality is an important determinant of forage selection. Studies at the Kostroma Moose Farm indicate that the nutritive value (protein, mineral, energy, and carotene contents) of deciduous foliage equals that of legumes. Moose feed selectively during autumn utilising residual green vegetation protected from frosts which remain about 60% digestible with a crude protein content of 12% (Renecker & Hudson, 1988). There is also a large resource of fallen leaves occasionally used by moose which supplies up to 2400 kg/hectare in aspen (*Populus tremuloides*) forests (Renecker & Hudson, 1986b).

Under intensive management and high stocking rates, damage to trees can be considerable. During late winter and early spring, moose strip bark from aspen and balsam poplar (*Populus balsamifera*) which has a digestibility of 65% (Renecker & Hudson, 1985). However, experience at Kostroma suggests that this problem could be totally eliminated by controlled grazing.

Growth and body composition

Weights of single neonates range between 11–16 kg with twins averaging about 10 kg (Coady, 1983). Hand-reared calves grow at an average rate of 0.7 kg/day for the first four months (Welch *et al.*, 1985). Wild moose in the Pechora-Illychsky Reserve are 117–124 kg at one year whereas farm-raised individuals reach 150 kg. Wild moose reach mature weights at five or six years of age with cows weighing about 350 kg and bulls 480 kg. On the Pechora-Illychsky Reserve farm, adult cows weigh about 350–460 kg and adult bulls 480 kg.

During spring and summer, when high quality forage is abundant, weight gains exceed 1.3 kg/day, but in winter, losses may be 0.8 kg/day (Renecker & Hudson, 1985). Rutting males may lose 1.3 kg/day (Renecker & Hudson, 1986a) for a cumulative total of 20% of pre-rut weight.

Meat yields of tame moose in the Soviet Union range from 50–70% of live weight. In Elk Island National Park, carcase weights for adult males and females are 190 and 175 kg, respectively, with a dressing percentage of approximately 50% (Hawley, 1985). Since moose have a greater proportion of muscle tissue in the front quarters compared with cattle (Berg & Butterfield, 1976), animals killed with inaccurate shots can have a low meat yield. Trimming losses of traumatised tissue can range from 0–26% with an overall average of 3.9% depending on whether killed by a head or forequarter shot (Hawley, Sylvén & Wilhelmson, 1983). Carcases yield about 74% saleable cuts (Hawley *et al.*, 1983).

Moose meat is typically lean (Zhitenko, 1973) and rich in vitamins and trace minerals (Table 20.4). Protein, fat, and energy contents of 25.0%, 0.9%, and 5.2 J/g, respectively, compare to 16.6%, 41.0%, and 17.9 J/g, respectively, for prime grade beef (Shaw, 1985:169).

Reproduction and lactation

Like other northern wild ruminants, moose are seasonally polyoestrus. The rut peaks during September and October (Lent, 1974). Oestrus lasts less than 24 hours, but the cow may be receptive for 7–12 days (Knorre, 1961). The cycle is 21 days with the peak breeding season falling within one oestrus period, although late conceptions have been documented (Moisan, 1955; Coady, 1974; Renecker, 1987b). The gestation period is about 240–260 days with parturition occurring from mid May to early June.

In wild populations, moose normally lactate for four months until the autumn rut. Wild cows probably produce from 75–200 litres of milk over a lactation cycle (Knorre, 1961; Filonov, 1983) compared with 300 (primiparous cows) to 500 litres (older cows) for tame individuals over a six month lactation at the Pechora-Illychsky Reserve. The maximum volume produced per milking was 3 litres with a daily volume of seven litres from five milking bouts. Lower yields were obtained in preliminary studies in Alberta

Table 20.4. *Vitamin and microelement content in moose meat.*[a]

Item (mg)	Young	Adult	Male	Female	Exhausted animal	Refrigerated 3-months
Vitamin						
A	22.9	17.6			12.7	16.5
B1	1.05	0.88			0.39	0.64
B2	1.0	0.82			0.33	0.52
B6	0.75	0.69			0.39	0.59
B12	5.4	5.3			4.2	4.3
PP	18.5	18.0			15.2	17.8
Microelements						
Zn	15.8		18.26	17.24		
Cu	2.45		2.9	2.45		
Mn	0.110		0.095	1.104		
Co	1.83		1.75	2.10		
Mo	0.03		0.063	0.056		
Fe	67.4		70.3	75.6		

[a]Zhitenko (1973).

(Renecker, 1987b). During the 20 year operation of the Kostroma Farm, 13,200 litres of milk were produced.

Moose milk is high in fat, protein, salts, and vitamins and relatively low in lactose (Table 20.5). It has been used for its supposed prophylactic and medicinal properties in the treatment of gastrointestinal disorders, disturbances of mineral metabolism, avitaminoses in children, and recovery of weakened people.

Disease susceptibility

Under intensive management, susceptibility to disease is a main constraint. Perhaps the most frequent problem is chronic scours which can result either from nutritional disorders or from comparatively minor loads of gastrointestinal parasites such as *Strongylus, Trichuris, Eimeria,* and *Monieza.*

Outbreaks of winter ticks can seriously influence survival. Heavy tick infestations and resulting hair loss have been identified in North America (Samuel & Barker, 1979; McLaughlin & Addison, 1986; Welch, 1988). When immune defences are overtaxed and the body weakened by this ectoparasite, secondary infections such as pneumonia can occur.

Table 20.5. *Comparative composition of milk from moose.*

	Soviet Union		North America	
	Moose[a]	Kholmogor cow	Holstein cow[b]	Moose
Total solids (%)	24.6	13.0	11.3	24.5–32.4[b,d]
Fat (%)	2.5	1.3	2.5–3.6	4–11.7[b,d]
Protein (%)	10.3	3.2	3.1–3.2	9.2–18.8[b,d]
Caseins (%)	8.0	2.7		
Albumin and globulin (%)	2.3	0.5		
Lactose (%)			4.3–5.0	0.5–6.8[b,d]
Energy (MJ/kg)			2.4–3.0	4.1–7.4[b,d]
Ash (%)	1.8	0.7		1.6–2.0[c,d]
Vitamin A (mg/l)	0.6	0.02		
Carotene (mg/l)	0.4	0.01		

[a]Kozhukhov (1973).
[b]Renecker (1987b).
[c]Cook, Rausch & Baker (1970).
[d]Franzmann, Arneson & Ullrey (1975).

Moose appear extremely susceptible to malignant catarrhal fever, a viral disease common in domestic sheep and to a lesser extent cattle. In moose, pathological symptoms are elevated body temperature, haemorrhagic enteritis, blindness, and death in 6–7 days.

Economic opportunities

Uses

Sport hunting is the most valuable benefit from wild populations in North America where hunters spend about US$ 110 million each year (Bisset, 1987). Total aesthetic and recreational values are estimated at US$ 360 million (Bisset, 1987). But moose additionally provide meat, hides, and even draught.

Meat makes the greatest economic contribution. Moose meat selling for US$ 7/kg carcase on the domestic Swedish market (Hawley *et al.*, 1983) earns about US$ 150 million annually. Moose meat cannot be sold in North America, but using a replacement value of beef (US$ 3/kg) a product value of US$ 38.9 million could be placed on the regulated harvest.

Moose hides are strong, flexible, and durable. For native people, hide garments can attain good prices especially when trimmed with native artwork. For example, a pair of mittens could sell for US$ 38–56, a leather vest for US$ 190, and a jacket which could be made from a large hide for US$ 375–550. At US$ 9/hide from adult moose (Hawley *et al.*, 1983), this could have considerable potential in northern communities with few other resource-based industries.

Considerable work in the Soviet Union has been conducted on use of moose for riding and pulling sleds. When suitably trained, moose perform as well as horses, carrying packs weighing 80–120 kg for distances of 30 km/day or pulling sleds weighing 400–500 kg. Although moose are easily heat stressed and can only work under cool conditions, they easily negotiate forest areas with swamps and deadfall, do not stray far at night, and readily return to a call.

Production costs

Specialised needs and the high cost of suitable rations greatly add to production costs in intensive systems. Fences should be strong and at least 2.1 m high with a top rail which increases installation costs to US$ 6100–9200/km. Cost of labour and solid feed during hand-rearing of individual calves averages about US$ 1.50/day over the first year. Costs of milk and veterinary supplies are additional. Feed and veterinary supplies and services at the Ministik Wildlife Research Station have been as high as US$

640/head/year, but average US$ 250/head/year. Despite the high production costs, there is a strong demand for tame moose with individual animals valued at US$ 5250–7400.

Conclusions

The tractability and favourable disposition make moose likely candidates for domestication. However, they are sensitive to diseases, and expensive to feed and fence. The system of free-grazing implemented in the Soviet Union has great merit and seems to minimise problems of disease which plague intensive containment systems. However, straying losses are high when animals range freely during summer. Opportunities also exist for moose production on extensively managed mixed-species ranches especially for native people with a large contiguous land base. Nevertheless, the greatest immediate potential seems to be improved management of wild stocks achieved through selective harvests and habitat manipulation (Lavsund, 1981; Stewart, 1983).

References

Addison, E. M., MacLaughlin, R. F. & Fraser, D. J. H. (1983). Capture and early care of moose calves. *Alces*, 18, 246–70.

Bailey, T. N. (1978). Moose populations on the Kenai National Moose Range. *Proceedings of the North American Moose Conference and Workshop*, 14, 1–20.

Ballard, W. B., Miller, S. M. & Whitman, J. S. (1986). Modeling a southcentral Alaskan moose population. *Alces*, 22, 201–44.

Berg, R. T. & Butterfield, R. M. (1976). *New Concepts of Cattle Growth*. Sydney: Sydney University Press.

Bergerud, A. T., Wyett, W. & Snider, B. (1983). The role of wolf predation in limiting a moose population. *Journal of Wildlife Management*, 47, 977–88.

Bisset, A. R. (1987). The economic importance of moose (*Alces alces*) in North America. In *Proceedings of 2nd International Moose Symposium*. Uppsala: Swedish Wildlife Research, Supplement (in press).

Blood, D. A. (1973). Variation in Reproduction and Productivity of an enclosed herd of moose (*Alces alces*). *Proceedings XI International Congress of Game Biologists*, pp. 59–66. Stockholm, Sweden.

Bobek, B. & Morow, K. (1987). Present status of moose in Poland. In *Proceedings of 2nd International Moose Symposium*. Uppsala: Swedish Wildlife Research, Supplement (in press).

Bodden, K. R. (1981). *The Economic Use by Native Peoples of the Resources of the Slave River Alberta Native Community of Fort Resolution*. MA Thesis. Edmonton: University of Alberta.

Cederlund, G. & Markgren, G. (1987). The development of the Swedish moose population 1970–1983. In *Proceedings of 2nd International Moose Symposium*. Uppsala: Swedish Wildlife Research, Supplement (in press).

Coady, J. W. (1974). Late pregnancy of a moose in Alaska. *Journal of Wildlife Management*, 38, 571–2.

Coady, J. W. (1982). Moose *Alces alces*. In *Wild Mammals of North America*, ed. J. A.
Chapman & G. A. Feldhamer, pp. 902–22. Baltimore: Johns Hopkins University Press.

Cook, H. W., Rausch, R. A. & Baker, B. E. (1970). Moose (*Alces alces*) milk. Gross
composition, fatty acid, and mineral constitution. *Canadian Journal of Zoology*, 48,
213–15.

Crête, M. & Joly, R. (1981). Resultats des deux premieres annees d'un plan quinquennal
d'inventaire aerien pour la gestion de l'original au Quebec. *Alces*, 17, 15–29.

Crête, M. & Messier, F. (1984). Response of moose to wolf removal in southwestern
Quebec. *Alces*, 20, 107–28.

Dodds, D. G. (1959). Feeding and growth of a captive moose calf. *Journal of Wildlife
Management*, 23, 231–2.

Edwards, R. Y. & Ritcey, R. W. (1956). The migration of a moose herd. *Journal of
Mammalogy*, 37, 486–94.

Feit, H. A. (1987). North American native management and use of moose populations. In
Proceedings of 2nd International Moose Symposium. Uppsala: Swedish Wildlife Research,
Supplement (in press).

Filonov, C. P. & Zykov, C. D. (1974). Dynamics of moose populations in the forest zone of
the European part of the USSR and in the Urals. *Naturaliste canadien*, 101, 605–29.

Filonov, K. P. (1983). *Moose*. Moscow: Lesnaya Promyshlennost (in Russian).

Formozov, A. N. (1946). *Snow Cover as an Integral Factor of the Environment and its
Importance in the Ecology of Mammals and Birds* (translated from Russian). Publication
1, Boreal Institute. Edmonton: University of Alberta.

Franzmann, A. W., Arneson, P. D. & Ullrey, D. E. (1975). Composition of milk from
Alaskan moose in relation to other North American wild ruminants. *Journal of Zoo
Animal Medicine*, 6, 12–14.

Franzmann, A. W. & Schwartz, C. C. (1986). Black bear predation on moose calves in
highly productive versus marginal moose habitats on the Kenai Peninsula, Alaska. *Alces*,
22, 139–54.

Geist, V. (1974). On the relationship of social evolution and ecology in ungulates. *American
Zoologist*, 14, 205–20.

Goulet, L. A. (1985). Winter habitat selection by moose in northern British Columbia.
Alces, 21, 103–25.

Hauge, T. M. & Keith, L. B. (1981). Dynamics of a moose population in northeastern
Alberta. *Journal of Wildlife Management*, 45, 573–97.

Hawley, A. (1985). Commercial meat production from wild cervids. In *Biology of Deer
Production*, ed. P. F. Fennessy & K. R. Drew, pp. 327–38. Wellington: Royal Society of
New Zealand.

Hawley, A. W. L., Sylvén, S. & Wilhelmson, M. (1983). Commercial moose meat
production. *Livestock Production Science*, 10, 507–16.

Jezierski, W. & Lewandowski, K. (1987). The range expansion of the moose in northeastern Poland. In *Proceedings of 2nd International Moose Symposium*. Uppsala: Swedish
Wildlife Research, Supplement (in press).

Karns, P. D. (1982). Twenty-plus years of aerial moose census in Minnesota. *Alces*, 18,
186–96.

Kelsall, J. P. (1987). The distribution and status of moose (*Alces alces*) in North America.
In *Proceedings of 2nd International Moose Symposium*. Uppsala: Swedish Wildlife
Research, Supplement (in press).

Kelsall, J. P. & Prescott, W. (1971). *Moose and Deer Behaviour in Snow in Fundy National
Park, New Brunswick*. Wildlife Series Number 15. Ottawa: Canadian Wildlife Service.

Kelsall, J. P. & Telfer, E. S. (1974). Biogeography of moose with particular reference to
western Canada. *Naturaliste canadien*, 101, 117–30.

Kistchinski, A. A. (1974). The moose of north-eastern Siberia. *Naturaliste canadien*, 101, 179–84.

Knorre, E. P. (1939). Results in two-year experiments on moose domestication. In *Nauchno-metodicheskiye zapiski Glavnogo upravleniya po zapovendnikam*, No. 4 (in Russian).

Knorre, E. P. (1959). Moose ecology. In *Trudy Pechora-Illychskogo zapovednika*, Vol. 7 (in Russian).

Knorre, E. P. (1961). Advances in and prospects of moose domestication. In *Pechora-Illychskogo Zapovednika*, Vol. 9 (in Russian).

Knorre, E. P. (1973). History and advances in experimental studies on moose domestication. In *Odomashnivaniye Losya*. Moscow: Nauka (in Russian).

Kozhukhov, M. V. (1973). Results of the 20-year experimental studies on moose domestication in the Pechora-Illychsky Reserve. In *Odomashnivaniye Losya*. Moscow: Nauka (in Russian).

Krefting, L. W. (1974). Moose distribution and habitat selection in North Central North America. *Naturaliste canadien*, 101, 117–30.

Krzywinski, A., Niedbalska, A. & Krzywinska, K. (1987). Collection and freezing the semen of bull moose. In *Proceedings of 2nd International Moose Symposium*. Uppsala: Swedish Wildlife Research, Supplement (in press).

Landowski, V. J. (1969). Artificial feeding and development of young moose *Alces alces* (L. 1958). *Der Zoologische Garten*, 36, 327–36 (in German).

Larsen, D. G. (1982). Moose inventory in the southeast Yukon. *Alces*, 18, 142–67.

Lautenschlager, R. A. & Crawford, H. S. (1983). Halter training moose. *Wildlife Society Bulletin*, 11, 187–9.

Lavsund, S. (1981). Moose as a problem in Swedish forestry. *Alces*, 17, 165–78.

Lent, P. C. (1974). A review of rutting behavior in moose. *Naturaliste canadien*, 101, 307–23.

Lynch, G. (1975). Best timing of moose surveys in Alberta. *Proceedings of the North American Moose Conference and Workshop*, 11, 141–51.

Manteufel, P. A. (1935). Moose domestication. In *Priroda isocialisticheskoye khozyaystvo*, Vol. 7 (in Russian).

Markgren, G. (1966). A study of hand reared moose calves. *Viltrevy*, 4, 1–35.

Markgren, G. (1974). The moose of Fennoscandia. *Naturaliste canadien*, 101, 185–94.

McLaughlin, R. F. & Addison, E. M. (1986). Tick (*Dermacentor albipictus*)-induced winter hair-loss in captive moose (*Alces alces*). *Journal of Wildlife Diseases*, 22, 502–10.

Mercer, W. E. & Strapp, M. (1978). Moose management in Newfoundland 1972–1977. *Proceedings of the North American Moose Conference and Workshop*, 14, 227–46.

Middendorff, A. Th. (1853). *Reise in den aussersten Norden und Osten Sibiriens*. St Petersbourg, Vol. II, No. 2.

Moisan, G. (1955). Late breeding in moose (*Alces alces*). *Journal of Mammalogy*, 37, 300.

Moodie, D. W. & Kaye, B. (1976). Taming and domesticating the native animals of Rupert's Land. *Beaver*, 307 (winter), 10–19.

Nygren, T. (1987). The history of moose in Finland. In *Proceedings of 2nd International Moose Symposium*. Uppsala: Swedish Wildlife Research, Supplement (in press).

Peek, J. M. (1974). Initial response of moose to a forest fire in northeastern Minnesota. *American Midland Naturalist*, 91, 435–8.

Peterson, R. O. (1977). *Wolf Ecology and Prey Relationships on Isle Royale*. National Park Service Scientific Monograph Series Number 11.

Pimlott, D. H. (1959). Reproduction and productivity of Newfoundland moose. *Journal of Wildlife Management*, 231, 381–401.

Regelin, W. L., Schwartz, C. C. & Franzmann, A. W. (1979). Raising, training, and maintaining moose (*Alces alces*) for nutritional studies. *Proceedings XIV International Congress of Game Biologists*, pp. 425–8. Dublin, Ireland.

Renecker, L. A. (1987a). *Bioenergetics and Behavior of Moose (*Alces alces*) in the Aspen-Dominated Boreal Forest*. Ph.D. Thesis. Edmonton: University of Alberta.

Renecker, L. A. (1987b). The composition of moose milk following late parturition. *Acta Theriologica*, 32, 117–21.

Renecker, L. A. & Hudson, R. J. (1985). Estimation of dry matter intake of free-ranging moose. *Journal of Wildlife Management*, 49, 785–92.

Renecker, L. A. & Hudson, R. J. (1986a). Seasonal energy expenditures and thermoregulatory responses of moose. *Canadian Journal of Zoology*, 64, 322–7.

Renecker, L. A. & Hudson, R. J. (1986b). Seasonal foraging rates of free-ranging moose. *Journal of Wildlife Management*, 50, 143–7.

Renecker, L. A. & Hudson, R. J. (1988). Seasonal quality of forages used by moose in the aspen-dominated boreal forest, central Alberta. *Holarctic Ecology*, 11, 111–18.

Saether, B-E. & Haagenrud, H. (1985). Life history of the moose *Alces alces*: relationship between growth and reproduction. *Holarctic Ecology*, 8, 100–6.

Samuel, W. M. & Barker, M. J. (1979). The winter tick, *Dermacentor albipictus* (Packard, 1869) on moose *Alces alces* (L.) of central Alberta. *Proceedings of the North American Moose Conference and Workshop*, 15, 303–48.

Schwartz, C. C., Regelin, W. L. & Franzmann, A. W. (1985). Suitability of a formulated ration for moose. *Journal of Wildlife Management*, 49, 131–41.

Shaw, J. H. (1985). *Introduction to Wildlife Management*. New York: McGraw-Hill.

Simkin, D. W. (1965). Reproduction and productivity of moose in northwestern Ontario. *Journal of Wildlife Management*, 29, 740–50.

Stewart, R. R. (1983). Sex- and age-selective harvest strategy for moose management in Saskatchewan. In *Game Harvest Management*, ed. S. L. Beasom & S. F. Roberson, pp. 229–38. Kingsville: Caesar Kleberg Wildlife Research Institute.

Syroechkovsky, E. E. & Rogacheva, E. V. (1974). Moose in the Asiatic part of the USSR. *Naturaliste canadien*, 101, 595–604.

Telfer, E. S. (1970). Winter habitat selection by moose and white-tailed deer. *Journal of Wildlife Management*, 34, 553–9.

Telfer, E. S. (1984). Circumpolar distribution and habitat requirements of moose (*Alces alces*). In *Northern Ecology and Resource Management*, ed. R. Olson, R. Hastings & F. Geddes, pp. 145–82. Edmonton: University of Alberta Press.

Tomm, H. O., Beck, J. A. & Hudson, R. J. (1981). Response of wild ungulates to logging practices in Alberta. *Canadian Journal of Forest Science*, 11, 606–14.

Turkin, N. V. & Satunin, K. A. (1902). *Mammals of Russia*. St Petersbourg (in Russian).

Welch, D. A. (1988). *Thermoregulation and Maintenance Energy Expenditures of Moose (*Alces alces*) Infested with the Winter Tick (*Dermacentor albipictus*)*. M.Sc. Thesis. Edmonton: University of Alberta.

Welch, D. A., Drew, M. L. & Samuel, W. M. (1985). Techniques for rearing moose calves with resulting weight gains and survival. *Alces*, 21, 475–91.

Wilhelmson, M. & Sylvén, S. (1979). The Swedish moose population explosion; precautions, limiting factors and regulation for maximum meat production. *Proceedings of North American Moose Conference and Workshop*, 15, 19–31.

Yazan, Y. & Knorre, E. P. (1964). Domesticating elk in a Russian national park. *Oryx*, 7, 301–4.

Zhitenko, P. V. (1973). Nutritive value of moose meat. In *Odomashnivaniye Losya*. Moscow: Nauka (in Russian).

21

Qiviut production from muskoxen

ROBERT G. WHITE, B. ANN TIPLADY &
PAM GROVES

Abstract

This chapter briefly reviews agricultural projects dealing with production of
qiviut, the insulative underwool of muskoxen (*Ovibos moschatus*). Qiviut is
extremely fine (10–16 μm in diameter) and commands premium prices of over
US$ 220/kg. The fibre is naturally shed and well-fed adults produce 2.6–3.5 kg
annually. Several factors regulate productivity of muskoxen, including seasonal
body weight dynamics, timing of rut and calving, food requirements, and health
and veterinary care. Management systems for qiviut and meat production are
currently based on either wild harvest or intensive husbandry. A third system of
close herding, not dependent on fences, has been advocated but remains untested.
A thorough economic analysis of these systems is lacking, but current expectations are that intensive herding systems must depend on supplementary revenue
from tourism. Future agricultural potential is limited by the availability of
animals, their vulnerability to post-natal infection, and lack of knowledge of the
potential qiviut market. Harvest of wild muskoxen provides significant economic
benefits for Alaskan and Canadian villages. Control over this form of use is
essential for managing animal numbers and distributions on native ranges.

Agricultural evaluations

Muskoxen were first evaluated as an agricultural animal in the
1930s when 33 animals from Eastern Greenland were shipped to College,
Alaska, by the US Biological Survey (Palmer, 1944). That project was
terminated after 2 years. In the early 1960s, attempts were made in Norway,
Canada, and Alaska to establish an agricultural industry based on marketing
of qiviut, the highly valued underwool (Teal, 1958; Wilkinson & Teal, 1984).
Although projects in Norway and Canada have been abandoned, the
endeavour in Alaska remains in operation at Palmer. This herd is descended
from a nucleus of 33 animals obtained from Nunivak Island in 1964 and
raised at the University of Alaska, Fairbanks. The muskox herd became too

large for the Fairbanks site and was transferred to a tundra range near the mouth of the Yukon River near the village of Unalakleet in 1975–76. Within a few years, further progress at this site was precluded by overgrazing (McKendrick, 1981) and the high costs of extending the range or providing supplemental feed in this remote location. The animals were moved to the present site at Palmer, a former dairy farm which appears to have high potential for the successful continuation of the project as it is well drained, hay production is assured, and snow depth is minimal due to moderate winter winds.

Domestication projects in Norway and Canada were abandoned mainly because of the continuous taming necessary and high costs of maintaining animals (Wilkinson & Teal, 1984). In addition, the system for marketing qiviut was not as well developed as in Alaska. Economic factors are cited as the main limitations to future expansion of muskox domestication (Young & Greer, 1975). However, a thorough economic analysis is required to account for changing economic conditions.

In Alaska, qiviut is carefully harvested in spring and processed into a fibre of uniform quality. The yarn is then distributed to native knitters who produce fine quality garments with patterns based on traditional designs. Despite a relatively long period since the project was established, approximately 25 years, we consider the project still experimental as no suitable location and final system of husbandry and qiviut production have stood the test of time.

Many zoos have captive muskoxen and research herds are established in Alaska (Large Animal Research Station, Institute of Arctic Biology, University of Alaska – Fairbanks), Norway (Department of Arctic Biology, University of Tromsø), and Canada (Western College of Veterinary Medicine, University of Saskatchewan). Both the Alaskan and Norwegian animals are of Greenlandic stock whereas those in Canada are from Banks Island. This paper reviews the production of qiviut as a fibre and the underpinning food requirements, and health and veterinary care that must be considered for intensive husbandry in comparison with harvesting wild populations.

Qiviut

Qiviut is a fine underwool that grows under long guard hairs which protect it from the weather. The fibre, which serves to maintain an air-filled space near the skin (von Bergen, 1932), is produced by secondary follicles and commences growth annually in late summer. The fibres are 10–16 µm in diameter (von Bergen, 1932) and grow to a length of approximately 4–8 cm in adults (Wilkinson, 1975) (Fig. 21.1). The combination of qiviut and guard hairs provides muskoxen with excellent insulation against the rigours of

Fig. 21.1. Above; muskox qiviut fibre (× 1000) showing surface scales responsible for good yarn formation. (Photograph by Mary Ann Borchert.) Below; large guard hair with a qiviut fibre for size comparison (× 128). (Photograph by Fran Reed.)

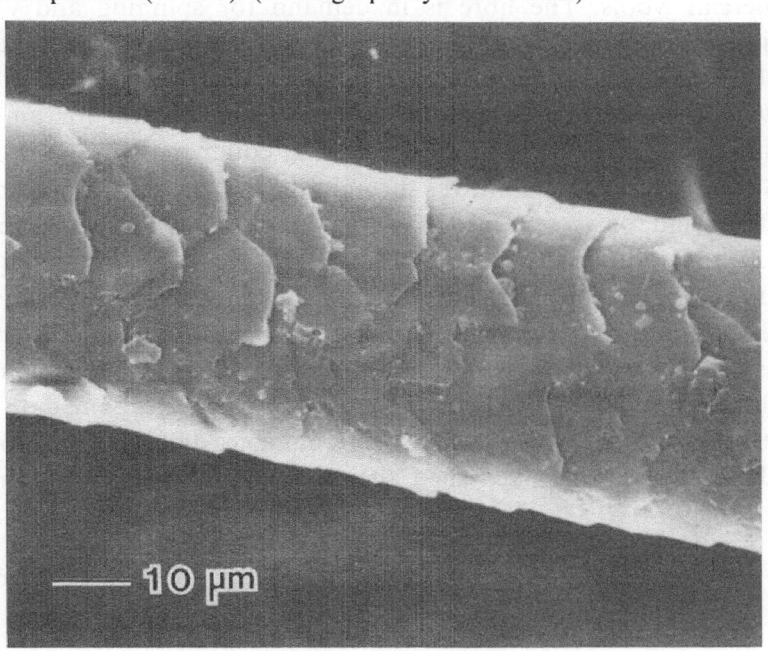

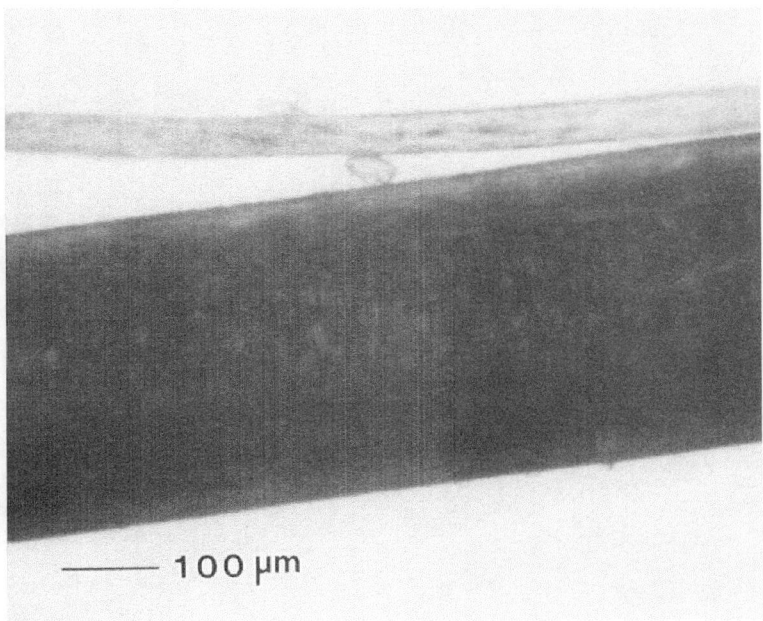

northern environments (Irving, 1972; Blix *et al.*, 1984). Qiviut is considered to be more insulative than superfine merino wool, and is prized for human wear. Its spinning quality is higher than alpaca and exceeds that of the finest of commercial wools. The fibre is in demand for spinning and weaving by individuals and guilds in rural and urban communities and currently sells for over US$ 220/kg.

Qiviut is shed each spring during a two-week period for an individual animal. Under normal conditions, shedding commences in the shoulder and rump areas and this wool moves through the long guard hairs to be released into the wind or is removed in large lumps against shrubs and bushes (Fig. 21.2). Traditionally, northern native peoples have collected qiviut from the tundra for use in preparation of fibres or as a simple insulative stuffing inside mittens and footwear. Males and non-pregnant females start shedding earliest, 15–20 April in the Fairbanks area. Pregnant females commence shedding approximately 5–10 days after calving. Since calving date is variable (Tener, 1965), these females may shed from as early as 10 May to as late as mid June. Newborn calves shed their fetal underwool approximately six

Fig. 21.2. Muskox shedding qiviut in early summer. (Photograph by Pam Groves.)

weeks after birth. Actual dates of calving and qiviut shedding appear related to both latitude and temperature; both occur later in more northern and colder regions.

Qiviut yields increase linearly with age to approximately three years in both males and females. Yearlings of either sex produce just over 1 kg annually, whereas adult females produce approximately 2.6 and adult males 3.5 kg (Table 21.1, Fig. 21.3).

Table 21.1. *Qiviut yield and production efficiency in relation to food intake.*

Cohort	Body weight (kg)	Annual (kg)	yield (g/kg BW)[b]	Efficiency[a] (g qiviut/kg DM[c] intake)
Yearling	119	1.13	9.6	1.04
Female	210	2.61	12.4	2.02
Male	250	3.52	14.5	2.31

[a]See Table 21.2 for annual food intake.
[b]Body weight.
[c]Feed dry matter.

Fig. 21.3. Yield of qiviut in relation to age in captive muskoxen at the Large Animal Research Station, University of Alaska, Fairbanks.

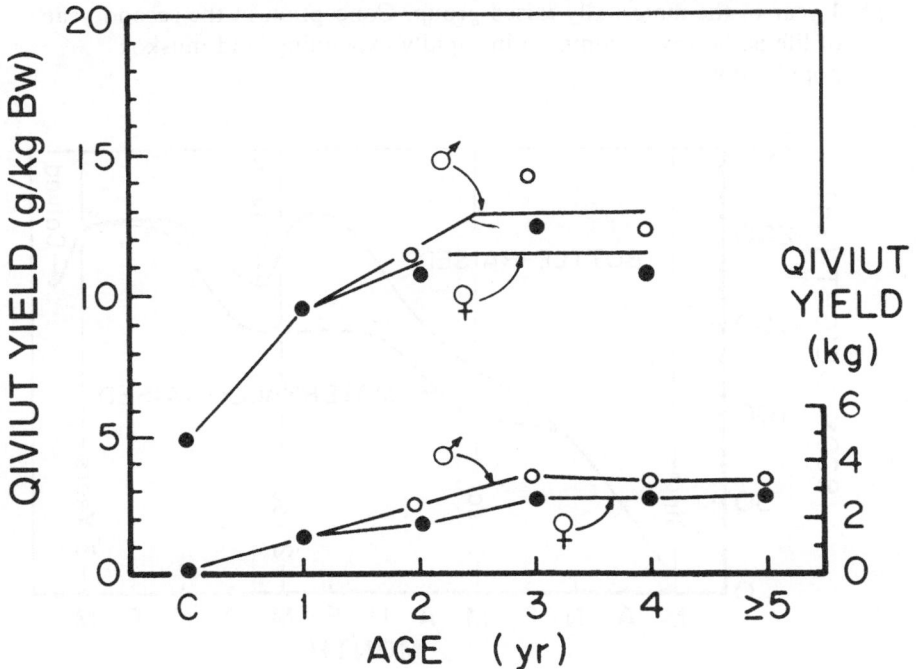

Under intensive systems, qiviut is usually collected by enclosing the animal in a small stall or crush and using long-toothed combs to remove the already loosened fibre. Attempts to remove tightly held qiviut cause the animal to become considerably agitated. Usually all qiviut is removed from one muskox in two or three combing sessions.

Productivity

Animal growth curves and adaptations

Muskoxen grow with annual cyclicity from birth (Tener, 1965; Hubert, 1977; White *et al.*, 1981). Summer growth of calves that are maternally raised averages 400 g/day and the calves usually enter a winter growth stasis in late October or early November, and commence regrowth the following May (Fig. 21.4). Winter growth stasis is variable in calves and little is known about controlling factors. For example, bottle-raised calves may continue to grow almost 100 g/day throughout winter, and this leads to a sufficient weight gain so that females commence breeding activity at approxi-

Fig. 21.4. Representative growth curves of captive muskox females showing the effect of bottle-raising in comparison with maternal rearing. Bottle raised and maternally raised calves were weaned in mid winter. Under these conditions, breeding age in females was delayed 1 year in the maternally raised group. Conception in the second year of life is, however, common in rapidly expanding wild muskox populations.

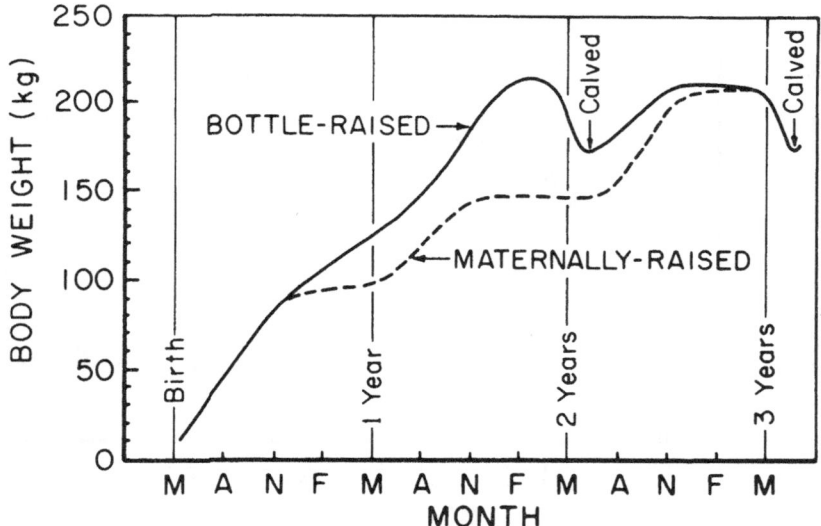

mately 16 months of age (Fig. 21.4). Under conditions of complete winter growth cessation in female calves, breeding usually commences a year later, at 28 months of age. In the wild, growth rates in summer can be low and when combined with winter food limitation, may delay first reproduction until the fourth year of life.

Males are larger than females (Table 21.1) and although this differential growth may commence in calves, it is most apparent in the second summer of life. Summer growth in males continues for several years, whereas adult females achieve adult size before they first calved (Figs. 21.4, 21.5).

Females do not regain body reserves until mid August, when rapid weight gain begins (Fig. 21.5). In central Alaska, this growth occurs when plant quality may be declining. Therefore, autumn may be very important for controlling the oestrus cycle and reproductive potential. If calves are artificially weaned from their mothers, females can replenish body reserves earlier in the summer and this has been advocated as a method of ensuring annual reproduction in captive herds (Wilkinson & Teal, 1984).

As with time of birth, date of natural weaning also varies from year to year. In the research herd at the University of Alaska, weaning has taken place from as early as 20 December to as late as 10 March. Males tend to be weaned earlier than females, though this difference is not significant.

The growth curves discussed above will be modified by the food supply and

Fig. 21.5. Body weight trends of four female muskoxen at the Large Animal Research Station, University of Alaska, Fairbanks. Important findings are that lactating females maintain body weight during most of the summer (e.g., 1984) and the main weight regain period is limited to the rutting period (mid August to early October).

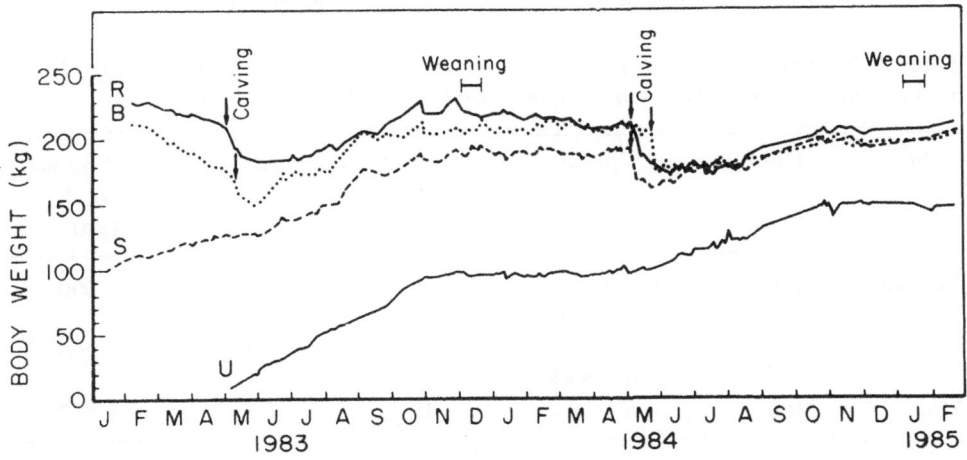

we emphasise that the curves shown in Figs. 21.4 and 21.5 are for well fed animals. Suboptimal feeding will modify this annual pattern of weight change.

Food requirements

Extensive studies at the University of Alaska and the University of Saskatchewan have indicated that summer food requirements using good quality hay or pelleted ration, for yearlings, females, and males are approximately 4, 5, and 6 kg/day, respectively (Chaplin, 1984; White *et al.*, 1984). In contrast, winter requirements are approximately two kg/day for all animals (Table 21.2). Assuming a summer of five months and a winter of seven, the annual food intake would be approximately 1000, 1300, and 1500 kg for the yearlings, females, and males, respectively. Our early studies indicated that the two kg/day requirement in winter is voluntary and attempts to make the animals eat more than this amount of food are unsuccessful unless the animals have suffered severe food restriction before the onset of winter. Under these conditions, compensatory growth can occur.

Muskoxen accept a wide range of hay and pelleted rations (Palmer, 1944). They tolerate legumes such as alfalfa and northern hays grown under agricultural conditions. In addition, they will harvest a wide range of shrubs and leaves such as those from willows (*Salix*), birch (*Betula*), dwarf berry shrubs (*Vaccinium, Empetrum*) and even alder (*Alnus*) (Tener, 1965; Hubert, 1977; Jingfors, 1980; McKendrick, 1981; Robus, 1981).

In winter, muskoxen browse on woody material and indeed prefer to supplement a hay diet with some browse. Under fairly natural conditions, we have found that female muskoxen that raise their own calves can reproduce annually on an all hay diet provided that the quality remains high and that they have access to grazing on pasture during the summer.

Table 21.2. *Daily and annual food requirements of muskoxen held in captivity. Foods used include good quality brome hay and commercial pelleted rations.*

Cohort	Daily Summer Intake		Daily winter intake		Annual total
	$(g/kg^{0.75})$	(kg)	$(g/kg^{0.75})$	(kg)	(kg)
Yearling	110	4.0 (3.5-4.9)	45	2.0	1,090
Female	102	5.0 (4.8-5.6)	38	2.1	1,290
Male	100	6.1	35	2.3	1,522

Muskoxen are highly adapted to restricted diets in winter and prefer snow over free water. Mixed diets based on hay will supply the essential nutrients and are preferable to high protein concentrates or pelleted rations fed as the sole diet.

Providing a diet which is too rich (i.e., excessive grain) can be detrimental. Calving problems have been associated with overweight cows. These problems, which range from difficult deliveries to stillbirths and retained placentas, are often fatal for the cows. Laminitis has also been related to excessive nutrition (Dieterich & Fowler, 1986). Detailed information on metabolic rates and water requirements of muskoxen is contained in the *Proceedings of the First International Muskox Symposium* (Klein, White & Keller, 1984).

The rate of qiviut production is in the order of 1–2.3 g of qiviut/kg dry matter eaten on an annual basis (Table 21.1). In economic terms for Alaska, this equates to US$ 2.20–5.06 of qiviut produced per US$ 0.42 of food consumed.

Health and veterinary care

Under intensive husbandry, normal veterinary care given domestic animals should be practised (Dieterich & Fowler, 1986). In particular, females should be inoculated *pre-partum* against *Escherichia coli* since calves are extremely prone to infection by various soil and faecal-bound organisms during the first two weeks of life. The calves may also be inoculated immediately after birth. Provided that a small amount of fortified pelleted ration is used, we have found no need for injections of either trace minerals or vitamins. An exception may well occur where fodder is deficient in selenium.

Captive animals should be treated for parasitic infections on a regular schedule. Standard worming medications effectively control stomach, intestinal, and lung nematodes. Caution must be used if a heavy lungworm infection exists as the killing of a large number of parasites by the drug can cause anaphylaxis. Coccidia are also found in muskoxen but commonly only cause problems in young animals at times of stress. Outbreaks of coccidiosis cause scours which can be life threatening to a young calf. Therefore, at stressful times, a preventative dose of a sulfa drug is indicated.

Central nervous system infections, often caused by *Corynebacterium pyogenes*, have been observed in captive muskoxen. Despite extensive broad-spectrum antibiotic treatment, these infections are usually fatal. They may be related to facial ulcerations caused by plant awns or other head wounds and tend to affect animals under stress.

Muskoxen have been affected by outbreaks of contagious ecthyma. Healthy, unstressed animals develop only minor lesions which cause no severe health problems. Most affected animals are young, so immunity is probably developed after exposure. Good herd health management and avoidance of stress should prevent major outbreaks.

Muskoxen are very susceptible to capture myopathy, so care should be taken when handling and transporting animals. To avoid excessive stress, it is often advisable to chemically restrain wilder individuals when they must be handled. Xylazine combined with etorphine has been used successfully to sedate muskoxen. Recommended dosages for calm, captive animals are: yearlings, 2.5 mg etorphine, 20 mg xylazine; two to three year olds, 3.0 mg etorphine, 25 mg xylazine; large cows, 3.5 to 4.0 mg etorphine, 25 mg xylazine; bulls 4.5 mg etorphine, 25 mg xylazine (Dieterich & Fowler, 1986). An excited animal requires a larger dose. The animal's ability to thermoregulate is compromised while restrained. Caution must be used to prevent overheating especially if the muskox has exerted itself during the handling procedure.

Management systems

At least three systems for muskox production have been proposed: (1) subsistence and commercial harvest of wild populations, (2) close herding of unfenced females raising their own young and using restraint to harvest qiviut, and (3) intensive husbandry on fenced pastures with various methods of taming animals by early weaning.

Wild harvest

In various forms, this practice is carried out in the high arctic islands of Canada, Greenland, and various sites in Alaska. The main products, meat, hides, and qiviut, are used either commercially or by local communities (Chapter 5). Collection of qiviut from shrubs also occurs, providing material for local native goods. Provided that such systems are well managed, the income for local populations is high. However, no estimate is available of the economics of this system. Such an economic analysis should include costs of the considerable efforts that must be made by management agencies to monitor essential and critical ranges and to estimate numbers and productivity of the hunted populations.

Close herding

Close herding techniques involve grazing animals on summer ranges and providing a winter supplement of hay and pelleted rations. In some areas, alfalfa may be the basis of the hay supply. The systems for herding reindeer employed in Alaska might be an appropriate model. The animals in this situation would not be tame. Techniques for rounding up and restraining the animals at harvest time need to be developed if this system is used. Veterinary care for such a herd would be minimal. Annual vaccinations and worming medications could be administered at the time of qiviut collection. The economics of this system are also unknown and need to be investigated.

Farming
An intensive husbandry system has been under development since 1964 with the domestic herd at Palmer. The production of tame animals has been achieved by early weaning of the calves and regular handling throughout their lives.

Weaning has been conducted at ages ranging from birth to nine months. Calves weaned at birth become extremely tame, although when mature the males of this group may become dangerous. Because of the frequent feedings necessary, weaning at birth has a heavy labour requirement. Calves weaned at nine months are extremely difficult to tame and many will remain difficult to handle despite frequent handling. Between these two extremes, a number of factors must be balanced to determine the best age for weaning. These include nutritional benefits derived from natural nursing, stress to the calf, ease of taming, and the time required and available for the weaning/taming process. Calves weaned at 2–3 months usually tame quite rapidly and suffer minimum stress. When fed a bottle of milk daily throughout their first winter, such calves mature into healthy, easily handled adults.

If animals are accustomed to intensive handling and close contact with humans, the qiviut harvest can be conducted with minimum stress to both animals and people. It has been found that handling during the first year will facilitate future handling. This early handling can range from bottle-feeding, grooming, and weighing to halter and harness breaking. Regular handling such as biweekly weighings and daily contact during feeding will maintain the tameness of adult muskoxen.

Another important aspect of domestication is controlled breeding which prevents inbreeding and allows genetic selection for high qiviut yield and temperament. However, the heritabilities of these traits are unknown so it will take many more generations for such selection to show definitive positive results. With a controlled breeding programme, facilities must be provided to keep cows and bulls separate. Particularly sturdy fences are required to contain rutting bulls with harems. Non-rutting bulls can be handled by experienced people but they do fight amongst themselves. Castration is an effective means to control fighting and facilitate handling. Bulls not required for the breeding programme can be castrated between six months and three years of age. These steers, as qiviut producers, can remain valuable members of the herd.

It has been the practice to dehorn cows and steers during their first winter. The horns are not needed for defence and dehorned animals have a decreased ability to harm each other or humans. Severe head wounds resulted when dehorned bulls engaged in fights during rut. Therefore, the potentially dangerous hooks should be trimmed retaining the boss for protection during fights. Overtrimming may result in the eventual loss of the entire horn, so it is best to occasionally trim small amounts of the horn as the animal grows and make final adjustments when the bull is fully mature.

An additional source of income for the domestication project at Palmer is the development of an educational tourist facility. Visitors pay an admission fee and are given a guided tour to feed the muskoxen through the fence and view photographic exhibits on the muskox life cycle and qiviut production. The initial response has been favourable and it appears to have potential to generate substantial income and thereby contribute to the economic success of the project.

Approximate forage production and minimal range requirements for farming muskoxen in central Alaska are shown in Table 21.3. Provided that approximately 32% of the land is used for hay to feed animals during winter, then approximately 10 adult equivalents/ha can be maintained with minimal dependence on commercial supplements. These results are compared with stocking rates reported by McKendrick (1981) for the muskox farm at Unalakleet. At that time, lack of hay production and supplements led to

Table 21.3. *Approximate forage production and minimal range requirements for intensive husbandry of muskoxen in central Alaska.*

Forage type	System	Hay production (kg/ha)	Range proportion[a]	Range requirement[b] Annual (ha/adult)	(Animal days/ha)
Brome pasture	Farm at 32% for hay	6900	2.17	0.08	4560
Mixed native pasture[c]	Free-ranging	4000–7000	2.63–1.68	0.01–0.02	61–101

[a]Summer/Winter (units: adults/ha).
[b]Expressed as non-lactating cow equivalents.
[c]Unalakleet Muskox Farm: McKendrick (1981).

serious overgrazing. The productive agricultural lands of the Tanana and Matanuska Valleys of Alaska hold more potential for intensive farming, whereas native ranges are better suited to wild harvest systems or possibly close herding. Tables 21.1–21.3 provide a preliminary basis for economic analysis, although the demand for qiviut, hide, and meat products is not well-established.

Prospects

There is an agricultural potential for muskoxen but it is currently limited more by the supply of animals than by knowledge of husbandry requirements. Systems based on agricultural lands require proper fencing and handling facilities. The success of such ventures will depend on the extent to which raw qiviut can maintain a premium price and thereby justify the capital expenditure required for the facilities. Production of high quality knitted products, as marketed by the Musk Ox Producers Cooperative, is a model that has maintained a high market price over a 25-year period.

Socioeconomic benefits of wild harvest systems vary regionally, and the marketing of meat products is limited by inspection requirements of field-killed animals. Certainly, there is increasing subsistence use as muskox populations reach unprecedented numbers in the north. An understanding of the basis of this population outbreak is needed, as a 'bust' may well follow this 'boom' portion of the population cycle, thereby depriving villages of this resource. An additional concern on native ranges is the extent that muskox may compete with other subsistence resources, such as moose (*Alces alces*), reindeer (*Rangifer tarandus*), and caribou (*R. tarandus*).

References

Blix, A. S., Grau, H. J., Markussen, K. A. & White, R. G. (1984). Modes of thermal protection in newborn muskoxen (*Ovibos moschatus*). *Acta Physiologica Scandinavica*, 122, 443–53.

Chaplin, R. (1984). Nutrition, growth and digestibility in captive muskox calves. In *Proceedings of the First International Muskox Symposium*, Biological Papers of the University of Alaska, Special Report No. 4, ed. D. R. Klein, R. G. White & S. Keller, pp. 195. Fairbanks: University of Alaska.

Dieterich, R. A. & Fowler, M. E. (1986). Musk-oxen. In *Zoo and Wild Animal Medicine*, ed. M. E. Fowler, pp. 996–8. Philadelphia: W. B. Saunders Co.

Hubert, B. A. (1977). Estimated productivity of muskoxen, Truelove Lowland. In *Truelove Lowland, Devon Island, Canada: A High Arctic Ecosystem*, ed. L. C. Bliss, pp. 467–91. Edmonton: University of Alberta Press.

Irving, L. (1972). *Arctic Life of Birds and Mammals*. New York: Springer–Verlag.

Jingfors, K. T. (1980). *Habitat Relationships and Activity Patterns of a Reintroduced Alaskan Muskox Population*. M.Sc. Thesis. Fairbanks: University of Alaska.

Klein, D. R., White, R. G. & Keller, S. (ed.) (1984). *Proceedings of the First International Muskox Symposium*, Biological Papers of the University of Alaska Special Report No. 4. Fairbanks: University of Alaska.

McKendrick, J. D. (1981). Response of arctic tundra to intensive muskox grazing. *Agroborealis*, 13, 49–65.

Palmer, L. J. (1944). Food requirements of some Alaskan game animals. *Journal of Mammalogy*, 25, 49–54.

Robus, M. A. (1981). *Muskox Habitat and Use Patterns in Northeastern Alaska*. M.Sc. Thesis. Fairbanks: University of Alaska.

Teal, J. J. Jr (1958). Golden fleece of the arctic. *Atlantic Monthly*, 201, 76.

Tener, J. S. (1965). *Muskoxen in Canada*. Canadian Wildlife Service, Monograph, No. 2. Ottawa: Queen's Printer.

von Bergen, W. (1932). The possibilities of muskox wool as a textile fibre. *The Melliland Textile Monthly*, 47, 1–15.

White, R. G., Bunnell, F. L., Gaare, E., Skogland, T. & Hubert, B. (1981). Ungulates on arctic ranges. In *Tundra Ecosystems: A Comparative Analysis*, IBP Vol. 25, ed. L. C. Bliss, O. W. Heal & J. J. Moore, pp. 397–483. Cambridge: Cambridge University Press.

White, R. G., Holleman, D. F., Wheat, P., Tallas, P. G., Jourdan, M. & Henrichsen, P. (1984). Seasonal changes in voluntary intake and digestibility of diets by captive muskoxen. In *Proceedings of the First International Muskox Symposium*, Biological Papers of the University of Alaska Special Report No. 4, ed. D. R. Klein, R. G. White & S. Keller, pp. 193–4. Fairbanks: University of Alaska.

Wilkinson, P. F. (1975). The length and diameter of the coat fibres of the muskox. *Journal of Zoology (London)*, 177, 363–75.

Wilkinson, P. F. & Teal, P. N. (1984). The muskox domestication project: an overview and evaluation. In *Proceedings of the First International Muskox Symposium*, Biological Papers of the University of Alaska Special Report No. 4, ed. D. R. Klein, R. G. White & S. Keller, pp. 164–6. Fairbanks: University of Alaska.

Young, B. A. & Greer, L. (1975). *Musk Oxen: The Question of Domestication*. Edmonton: Department of Animal Science, University of Alberta.

22

Musk production from musk deer

MICHAEL J. B. GREEN

Abstract

Musk from male musk deer (*Moschus* spp.) is one of the most valuable animal products in the world. Over-exploitation, combined with habitat destruction, has been responsible for the widespread decline of the musk deer. Attempts in China to raise musk deer in captivity are reviewed but the available information is inadequate for assessing the economic viability of such farms. Musk deer are difficult to breed in captivity and their solitary habits and territorial behaviour are a handicap to intensive husbandry. An alternative of harvesting musk from free-ranging populations is discussed. If developed at rural levels, this strategy could help conserve musk deer and their habitats, provided that commercial operations are accompanied by proper protection of musk deer in conservation areas.

Introduction

The male musk deer has long been valued for its musk, a secretion of the preputial gland. Not only is musk one of the oldest and most esteemed raw materials used in perfumery because of its fixative and scent attributes (Parry, 1925), but pharmacological properties have long been ascribed to it (Pereira, 1857).

Musk deer (Fig. 22.1) are distributed sporadically throughout mountain forests of eastern Asia, from the Arctic Circle to China and the Himalayan region (Green, 1986). At least three species are recognised: Siberian musk deer (*M. moschiferus*) in the eastern Soviet Union, Korea, and northern China; dwarf musk deer (*M. berezovskii*) in southern China and northern Vietnam; and alpine, or Himalayan, musk deer (*M. chrysogaster*) in western China and the Himalayan region (Grubb, 1982).

Traditionally, musk deer are killed to excise the musk gland or *pod* as it is known in the trade. Severe hunting pressures exacerbated by extensive habitat destruction are responsible for the widespread decline of the musk

deer throughout much of its Palaearctic distribution. This decline, documented for the Himalayan population (Green, 1986), is responsible for the
increasing scarcity of musk, the price of which has progressively risen from
a quarter of its weight in gold in the 1850s to three times its weight in gold
by the 1970s, making musk one of the world's most valuable animal
products (Green, 1986). Currently, the status of the Himalayan musk deer is
threatened (IUCN, 1974), and commercial trade in Himalayan musk is
prohibited under the Convention on International Trade in Endangered
Species of Wild Fauna and Flora (CITES). In contrast, musk from Chinese
and Soviet populations may be traded under CITES subject to strict
regulation.

As early as 1919, Clements (cited in Parry, 1925) suggested that musk deer

Fig. 22.1. Male musk deer (photograph by the author).

should be reared in semi-captivity for the production of musk which, as subsequently noted by Flerov (1952), can be extracted through the external orifice of the gland without killing the animal. This technique has been adopted by the Chinese who, in the face of diminishing supplies of musk from wild animals, began to domesticate the musk deer in 1958 as part of a national programme to utilise animal (and plant) resources in traditional *materia medica* (Zhang, Dang & Li, 1979; Zhang, 1983). In India, small experimental farms have been established in Himachal Pradesh at Kufri (Jain, 1980), and in Uttar Pradesh at Kanchula Kharak and Meroli (Green, 1985), but these do not operate commercially. In Europe, the United States, and New Zealand, there is considerable interest in farming musk deer but, currently, there is no captive stock.

Structure, scale, and economy of musk production

The number of farms in China is increasing rapidly due to the high demand for musk. Sichuan Province, which reputedly produces half of the country's musk, has 21 communal farms in addition to four state farms established in 1958 (Bista, Shrestha & Kattel, 1979). About 1000 musk deer were raised on farms in Sichuan in 1984, half of which were state-run and the rest commune-run, which suggests that currently some 2000 or more animals are farmed in China. Other provinces in which farms have been established are Qinghai (Zheng, 1980), Shaanxi (Zhang *et al.*, 1979), Shanxi (Zhou, 1965), Anhui (Anon., 1975; Hung, 1975) and Quangxi Autonomous Region (Xiang, 1974).

Even if the amount of musk produced from farmed animals were known, this would not be representative of the trade since most Chinese musk originates from the wild. Although the musk deer is nationally protected in China, this does not have any legislative basis. In fact, the demand for musk is so great that commune leaders authorise the collection of musk and actively encourage hunting by supplying free ammunition. The commune at A'mne Machin in Qinghai Province, for example, sells Y 100,000 of musk from wild animals each year (Rowell, 1983). This represents an annual kill of at least 250 males. Similarly, the present glut of Chinese musk on the market in Hong Kong, centre of the international traffic in musk, must originate from wild animals since it is in pod form (Green, 1986).

Although not much is known about the scale of the trade within China, 215–300 kg of Chinese musk was annually imported by Japan, the world's largest consumer of musk (Green, 1986), during 1981–85 according to official Japanese trade statistics. This represents over 50% of the current international trade in musk, which is only a fraction of what was exported from

China and the Indian subcontinent (about 1400 kg/annum) at the turn of this century (Green, 1988).

Trading in musk has always been highly profitable, given its high market value. Although most profit is gained at intermediate levels, musk contributes significantly to the economy of rural communities in remote areas with few alternative sources of livelihood. About 50 g of musk, which on average represents that obtained from two males (Green, 1988), will provide a family living in a remote region of Nepal with a year's income (Blower, 1974). In West Nepal, hunting musk deer provided the 60 families of Dalphu Village with 20% of their annual cash income (Jackson, 1979). Similarly, the sale of musk acquired from hunting provided the 800-member commune at A'mne Machin, in western China, with 26% of its annual income (Rowell, 1983).

The economics of farming musk deer in China are unquantified. State farms, subsidised by the government, currently operate at a loss (Bista *et al.*, 1979). High maintenance costs may reflect the difficulty of raising musk deer in captivity. Another factor may be the reputed inferior quality of musk from captive males, which are confined to wooden cages to prevent fighting (Green & Taylor, 1986).

The price of musk in China appears low compared with its value elsewhere. In 1979, Chinese musk was worth Y 6/g (US$ 3860/kg) (Bista *et al.*, 1979), whereas the wholesale price of Himalayan musk was four times higher at Rs 125/g (US$ 15,380/kg)(Green, 1988). At that time, the international trade price was US$ 24,000/kg, based on Japanese import statistics, and granular musk sold for up to US$ 45,000/kg. Such large differences in price still exist. If the price of musk fixed by the Chinese Government were on a par with overseas prices, Chinese farming operations would be more profitable.

Management and extraction of musk

Zhang *et al.* (1979) provide a detailed account of Chinese experience in farming musk deer. *Moschus berezovskii* has proved easier to domesticate than *M. chrysogaster (sifanicus)* (Bista *et al.*, 1979), the latter species being farmed only in a few locations such as at Dong Guo Forestry Farm in Jian Zha County, Qinghai (Zheng, 1980). Most success with *M. berezovskii* has been achieved in Sichuan Province where several of the state farms were visited by a Nepalese delegation in 1979 (Bista *et al.*, 1979). Some of these farms are used for breeding while others maintain only males for musk extraction. Musk deer are kept in large enclosures, each containing 10–15 animals with a ratio of 3–7 females to one male, but at the Quang Xian extraction farm males are confined to small wooden cages of less than a cubic metre.

Various rations have been prescribed, comprising a variety of brans, pulses, vegetables, fruits, and woody plant leaves, supplemented with minerals (salt, calcium, potassium, and iodine) and vitamins as necessary (Zhou, 1965; Anon., 1974, 1975; Bista *et al.*, 1979; Zhang, 1983). Rations need to be high in protein and fermentable carbohydrates and low in fibre because the musk deer is essentially a concentrate feeder, albeit with an ability to adapt to poorer diets when high quality food is unavailable (Green, 1987b).

At Ma Er Kang, China's first musk deer farm, musk is extracted from males in late winter (Anon., 1974). The musk is removed using a spatula, which is inserted into the musk sac, via the external orifice, while the animal is manually restrained. The procedure takes several minutes after which the opening of the musk sac is treated with antiseptic cream. The extracted musk is dried, in a dessicator if available, weighed, and stored in an airtight container.

Reproduction and production of musk

Musk deer are seasonal breeders. The rut occurs between November and early January (Bista *et al.*, 1979; Bannikov, Ustinov & Lobanov, 1980; Green, 1985) and the young are born in May–June (Zheng & Pi, 1979; Bannikov *et al.*, 1980; Green, 1985) following a gestation period of 178–198 days (Table 22.1). There appears to be a trend of increasing length of gestation and heavier birth weight with increasing size of species. *Moschus berezovskii*, the smallest species, has the shortest gestation period and lowest birth weight. Conversely, *M. chrysogaster* is the largest species and has the longest gestation period and highest birth weight.

Mean litter size varies between one and two. It is appreciably lower for *M. chrysogaster* than for either *M. moschiferus* or *M. berezovskii* in both captive and wild populations (Table 22.2), but whether differences are species specific or governed by environmental conditions is uncertain. The young grow rapidly, attaining most of their adult body weight by the age of six months (Green, 1985). Musk deer become sexually mature by 18 months of age (Egorov, 1965; Lobanov, 1970; Anon., 1974; Bannikov *et al.*, 1980) but females are apparently capable of breeding in their first year (Hodgson, cited in Jerdon, 1867). In captivity, musk deer have been known to live up to 20 years (Zhang, 1983).

Males secrete musk from an age of 12–18 months onwards. Peak production, marked by the visible swelling of the musk sac and scrotum and by a loss in appetite, occurs in May–July, prior to the autumn rut. Yellow, milky musk drains from a single layer of columnar epithelial cells lining the vesicles of the gland via ducts into the neck of the centrally situated sac. Here, over a period of 30 days or more, it matures into a powerfully scented, granular, red-brown substance (Zhang *et al.*, 1979; Bi *et al.*, 1980). The synthesis of musk is negligible in castrated males, suggesting that it is regulated by androgens from the testes (Zhang *et al.*, 1979). There is some evidence that musk is used to scent the urine (Green, 1987a). Most musk is produced from animals

Table 22.1. *The range in length of gestation and birth weight for different species of musk deer. Sample size is given in parentheses.*

Species /region[a]	Population status	Gestation (days)	Birth weight (g)	Authority
M. moschiferus				
Yakutia	wild	185–195 (?)	no data	Shaposhnikov (1956)
E. Sayan	wild	no data	635 (1)	Egorov (1965)
E. Sayan	wild	no data	460–500 (?)	Bannikov et al. (1980)
M. berezovskii				
Shensi	captive	178–189 (?)	no data	Zhang et al. (1979)
Sichuan	captive	179–187 (?)	350–558 (15)	Bista et al. (1979)
Sichuan[a]	captive	178–192 (?)	455–604 (?)	Anon. (1974)
M. chrysogaster (sifanicus)				
Qinghai	wild	no data	700–750 (2)[b]	Zheng (1980)
Himalaya	captive	170 (1)	no data	Hodgson (1831)[c]
Himalaya	captive	196–198 (2)	600 (2)	Green (1985)

[a]Provinces include Sichuan, Shaansi, and Anhui.
[b]Weight at three days old.
[c]Hodgson's record of 170 days is exceptionally short compared with other data. It is based on the birth of a musk deer in captivity in June following mating in January, as witnessed by the keepers. Possibly mating occurred earlier but was not noticed.

Table 22.2. *The mean litter size (l.s.) for different species of musk deer.*

Species -region	Population status	Pregnant females	Embryos/offspring			Mean l.s.	Authority
			Triplet	Twin	Single		
M. moschiferus							
Yakutia	wild	9	0	6	3	1.7	Egorov (1965)
E. Sayan	wild	76	?	?	?	1.8	Lobanov (1970)
E. Sayan	wild	90	12	50	28	1.8	Bannikov et al. (1980)
M. berezovskii							
Sichuan	captive	12	0	2	10	1.2	Bista et al. (1979)
Sichuan	captive	204	1	137	59	1.6	Anon. (1974)
M. chrysogaster (sifanicus)							
Qinghai	wild	12	0	2	10	1.2	Zheng & Pi (1979)
Himalaya	wild	7	0	1	6	1.1	Green (1985)

between 3 and 8 years of age (Bista *et al.*, 1979). Males in captivity produce little musk by the age of 14 years but the ability to secrete the substance still persists at 20 years of age (Zhang, 1983). About 18 g (10 g dry weight) of musk is harvested annually from captive males (Anon., 1974, 1975). This amount is comparable to yields obtained from males killed in the wild (Green, 1988).

Constraints

During initial attempts by the Chinese to build up captive stocks with musk deer from the wild, the mortality rate was 60–70%, with many animals dying from gastroenteritis primarily as a result of poor husbandry (Bista *et al.*, 1979). Newly captured fawns, which are preferred to adults because they are easier to tame, are particularly prone to such infection unless preventative measures are taken (Zheng, 1980). The other commonly fatal disease, to which young are very susceptible, is pneumonia (Bista *et al.*, 1979).

The musk deer has been a relatively difficult species to breed in captivity. Of 32 recorded births in zoos worldwide between 1959 and 1980, only 17 (53%) survived (Green, 1985). In China, survival of young has improved from 50% to over 90% at Foziling Farm in Anhui (Anon., 1975), and at Ma Er Kang Farm in Sichuan it averaged 74.4% ($n = 336$) during the period 1959–73 (Anon., 1974).

In Sichuan's state farms, advances in domesticating and breeding musk deer were sufficient for the capture of wild animals to be discontinued after 1965 (Bista *et al.*, 1979). Such progress is not widespread in China, as most captive stocks still need to be replenished with animals from the wild (Green & Taylor, 1986).

The musk deer's solitary habits (Green, 1985) tend to preclude intensive husbandry. Males are probably territorial and cannot be raised together in confined spaces without risk of injury from fighting. The Chinese practice of isolating males in small cages is not only inhumane but cannot be conducive to their productivity. Such treatment may be responsible for the reputed inferior quality of musk from farmed animals.

The future

Existing information on farming musk deer in China is inadequate for assessing economic viability. Even with improvements in rearing and breeding musk deer, there remains the problem of raising males at high densities without resorting to practices that are counter to animal welfare. Furthermore, as musk deer are easily stressed, extraction of musk needs to be performed under anaesthesia unless animals are tame.

Alternatively, the problems and costs of maintaining musk deer in captivity could be avoided by harvesting musk from free-ranging animals on the basis of either capturing live animals, then releasing them after extracting the musk (Green, 1978, 1986), or culling them as is practised in the Soviet Union (Bannikov *et al.*, 1980). A major constraint to the former is developing a suitable method of capture (see Green, 1985). Rural development schemes to harvest musk from wild or ranched animals would provide local communities with the motivation to protect not only the musk deer but also its habitat (Green, 1986; Green & Taylor, 1986).

Pure musk is virtually unobtainable due to the widespread and centuries-old practice of adulteration (Green, 1988). Some simple assay of its purity needs to be developed. In Japan, the leading importer of musk measures the muscone content by gas-liquid chromatography, but a less sophisticated technique is needed for ready application in the field.

In view of the highly lucrative nature of the musk trade, commercial operations to harvest musk from captive or wild animals must be accompanied by the proper protection of the musk deer in conservation areas (Green, 1986). At present, wild populations of musk deer in China (and elsewhere) are threatened by hunting. The extent to which this is occurring under the cover of farming enterprises in China is not known, but farming provides an infrastructure whereby musk obtained from wild animals can be traded with reduced risk of apprehension.

References

Anon. (1974). Feeding musk deer in captivity and collecting musk from the live animal. *Dongwuxue Zahzi, China*, 1974(2), 1–14.

Anon. (1975). Preliminary experience in raising the survival rate of musk deer. *Dongwuxue Zahzi, China*, 1975(1), 17–9.

Bannikov, A. G., Ustinov, S. K. & Lobanov, P. N. (1980). *The Musk Deer Moschus moschiferus in the USSR*. Gland: IUCN. Unpublished report.

Bi, S. Z., Yan, Y. H., Qeing, Z. X., Sheng, P. T., Wu, Y. M., Chen, C. F., Xu, H. J., Yang, G. K., Yin, T. B. & Lu, Y. J. (1980). Dissection and analysis of the musk gland of *M. moschiferus* and a preliminary investigation into its histology. *The Protection and Use of Wild Animals, China*, 1, 14–19.

Bista, R. B., Shrestha, M. N. & Kattel, B. (1979). Domestication of the dwarf musk deer (*Moschus berezovskii*) in China. Kathmandu: National Parks and Wildlife Conservation Office. Unpublished report.

Blower, J. (1974). *Note on Trade in Musk, Nepal*. Morges: WWF/IUCN. Unpublished report.

Egorov, O. V. (1965). *Wild Ungulates of Yakutia*. Moscow: Nauka. (Translated from Russian by Israel Program for Scientific Translations, Jerusalem.)

Flerov, C. C. (1952). *Fauna of the USSR, I. Mammals: Musk Deer and Deer*. Moscow: USSR Academy of Sciences. (Translated from Russian by Israel Program for Scientific Translations, Jerusalem.)

Green, M. J. B. (1978). Himalayan musk deer (*Moschus moschiferus moschiferus*). In *Threatened Deer*, pp. 56–64. Morges: IUCN.

Green, M. J. B. (1985). *Aspects of the Ecology of the Himalayan Musk Deer*. Ph.D. thesis. University of Cambridge.

Green, M. J. B. (1986). The distribution, status and conservation of the Himalayan musk deer (*Moschus chrysogaster*). *Biological Conservation*, 35, 347–75.

Green, M. J. B. (1987a). Scent-marking in the Himalayan musk deer *Moschus moschiferus*. *Journal of Zoology (London)* B, 1, 721–37.

Green, M. J. B. (1987b). Diet composition and quality in Himalayan musk deer based on fecal analysis. *Journal of Wildlife Management*, 51, 880–92.

Green, M. J. B. (1988). The musk trade, with particular reference to its impact on the Himalayan population of *Moschus chrysogaster*. In *Conservation in Developing Countries*. Bombay: Bombay Natural History Society (in press).

Green, M. J. B. & Taylor, R. (1986). The musk connection. *New Scientist*, 110 (1514), 56–8.

Grubb, P. (1982). The systematics of Sino-Himalayan musk deer (*Moschus*), with particular reference to the species described by B. H. Hodgson. *Saugetierkundliche Mitteilungen*, 30, 127–35.

Hodgson, B. A. (1831). Contributions in natural history (the musk deer and *Cervus jarai*). *Gleanings in Science*, 3, 320–24.

Hung, A. N. (1975). Domesticating musk deer. *China Reconstructs*, 24 (3), 47–8.

IUCN (1974). *Red Data Book: Mammalia*. Morges: IUCN.

Jackson, R. (1979). Aboriginal hunting in West Nepal with reference to musk deer *Moschus moschiferus moschiferus* and snow leopard *Panthera uncia*. *Biological Conservation*, 16, 63–72.

Jain, M. S. (1980). Observations on birth of a musk deer fawn. *Journal of the Bombay Natural History Society*, 77, 497–8.

Jerdon, T. C. (1867). *The Mammals of India: A Natural History of All the Animals Known to Inhabit Continental India*. Roorkee: Thomason College Press.

Lobanov, P. N. (1970). Characteristics of the distribution, structure and reproduction of the musk deer population in Eastern Sayan. *Ekologiya*, 6, 94–9.

Parry, E. J. (1925). *Parry's Cyclopedia of Perfumes*, Vol. 2, pp. 473–86. London: J. & A. Churchill.

Pereira, J. (1857). *The Economics of Materia Medica and Therapeutics*, 4th edn., Part 2, pp. 802–9. London: Longman, Brown, Green, Longmans & Roberts.

Rowell, G. (1983). China's wildlife lament. *International Wildlife*, 13 (6), 5–11.

Shaposhnikov, F. D. (1956). Material on the ecology of the musk deer in the north-eastern Altai. *Zoologicheskii Zhurnal*, 36, 1084–93.

Xiang, C. X. (1974). A study of the ecology of the musk deer and methods for its live capture in the Guangxi Zhuang People's Autonomous Region. *Dongwuxue Zahzi, China*, 1974(2), 9–10.

Zhang, B. (1983). Musk-deer: their capture, domestication and care according to Chinese experience and methods. *Unasylva*, 35, 16–24.

Zhang, B. L., Dang, F. M. & Li, B. S. (1979). *The Farming of Musk Deer*. Peking: Agricultural Publishing Company.

Zheng, S. W. (1980). The feeding and management of young wild musk deer. *The Protection and Use of Wildlife, China*, 1, 22–3.

Zheng, S. W. & Pi, N. L. (1979). A study of the ecology of *Moschus sifanicus*. *Acta Zoologicia Sinica*, 25, 176–86.

Zhou, G. Y. (1965). The experience of Ping Chuan Musk Deer Farm in Jin Nan. *Dongwuxue Zahzi, China*, 1965(4), 188–90.

SECTION H

Environmental and socioeconomic implications

RICHARD A. LUXMOORE

Having examined the various types of wildlife production systems currently practised and the history of some that preceded them, it remains to contemplate their future: what is the likelihood that they will be successful in the long term, and what will be the consequences if they are? The two questions are obviously interrelated, the solution of the first depending partially on the answer to the second.

In predicting whether a production system will persist, it is necessary to consider the biological potential of the species under the conditions of production, the technical feasibility of exploitation system, its economic viability, the ease with which it will be integrated into the human society, and the impact that it will have on the natural environment. All of these five parameters must be evaluated in relation to alternative, possibly competing, forms of land use. They represent a list of increasing difficulty of assessment, partially because the latter factors are more subjective but also because their effects become apparent over a longer time scale: the biological and technical aspects determine whether a system can succeed in the short-term; the economic and social viability, whether it can proceed beyond the experimental phase; and the environmental impact, whether it is a sustainable form of land use.

Obviously, many of these factors are affected by changing external circumstances, and this will intensify the problems of prediction. An industry which depends on supplying products for traditional medicine may suffer if modern medical treatments become more popular. A subsistence hunting culture may collapse if its practitioners develop a taste for a less arduous lifestyle. The social acceptability of killing animals for sport or commercial purposes may decline to the point at which it is prohibited. Conversely, agricultural over-production may release some land for less intensive uses.

Growing public appreciation of the aesthetic values of wildlife may favour similar changes. The ending of economic subsidies or overseas aid for conventional agriculture may make other production systems financially competitive. The realistic assessment of the extent of environmental degradation caused by current land uses may have a similar effect and force a return to less damaging forms.

The environmental impacts of different wildlife production systems are important not only for their own intrinsic value but also because they may be a potent factor in determining whether the systems gain acceptance. The implementation of wildlife production under sustainable systems of management can be seen as beneficial to conservation because it results in the maintenance of populations of the target species. It may also have significant benefits in terms of the preservation of large areas of habitat which are little modified from their natural state, and thus provide an economic argument for displacing more environmentally damaging types of land use. The less intensive forms of management are the best in this respect. However, the use of economic arguments to justify conservation in this way needs to be approached with caution as minor changes in the economic equation can lead to the opposite logical conclusion.

Wildlife production systems have been credited with the ability to produce large quantities of meat from marginal land without causing environmental degradation. Although they may not be able to live up to the more optimistic predictions, they are certainly capable of returning a financial profit under some conditions, and have clear advantages over other agricultural land uses, if not in preserving pristine natural habitat, at least in encouraging landscape diversity.

23

Impact on conservation

RICHARD A. LUXMOORE

Abstract

The World Conservation Strategy recognises that natural resources must be exploited to ensure human survival, but it seeks to ensure that they are used sustainably. The sustainable exploitation of wildlife, therefore, needs no justification, but it can further be used as an economic argument to slow or reverse the spread of more environmentally damaging forms of land use. This type of argument can be valuable but it relies on demonstrating that wildlife utilisation, either on its own or in conjunction with conventional agriculture, is economically the optimum strategy, and often the necessary data are lacking. The differing environmental impacts of various types of wildlife production systems are discussed, and it is concluded that the less intensive forms, such as controlled hunting or tourism, are generally of greater benefit to conservation than intensive forms. Most Western conservationists hold their views for essentially aesthetic reasons, but often they employ utilitarian arguments to justify conservation to others. This polarisation may be unnecessary, and a balance between aesthetics and utility may be more generally effective in putting conservation policies into practice.

Introduction

Many would regard exploitation and conservation as mutually exclusive, if not antonymous. If this were so, this chapter would be as short as it would be one-sided. The concept that the exploitation of wildlife could have anything other than an adverse impact on its conservation is often difficult to grasp: how can killing animals be good for them? But much of the difficulty stems from a natural tendency to oversimplify the problem. Some forms of exploitation are harmful; so, where doubt exists, it is simplest, for safety and ease of enforcement, to assume that they all are.

Responding to the need to define what is meant by conservation, and what should be done to achieve it, the International Union for Conservation of

Nature and Natural Resources, the United Nations Environment Program and the World Wildlife Fund published the World Conservation Strategy in 1980. This document sets out three main aims for conservation: to maintain essential ecological processes and life support systems; to preserve genetic diversity; and to ensure the sustainable utilisation of species and ecosystems. The third aim is very significant, and has at times been highly controversial as it acknowledged that three of the world's largest international environmental conservation organisations considered that the conservation movement should not merely tolerate, but should actively support the sustainable utilisation of wildlife.

If there were merely a choice between exploiting wildlife or leaving it alone, there would be little doubt which option would be preferable on conservation grounds. However, if this were the case, there would have been no need to write the World Conservation Strategy. Much of the world's natural environment has already been altered by man; much of the remaining land is under threat of alteration; and much of the altered habitat is in danger of further degradation. The total preservation of wildlife is neither attainable, nor necessarily desirable. In order to support a growing human population, the world's human carrying capacity must be raised from its primaeval state by environmental manipulation. Realistic, not idealistic, conservation acknowledges this and seeks to eliminate or minimise long-term degradation.

It is widely held that, in aboriginal subsistence cultures, the utilisation of wildlife is not only essential for their survival but is conducted in a way which is not environmentally damaging. In contrast, commercial exploitation of wildlife by developed societies is perceived to have a worse track record and public image. It is common knowledge that human predation was largely responsible for the extinction of some species, such as the Steller's sea cow (*Hydrodamalis gigas*) and the passenger pigeon (*Ectopistes migratorius*), and the near extinction of others, such as the North American bison (*Bison bison*) and several species of fur seal and large whales. Similarly, international demand for timber has been an important factor in the destruction of forests in South East Asia. However, these are examples of non-sustainable utilisation, and they are not the exclusive preserve of developed societies. It is now recognised that preindustrial societies were responsible for the deforestation of much of Europe and other parts of the world, including Easter Island. The Maoris are thought to have caused the extinction of the moa fauna of New Zealand, and prehistoric peoples may have been responsible for the disappearance of many of the large mammals from the New World. What is important in terms of conservation is therefore not the difference between utilisation and preservation, nor between commercial and subsistence use, but that between sustainable and non-sustainable use.

While acknowledging the principles of the World Conservation Strategy, many conservationists are uneasy with the promotion of the sustainable utilisation of wildlife, and view it as second best to the alternative of total preservation in national parks. Their chief practical justification is that even low levels of harvest may have subtle environmental effects. The preservation of some 'untouched' areas therefore serves as a safety net for our current ignorance.

Although sustainable utilisation is one of the principles of conservation and could be justified on these grounds alone, it can also have some very significant indirect benefits conserving habitats or providing a genetic refuge for target and non-target species alike. On the negative side, the direct effects of overexploitation are obvious but there are also more insidious problems: the encouragement of an exploitation ethic can be counterproductive to developing public sympathy for conservation. These direct and indirect, obvious and subtle impacts are the focus of this chapter.

Use it or lose it

Theoretical justification for conservation is of little use if it cannot be implemented in practice. Powerful arguments are needed to slow the spread of agricultural or industrial development. Most people would concede that some land should be set aside as national parks, immune from such pressures, but these usually constitute only a small percentage of the land area in each country; so conservation must play a clear role in unprotected areas. Even national parks, which may be theoretically inviolable, are vulnerable to the advances of hungry pastoralists and farmers, or of corporate entities wishing to mine mineral wealth within their borders.

Academic arguments stressing the long-term importance of preserving natural genetic diversity, or the aesthetic value of wildlife often carry little weight when ranged against the fear of imminent starvation or the financial muscle of industry. The balance can sometimes be tipped in favour of conservation by adding some economic weight through producing food or some other form of income from wildlife. This reasoning has given rise to the 'use it or lose it' philosophy: unless wildlife can be shown to be useful it will lose ground to the advance of rural development. Ehrenfeld (1976) has discussed the conservation of species which do not appear to have a resource value, and cautioned against inventing spurious economic arguments for this purpose. However, the distinction between resources and non-resources is not always clear and many of the same arguments apply to seemingly useful species.

Is wildlife a financial asset?

This might seem a foolish question in view of the foregoing chapters which describe the great potential and realised value of wildlife. However, to be consistently effective in displacing other land uses, wildlife production systems need to generate greater revenues from the same land, or to be otherwise economically preferable.

In the 1960s and 1970s, it was fashionable to promote wildlife exploitation as the ideal food production system in Africa (Talbot *et al.*, 1965). These arguments have since lost some of their force as most attempts to demonstrate meat productivity superior to domestic livestock have failed (Mentis, 1977). The argument has now shifted to more strictly economic grounds: game meat can be sold for higher prices than beef, either to luxury restaurants as in Kenya (Stelfox, 1984), or to European markets as in South Africa (see Chapter 2). Additional high income can be generated by the sale of breeding stock and, especially, hunting rights (Child & Nduku, 1986).

Except for intensive deer farming, there are few rigorous economic analyses of the profitability of wildlife production; often the income is given but the full costs, particularly of capital equipment, fencing, etc., are overlooked. It is even rarer to find a full consideration of the profitability of the alternative forms of land use, normally conventional agriculture, as these are often underpinned by government subsidies, either in the form of direct grants or the less obvious free veterinary and marketing advice. One analysis in Zimbabwe (cited by Child & Nduku, 1986) has shown that a game enterprise, based mainly on trophy hunting, generated Z$ 4.20/hectare compared with less than Z$ 3.58/hectare for the concurrent cattle enterprise. The ranch is now selling its cattle and concentrating exclusively on game. Most game farms in South Africa combine game with cattle or sheep (Luxmoore, 1985), since combined enterprises generate more income (Collinson, 1979; Conroy & Gaigher, 1982). It is thus not always a choice of game or domestic stock, but sometimes a choice of domestic stock on its own or in combination with game.

One significant economic advantage of simpler wildlife production systems is that they can be operated with minimum capital investment, and this is particularly important in times of high inflation. The gross income of a hunting operation may be far less than that produced by intensive agriculture on the same land. But the latter may require extensive land clearance, fencing, building, road maintenance, and purchase of stock. If the interest on the capital invested and the capital depreciation were considered, net income would be lowered.

Many of the other economic advantages of wildlife are difficult to quantify. Tours may often be the most important asset of wildlife, particularly in attracting foreign exchange. However, servicing foreign tourists often entails the import of certain goods, such as vehicles and luxury foods; so it is the net balance of foreign exchange which must be considered.

Hunting cultures often value the right to hunt far more than indicated by conventional economic analysis. This implies both that they would be prepared to tolerate higher reduction in agricultural income to preserve the ability to hunt or, conversely, that they would require greater financial compensation if the right were taken away. Conversely, pastoralists often view cattle as symbols of wealth and may strongly resist their replacement by wildlife, however much more profitable it might be.

With so many complex and interrelated factors, it is difficult to produce a convincing economic argument for wildlife conservation which cannot be refuted by those with opposing financial interests.

Implications of different wildlife production systems

This volume has described five major systems of wildlife production: subsistence hunting, commercial hunting, herding, ranching, and farming. All have different implications for conservation, and must be considered separately.

Intensive containment systems

Intensive containment systems might seem to provide the most security for species which had become very rare in the wild. After all, zoos are often a last resort for seriously endangered species, providing genetic refuge until offspring can later be used to repopulate the original habitat. But it is difficult to claim that commercial operations (unless zoos are included in this term) have so far been responsible for saving endangered mammals. There are several examples of threatened mammals being farmed commercially; they include the long-tailed chinchilla (*Chinchilla laniger*), the Himalayan musk deer (*Moschus chrysogaster*), farmed on a minute scale, and the Formosan sika deer (*Cervus nippon taiouanensis*), the last probably being extinct in the wild.

Captive breeding generally benefits only species involved since natural environments need not be preserved. In the early stages of captive breeding programmes, foundation stock are usually captured from the wild, and this may provide a small incentive for habitat conservation, but as farmed stocks build they tend to supply an increasing proportion of the demand. The long-term genetic integrity of captive populations also must be doubted, as they usually undergo planned or inadvertent genetic selection. One only needs to consider the wild ancestors of our domestic ungulates: the auroch (*Bos primigenius*) is extinct, the wild yak (*Bos gruniens*) and wild water buffalo (*Bubalus bubalis*) are endangered, and the banteng (*Bibos javanicus*) is vulnerable. The tarpan (*Equus caballus*) is extinct while another subspecies, Przewalski's horse, and the congeneric African wild ass (*Equus asinus*) are endangered. The wild bactrian camel (*Camelus bactrianus*) is vulnerable and the guanaco (*Llama guanicoe*) is much depleted. Several of the possible wild ancestors of sheep and goats, though their taxonomy is confused, are rare and much restricted in range. Only the wild boar (*Sus scrofa*) still thrives in the wild (IUCN, 1978; Mason, 1984). In general, the ancestors have not been preserved and their progeny are not suitable for reintroduction.

Despite this harsh assessment, intensive game farming may be better than non-sustainable crop agriculture and purposeful genetic selection may be better than inadvertent selection within small zoo populations. It remains a question of choices. Nevertheless, intensive containment systems are a special case, and the discussion in the rest of this chapter will therefore focus on less intensive systems.

Extensive systems

The less intensive forms of wildlife production systems have greater potential for conservation as they not only ensure the preservation of the exploited species but also may benefit some non-target species as well. In fenced ranching systems this effect may be minimal, as the habitat may be damaged by pasture improvement, and wildlife which compete for grazing may be eradicated. Unfenced systems which most closely approximate natural conditions offer the greatest potential for conservation. These can include commercial, recreational and subsistence hunting as well as commercial tourism. Their chief disadvantage is that they are more difficult to manage or police than the more intensive systems, and therefore risk overexploitation.

Has wildlife benefited from exploitation?

To answer this question in the affirmative, it is tempting to consider the exploited populations of wildlife which are flourishing in many parts of the world. The moose (*Alces alces*) is probably more numerous now in Scandinavia than it has ever been previously (Chapter 20). Other species of deer in Europe are also increasing in numbers or stable (Gill, 1987). In the Soviet Union, the saiga (*Saiga tatarica*) has reached a population of over 1 million, having previously been reduced to very low numbers (Chapter 9).

Most large mammals in North America are once more increasing in numbers or range, including some that are heavily hunted such as the white-tailed deer (*Odocoileus virginianus*). In South Africa, most of the large game species are increasing and some have been reintroduced to areas where they had been extirpated (Freudenberger, 1982; Luxmoore, 1985). Elephants are now culled in Zimbabwe to stabilise population growth, and other game species are gaining ground in agricultural areas (Child & Nduku, 1986). The capybara (*Hydrochaeris hydrochaeris*) of South America has increased its numbers in Venezuela where a substantial harvest is conducted annually. However, such examples prove little beyond the fact that the harvests are well regulated. They show that exploitation is not incompatible with conservation, and can take place once conservation has been effective. They do not necessarily demonstrate that exploitation has resulted in conservation.

In fact, most of these conservation successes owed little to the incentive of exploitation. The increase in the Scandinavian moose population is largely due to changes in sylvicultural practices. Protection of elephants in Zimbabwe, of saiga in the Soviet Union, of large game in North America, and of capybara in Venezuela was initially undertaken primarily because populations had fallen to dangerously low levels, and less because they were viewed as a potential resource. The growth of game farming in South Africa took place mostly before the lucrative export markets were developed, and Bigalke (1984) has pointed out that it was 'motivated by interest and enthusiasm in more cases than thought of food production and profit'. All of these examples of successful utilisation have occurred in relatively rich countries, which itself suggests that they may be attributable to surplus and leisure rather than dire necessity. Although the primary reasons for embarking on game conservation may have been aesthetic, the potential economic value of the game should not be underestimated in having affected the decision to conserve it. It is easier for a landowner to justify, even to himself, the decision to accept a loss of agricultural production if there is a slight, but unproven, possibility that wildlife may in future be profitable. Rigorous mathematical calculations are often not necessary. Where there is a spark of interest in wildlife conservation, the flimsiest of economic arguments may be sufficient to fan it into life.

Another clue to the success of many wildlife production systems lies in the word 'landowner'. Hardin's (1968) classic paper, 'The tragedy of the commons', eloquently explains why it is impossible to ensure sustainable use of a commonly owned resource except by powerful coercive measures. It is irrational to expect an individual to exercise restraint for a resource he does not own unless he is made to do so by an authority representing collective interests. The secret of success lies either in conferring effective ownership of wildlife on landowners or, where the land is publicly owned, in the existence of a powerful enforcement agency.

Hazards of commercial exploitation

Commercial exploitation can have several adverse biological effects. Some of the effects on the habitat have been discussed earlier in the consideration of the merits of the different production systems. Meat production systems tend to rely on a relatively small number of species, and these are usually the large, gregarious species which are simpler to harvest. They can therefore result in a reduction in species diversity and an imbalance in the grazing community. Financial pressure to maximise profits also can encourage overstocking, with ensuing habitat degradation even if wildlife rather than domestic livestock are used. Rowe-Rowe (1984) has shown that over half of the game farms in Natal have suffered from this. Natural climatic fluctuations pose a particular problem to the game farmer as in drought years the cattle farmer can either kill excess stock or provide supplementary feed. The game farmer is reluctant to reduce his stock as it may be expensive or impossible to replace them when favourable conditions return (Collinson, 1983; Chapter 15). These two problems are both less severe if trophy hunting is the main motive: it can be carried out at lower wildlife densities, causing less danger of overgrazing; and there is a greater incentive to maintain species diversity particularly of the rarer species which attract higher trophy prices.

Commercial pressures often demand the introduction of exotic species which can then escape to either interbreed with the local species or displace them by ecological competition. Numerous species of large game have been introduced into the southern United States where they now live ferally (Chapter 14). The red deer (*Cervus elaphus*) and the Himalayan tahr (*Hemitragus tahr*) were introduced to New Zealand for hunting (albeit not commercial) where they have caused considerable damage to the natural vegetation, and recently further species or races of deer have been introduced for farming, particularly wapiti (*C. elaphus nelsoni*) and Pere David's deer (*Elaphurus davidianus*). Fears of hybridisation have been expressed in South Africa where farmers have tried to introduce Hartmann's mountain zebra (*Equus zebra hartmannae*) and blesbok (*Damaliscus dorcas phillipsi*) where their close relatives (Cape mountain zebra (*Equus zebra zebra*) and bontebok (*Damaliscus dorcas dorcas*)) are indigenous. Introductions also threaten spread of disease, or movement of non-resistant populations into areas with endemic diseases.

Predators are a problem in most production systems. Even low levels of management or hunting often involve predator control; for example the hunting lobby has been vocal in demanding the reduction of wolf populations in Alaska (Skoog, 1983) and Canada (Carbyn, 1983). An exception is where the predator is itself hunted. Lion (*Panthera leo*) and leopard (*Panthera pardus*) are very valuable trophy animals, and Child & Nduku (1986) have described how this has allowed them to be tolerated, and possibly actively encouraged in parts of Zimbabwe. Commercial tourism is much more tolerant of predators, and indeed they may provide the greatest attraction (Thresher, 1981).

Apart from the direct biological problems, there are several more insidious pitfalls associated with the economic use of wildlife. If economic reasons are used as the major justification, then market forces may take control. There may then be no reason to conserve the species which are no longer of any utility. Logic would demand their removal as potential competitors. Even species which are superficially valuable may not prove to be so on closer analysis or if economic conditions change (Ehrenfeld, 1976). Clark (1973) has shown, by a rigorous economic analysis of the whaling industry, that a policy to maximise profits would result in the immediate harvesting of all the remaining whale stocks, and not in their indefinite sustainable harvest, as had previously been supposed. This conclusion was largely due to the slow replacement rate of whales in relation to international discount (inflation) rates. It is possible that an analysis of the exploitation of other slow-maturing species, such as rhinos and elephants, would have similar predictions.

International markets are notoriously fickle. The South African game industry relies heavily on the export of meat to Europe. Profits may be suddenly cut if the price drops, as happened in 1983 (Chapter 2), if imports are suddenly curtailed by changing veterinary regulations, spread of endemic disease in the exporting country, or political sanctions. Game viewing and trophy hunting can be disastrously affected by insurrection or political unrest. Numerous unpredictable events which are beyond the control of the wildlife producer can render his operation unprofitable. If this happens, is he to slaughter his wildlife and turn to some other activity as business practice would dictate?

A still more intractable problem concerns the public attitude to wildlife. Many of the early successes of conservation have been attributable to public concern for animal welfare or respect for wildlife. Educational campaigns have devoted much energy and money to disseminate these sentiments. It has been argued that promoting consumptive wildlife utilisation would reverse some of these gains and confuse the public. In reality, there is little evidence for this view, and it is countered by Sale's (1983) finding that 'economic dependence on wild resources promotes a close relationship between the way

· of life and social organisation of many tribes and the ecological factors governing the existence of wildlife'.

Another important consideration is the distribution of costs and benefits. Herdsmen prevented from grazing stock in a wildlife reserve, or hunters stopped from carrying out traditional pursuits are unlikely to conserve wildlife for safari outfitters selling wildlife to overseas visitors. More recent African wildlife utilisation schemes have attempted to direct financial returns to local communities. But, as Bell (1987) points out, it is often difficult to target the funds accurately enough, and unless the people affected are involved in the planning process they may, with justification, conclude that they are 'being treated as a nuisance which is being bribed to keep quiet'.

A hope of commercial wildlife production is that it would control prices on illegal markets by providing an alternative regulated legal supply. There may even be some substitution of products from abundant species for products of those that are endangered. However, the sword may be double-edged when poached game products are laundered through legal market mechanisms (Geist, 1985).

Conclusions

That exploitation has a legitimate role in wildlife conservation has been justified in the World Conservation Strategy, but it is a different matter to use exploitation as an argument to promote conservation. This chapter has attempted to examine some of the advantages and hazards of doing so. Aesthetic rather than utilitarian arguments for wildlife conservation are usually the driving force and foundation of Western conservation movements, but they are often presumptuously not thought suitable for influencing conservation in the Third World. Conversely, Western conservationists are portrayed as imposing their ideals on developing societies who can ill afford them, and utilitarian arguments are seen as the acceptable alternative under which the ideals can be disguised. Recent studies suggest that this polarisation may be mistaken and the subterfuge unnecessary, as the inhabitants of many developing countries may have an equally sophisticated awareness and appreciation of wildlife (Harcourt, Pennington & Webber, 1986). Sale (1983) has pointed out that 'there is a strong basis for an African conservation philosophy within traditional relationships between people and wildlife'. As in so many seemingly insoluble disputes, it is probable that a balance between aesthetic and utilitarian arguments may be the correct approach. Whatever the outcome, in admitting the importance of aesthetic ideals, I can do no better than echo Ehrenfeld's (1976) conclusion that we will have had 'the small, private satisfaction of having been honest for a while'.

References

Bell, R. H. V. (1987). Conservation with a human face: conflict and reconciliation in African land-use planning. In *Conservation in Africa: Policies and Practice*, ed. D. Anderson & R. Grove. Cambridge: Cambridge University Press.

Bigalke, R. C. (1984). Utilization of game. In *Proceedings of the July 1983 Workshop on the Conservation and Utilization of Wild Life on Private Land*, ed. P. R. K. Richardson & M. P. S. Berry, pp. 32–41. Pretoria: Southern Africa Wildlife Management Association.

Carbyn, L. N. (1983). Management of non-endangered wolf populations in Canada. *Acta Zoologica Fennica*, 174, 239–43.

Child, G. & Nduku, W. K. (1986). Wildlife and human welfare in Zimbabwe. Paper presented at the African Forestry Commission Working Party on Wildlife Management and National Parks, Bamako, Mali. Reprinted in IUCN/SSC Ethnozoology Specialist Group Newsletter No. 1.

Clark, C. W. (1973). The economics of overexploitation. *Science*, 181, 630–8.

Collinson, R. F. H. (1979). Production economics of impala. In *Beef and Game Management, Proceedings of the 3rd Hlabisa Soil Conservation Symposium*, pp. 90–103.

Collinson, R. F. H. (1983). Management for improved veld conditions and production on game ranches. *Proceedings of the 4th Hlabisa Soil Conservation Symposium*.

Conroy, A. M. & Gaigher, I. G. (1982). Venison, aquaculture and ostrich meat production: action 2003. *South African Journal of Animal Science*, 12, 219–333.

Ehrenfeld, D. W. (1976). The conservation of non-resources. *American Scientist*, 64, 648–56.

Freudenberger, D. (1982). Southern Africa. In *The Farming of Deer: World Trends and Modern Techniques*, ed. D. Yerex, pp. 31–44. Wellington: Agricultural Promotion Associates.

Geist, V. (1985). Game ranching: threat to wildlife conservation in North America. *Wildlife Society Bulletin*, 13, 594–8.

Gill, R. M. A. (1987). The current status and management of European Red Deer. Proceedings of the Red Deer Symposium, Confederation Internationale de la Chasse, 1986 (in press).

Harcourt, A. H., Pennington, H. & Webber, A. W. (1986). Public attitudes to wildlife and conservation. *Oryx*, 20, 152–4.

Hardin, G. (1968). The tragedy of the commons. *Science*, 162, 1243–8.

IUCN. (1978). Red Data Book, Vol. 1, Mammalia. Morges: IUCN.

Luxmoore, R. A. (1985). Game farming in South Africa as a force in conservation. *Oryx*, 19, 225–31.

Mason, I. L. (ed.) (1984). *Evolution of Domesticated Animals*. London: Longman.

Mentis, M. T. (1977). Stocking rates and carrying capacities for ungulates on African rangelands. *South African Journal of Wildlife Research*, 7, 89–98.

Rowe-Rowe, D. T. (1984). *Game Utilization on Private Land in Natal*. Pietermaritzburg: Natal Parks, Game and Fish Preservation Board.

Sale, J. B. (1983). *The Importance and Values of Wild Plants and Animals in Africa*. Gland: IUCN.

Skoog, R. O. (1983). Results of Alaska's attempts to increase prey by controlling wolves. *Acta Zoologica Fennica*, 174, 245–7.

Stelfox, J. B. (1984). Wildlife ranching. *Swara*, 7, 15–19.

Talbot, L. M., Payne, W. J. A., Ledger, H. P., Verdcourt, L. & Talbot, M. H. (1965). The meat production potential of wild animals in Africa. A review of biological knowledge. *Commonwealth Agricultural Bureau Technical Communication*, 16, 1–42.

Thresher, P. (1981). The present value of an Amboseli lion. *World Animal Review*, 40, 30–3.

24

Socioeconomic prospects and design constraints

ROBERT J. HUDSON & V. V. DEZHKIN

Abstract

Economic utilisation of wild herbivores has been argued on biotechnical, socioeconomic, and environmental grounds. Biotechnical arguments include environmental adaptations, complementary diets, high sustained productivity, and desirable carcase qualities of wild ungulates. Socioeconomic arguments include complementary patterns of inputs in relation to conventional animal agriculture, access to a broader range of markets, and provision of culturally consistent livelihoods for indigenous peoples. Environmental benefits may derive from maintenance of natural habitats, competition with illegal markets, and sometimes provision of an ultimate genetic refuge for rare species. However, the degree to which these benefits accrue depends on how wildlife production systems are designed. Ultimately, systems are constrained by the physiological, ecological, and behavioural needs of animals. But these often must be compromised in the interests of conflicting human needs such as a fixed domicile, nutritional and economic security, or competitive demands for labour and capital. The marketplace and public perceptions further limit options. Competition on international game meat markets has meant increasingly higher standards of quality, hygiene, and disease control. Generally, these tighter controls favour more intensive production systems which permit standardisation of carcase size and fatness as well as enable live delivery of animals to abattoirs for *ante-mortem* inspection and hygienic handling of meat.

Introduction

Whereas wildlife viewing and sport hunting provide solid economic justification for wildlife conservation, commercial meat production from wildlife remains more talked about than done (Eltringham, 1984; Kyle, 1987). Among the most ardent supporters of wildlife production (as well as its most bitter opponents) are people more interested in biology and conservation than agriculture and rural development. In fact, productivity and economic viability once seemed more of an excuse than a reason for commercial game production. Whereas biologists have been accused of overlooking practical realities, agriculturists have been slow to appreciate the advantages of wildlife utilisation and sometimes over-estimate the biotechnical and socioeconomic difficulties.

The controversy which surrounds the issue as well as obstacles thrown in its path often make game production much less competitive with other options for immediately improving conditions for rural people. Yet, where financial risks can be taken and longterm planning is possible, these obstacles usually can be overcome. Often, the necessary changes in legislation require justification on environmental grounds. However, subsequent growth of the industry is fuelled by its advantages as an agricultural strategy. To become an important form of land use, game production must be biotechnically feasible, economically rewarding, and culturally consistent, i.e., it must be a fundamentally rational activity. This chapter evaluates the agricultural potential of wildlife and explores constraints on the design of production systems with a view to understanding future transformations.

Prospects for economic utilisation

Despite considerable experimentation with new germ plasm in crop production, there has been surprisingly little effort to evaluate new agricultural animals. Of approximately 200 ungulate species, only about 35 have been used as domestic and semidomestic animals in the past and roughly a third persist as agricultural animals (Clutton-Brock, 1981; Mason, 1984). Through time, conventional livestock husbandry attained paramount economic importance, engendering the widespread view that livestock husbandry and game production were rigid alternatives. Animals were cast in traditional roles; some were for food and others for fun.

However, comparative research on domestic and wild ungulates, practical experience, and current economic successes cast doubt on the wisdom of such an absolute orientation to livestock as meat producers (Yerex, 1982; Fennessy & Drew, 1985). After all, there are few obvious biological barriers to domesticating new species; physical, ecological, and behavioural traits of wild and domestic ungulates show few consistent differences (Tennessen & Hudson, 1981). Even if there were, reasons probably are very different now than at the advent of the Neolithic when many of these domestications took place.

Today, the promise of wild herbivores as an economic resource relates to prospects of harnessing their high productivity, reducing material and energy subsidies, tapping expanded markets, and maintaining traditional livelihoods for indigenous peoples. The future of economic utilisation rests in the following biological and socioeconomic arguments (Field, 1979; Walker, 1979; Bigalke, 1986).

Biological arguments

The agricultural potential of wildlife was first argued from a biological standpoint because there was little experience on which to evaluate economic aspects (Darling, 1960; Dasmann, 1964; Talbot, 1966; Kay, 1970; Topps, 1975). The case was made that anatomical, physiological, and ecological adaptations of indigenous herbivores accounted for their high productivity on marginal lands relative to poorly adapted domestic stock, and that their favourable carcase characteristics preadapted them to their new economic role.

Environmental adaptations

Early work seems to have been motivated by the hope of demonstrating superior abilities of wild ungulates to digest coarse forages used sparingly by domestic stock. Although some wild species compared well with cattle or sheep on the basis of conventional measures of digestive efficiency, many did not (Gebczynska *et al.*, 1974; Arman & Hopcraft, 1975; Richmond, Hudson & Christopherson, 1977; Schaefer, Young & Chimwano, 1978; Foose, 1982). Anatomical and functional explanations for these differences were provided by Hofmann (1973, 1985) whose studies turned attention away from differences between domestic and wild to more important contrasts among feeding styles and broader questions of nutritional adaptation. Digestion coefficients used in conventional agricultural evaluations turn out to have qualified significance for range animals where the most important determinant of survival and productivity is daily digestible nutrient intake rather than completeness of digestion.

Another apparent advantage of wild ungulates is their ability to thrive in thermally extreme environments (Parker & Robbins, 1985) where shelter and/or water would have to be supplied for domestic livestock. Tropical ungulates adapt to heat loads by lowering metabolic rates, dissipating heat by panting or sweating, and storing heat by adaptive hyperthermia (Taylor, 1969a,b, 1970a,b). Closely related adaptations involve water and nitrogen economy since hot environments are often arid and water is needed both for evaporative cooling and to remove nitrogenous wastes (King *et al.*, 1975; King, 1979). Tropical ungulates also may respond to heat and aridity by merely limiting activity to cooler parts of the day or selecting succulent or hygroscopic foods (Taylor, 1968; Lewis, 1977).

Temperate and arctic species adapt to cold temperatures and strong seasonality with high thermal insulation and seasonal rhythms of energy metabolism (Christopherson, Hudson & Christophersen, 1979; Parker & Robbins, 1984; Kay, 1985; Falkow & Mercer, 1986; Renecker & Hudson, 1986). The mid-winter nadir of energy expenditures, inappetence, and low growth impulse presumably equips animals to cope with food shortage (Ringberg, 1979). However, the main feature is not that winter energy expenditures are low but rather that summer expenditures are comparatively high during the brief productivity pulse in northern latitudes (Hudson & Christopherson, 1985). The generally high metabolic rates of wild ruminants have even been cited as detracting from their efficiency as meat producers (Rogerson, 1968).

Whereas the significance of maintenance metabolism in the efficiency complex can be questioned (Hudson & Christopherson, 1985), the tendency of many wild animals to deposit mainly lean tissues (Ledger, 1968) clearly improves the apparent efficiency of growth. The energy content of fat is 39.7 kJ/g whereas lean tissue with associated water has an energy content of only about 5 kJ/g (see Price & White, 1985). Efficiency of utilisation of metabolisable energy for either maintenance or gain seems to vary little between wild and domestic ruminants (Simpson *et al.*, 1978; Robbins, 1983).

Undoubtedly more important than slight variation in bioenergetic efficiency is relative resistance to diseases such as trypanosomiasis, foot-and-mouth, theileriosis, or malignant catarrhal fever (Edwards & McDonnell, 1982). Some wild species do thrive where endemic diseases obviate husbandry of conventional livestock. But greater resistance may mean that wild ungulates serve as carriers (Hammond & Branagan, 1973; Tessaro, 1986). Therefore, game may be eliminated to protect livestock or veterinary restrictions may be placed on exports of live animals or products. Under intensive husbandry many wild ungulates are quite sensitive to common diseases of farm livestock.

Complementary resource use
The advantages of mixed-species grazing are widely recognised by pastoral people who keep an assortment of large and small stock (Dahl, 1981; Lamprey, 1983). On natural ranges offering a spectrum of forage types and a diverse landscape, a community of herbivores whether domestic or wild offers a way to balance grazing pressure and to ensure full but sustained utilisation of primary production. Distinctive distributions and foraging habits can be traced to such factors as digestive anatomy and physiology, relative independence of water and cover, gregariousness, and response to

disturbance and predators (Bell, 1971; Sinclair, 1985; McNaughton & Georgiadis, 1986).

Although mixed-species grazing may provide the opportunity to better manage rangelands (Owen-Smith & Cooper, 1985), game production does not necessarily guarantee it (Bigalke, 1986; Chapter 15). It is not difficult to demonstrate that mixed-species grazing results in broader utilisation of range forages (Taylor & Walker, 1978) despite the sizeable dietary overlap in heavily stocked natural communities (Hansen, Mugambi & Bauni, 1985). However, distributions of ranch game may be hard to control and if breeding stock is expensive or difficult to obtain, game farmers may be reluctant to destock during drought.

Sustained productivity

Productivity of grazing systems depends proximally on net primary productivity and ultimately on rainfall, soil fertility, and habitat structure. These relationships are best known for African grazing systems which have remained relatively intact. For eutrophic savannas, Coe, Cumming & Phillipson (1976) established consistent relationships between rainfall and herbivore biomass and productivity. Dystrophic savannas developed on nutrient-poor soils are generally much less productive and support a different herbivore guild (Bell, 1982; East, 1984). Of course, many other factors determine carrying capacity and confound such simple bivariate relationships (McNaughton & Georgiadis, 1986).

It is often claimed that carrying capacity for wildlife communities is many times greater than for domestic livestock (Talbot, 1966; Brown *et al.*, 1986) but this is not the general case particularly in areas of low rainfall (Lamprey, 1983). Nevertheless, communities of wild ungulates often do attain respectable biomass densities (Mentis, 1977).

The main problem with trying to compare wild and domestic herbivores is defining resource boundaries. Wild ungulates often move widely making it difficult to define the land base with any degree of precision. If domestic livestock receive supplements at any time of the year, it is similarly hard to include this contribution. Consequently, the difference between domestic and wild herbivores often is one of intensity of management. Ideally, comparisons should be based on responsiveness to increasing intensities of management (Fig. 24.1). Despite the high productivity of wild ungulates in natural grazing systems, the competitive edge may shift to domestic stock with intensive husbandry. But there are several exceptions such as the impressive productivity of farmed red deer on intensively managed pastures which often exceeds that of beef (Drew, 1985).

The contrast of domestic and wild probably means little without considering such factors as allometric scaling or digestive adaptation (Demment & van Soest, 1985; Hudson, 1985). Characteristics such as population density, rate of increase, growth rates, fecundity, and longevity tend to scale allometrically (Peters, 1983; Schmidt-Nielsen, 1984; Georgiadis, 1985). Therefore, differences between domestic and wild, or exotic and indigenous herbivores may be largely a matter of body size.

Carcase characteristics
Many ungulates seem almost preadapted as meat animals. The dressed carcase weight relative to live weight (dressing percentage) typically is high in wild ungulates (Ledger, 1968; von la Chevallerie, 1972; Berg & Butterfield, 1976). At least part of the reason is the relatively small digestive fill of wild selective/mixed feeders compared with cattle, the usual standard of comparison. The relatively large hind quarters which contain many of the best commercial cuts may offer a further advantage (Kay *et al.*, 1981; Wenham & Pennie, 1986).

Typically, game meats are lean with little marbling (intra-muscular fat), an attractive marketing feature for consumers concerned with calories and

Fig. 24.1. Productivities of wild and domestic herbivores in relation to intensity of management.

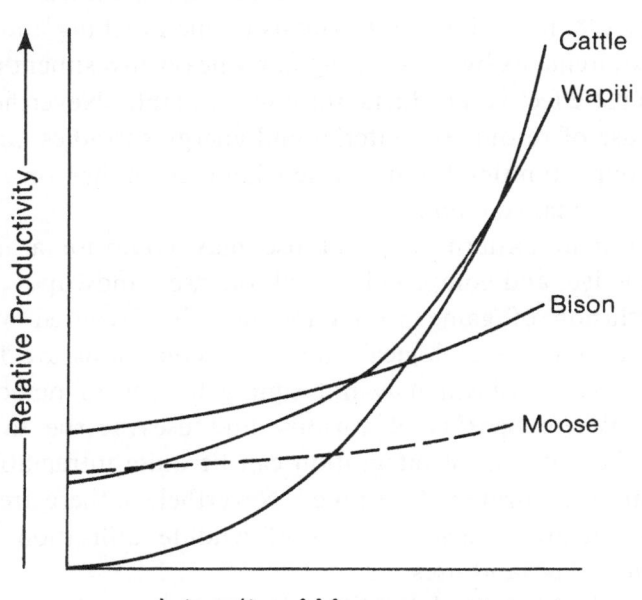

cholesterol (Hawley, Sylven & Wilhelmsen, 1983; Drew, 1985; Hawley, 1986). Fatty acid composition has been studied along a similar vein (Crawford *et al.*, 1970; Manley & Forss, 1979). Game meat is highly palatable (Forss *et al.*, 1979), attracting premium prices on gourmet markets.

Socioeconomic arguments

Biological arguments in a market economy are relevant largely to the extent that they influence input costs, product volumes, or market prices. Under some circumstances, wild ungulates do incur lower cash costs of production and venison usually attracts higher prices than conventional meats. However, game production on farms and ranches is viewed more realistically in the long-term as an opportunity to diversify agriculture in terms of both inputs and products.

Complementary inputs

Wildlife production is commonly perceived as a low-input system with less dependence on fossil fuels and agricultural chemicals than conventional livestock husbandry (Telfer & Scotter, 1975; Kyle, 1987). It can offer an appropriate, complementary, and more sustainable land use, but to what degree depends upon how game is produced. Even extensive systems of wildlife utilisation may require sizeable inputs although they may be hidden or externalised as in sport hunting. Commercial hunting requires either heavy investment in manpower or heavy dependence on off-track vehicles or aircraft (Hawley, 1985). Containment systems (game ranching and farming) reduce labour requirements by substituting fairly heavy investment in fencing although more cost-effective products are now available. Nevertheless, this complementary use of resources, material and energy subsidies, and labour offers excellent opportunities for integrated land use either on individual farms or within regional economies.

Regionally, optimal patterns of land use may comprise a mosaic of protective, productive, and compromise (multiple use) landscapes (Dezhkin, 1975, 1978). Inclusion of game production may be favoured over more traditional land uses when ecological and social conditions are fully considered. However, the problem may not simply be one of demonstrating competitiveness. Where benefits of various land uses accrue to different interest groups, the problem of integration can be quite intractable even if optimal mixes can be accurately determined. Nevertheless, there are excellent opportunities to integrate various forms of wildlife utilisation with one another as well as other land uses.

Recreational and commercial hunting often can be usefully combined.

After all, one of the most expensive aspects of commercial hunting is the harvest (Hawley, 1985). A widely used method for externalising these costs is to harness the volunteer efforts of sportsmen who surrender the carcase in return for the opportunity to hunt. The problem is more difficult in North America where wildlife is in public ownership and landowners generally are not allowed to benefit directly by marketing wildlife products (White, 1986). This has resulted in a variety of indirect charges and the introduction of exotic species which are not subject to the same regulations (Teer, Burger & Deknatel, 1983).

There is an opportunity and a need to integrate wildlife and forest management. Spectacular increases in populations of some wild ungulates can be traced to sylvicultural practices (Chapter 20). On the other hand, local over-abundance of game may inflict serious damage to forests. In Western Europe, forest management systems encourage integration by providing incentives through commercial hunting for sport or venison. In North America, most commercial forests are on public lands and forest companies are required to follow strict guidelines intended to protect other interests including wildlife. However, they are not allowed to profit from purposeful management of wildlife. Therefore, there is little incentive to do much more than required by government agencies.

Equally successful integration is possible on rangelands where grazing indigenous ungulates with or without domestic livestock leads to fuller sustained utilisation (Dezhkin & Menkova, 1981). This opportunity has been most fully captured in eastern and southern Africa where wildlife additionally provides a hedge against drought and much needed enterprise diversification (Chapter 15). Even on intensively managed pastures, roughage grazers such as cattle may improve conditions for more selective feeders such as red deer or sheep (Chapter 16).

Diversified market opportunities
A more important way that game production diversifies agriculture is by tapping new markets, particularly the international venison trade (Chapter 2). For many years, there has been discussion in rural development circles about whether it is best to provide low-cost protein for rural consumption or to serve specialty export markets generating foreign exchange which could be returned to provide better local health and educational services. In general, where projects are even moderately capital intensive, the most lucrative markets must be tapped. But this does not mean that subsistence production could not be usefully enhanced (Chapter 25).

Of course, specialty prices hold only as long as the product is in short supply. Profitablity will decline as world supplies meet market demands. Traditional European markets appear to be well served but opportunities in North America and the Pacific Rim have scarcely been touched. It may be difficult to determine how quickly this will occur but the diversified regional economies that result still will benefit from additional stability.

Culturally consistent livelihoods

Wildlife production often is seen as a culturally consistent livelihood for aboriginal peoples. Although sealing, whaling, and trapping by northern native people have suffered from the campaigns of environmental and animal rights activists, conservationists seem to be rallying to this cause and steps have been taken to involve subsistence and even commercial wildlife utilisation in the evolving World Conservation Strategy.

Since it is hard to be openly against economically disadvantaged people, opponents argue that it is simply no longer possible to allow subsistence hunting or that the trend toward commercialisation is not just a modern manifestation of traditional pursuits. However, there are ways to regulate wildlife utilisation without banning it and traditional man/animal relationships always have and always will evolve. The Arjemark (hunting/fishing) economy of the Sami people was transformed into reindeer husbandry in the 16th century (Chapter 11). Indians of western Canada soon transformed subsistence hunting to commercial procurement of meat to supply the fur trade in the 18th and 19th centuries (Chapter 13). For these people to engage in commercial game production or even lead its transformation in the modern world would hardly be inconsistent with their history.

Design constraints

Whereas a strong case can be made for new agricultural animals, there is less agreement about how their productive potential should be harnessed. A common view is that husbandry systems should be designed to meet the needs of animals rather than trying to force them into familiar systems designed for conventional livestock since, under such circumstances, ecological adaptations may be compromised. While this may be true, practical socioeconomic and legal constraints limit the options. There are circumstances where the needs of people are so pressing that species must be selected to suit traditional husbandry practices (King & Heath, 1975; King, Heath & Hill, 1977). The following sections outline the biological, ecological, managerial, economic, and social constraints which frame viable options for commercial wildife production.

Biological constraints

Many adaptations of wild ungulates are dissipated under intensive management where they lose opportunities to range widely and feed selectively (Lewis, 1977) so their ability to thrive in harsh environments is not a particularly useful indicator of their agricultural potential (Moss, 1975). For example, wild ungulates may owe their apparent disease resistance simply to light stocking densities; moose (*Alces alces*), bighorn sheep (*Ovis canadensis*), muskoxen (*Ovibos moschatus*), and caribou (*Rangifer tarandus*) are notably susceptible to disease when confined and may be better hunted, herded or ranched than farmed. Similarly, adaptations such as digestive specialisation need not preadapt them to life as semi-domestic animals; wild browsers may be complementary to range cattle (Owen-Smith & Cooper, 1985) but they are typically expensive to feed under intensive husbandry (Chapter 20).

Opportunities for intensifying management also may be constrained by behavioural traits such as emotionality, aggressiveness, or territoriality (Walther, 1984; Hemmer, 1985). Injuries are common among small flighty species such as fallow deer (*Dama dama*), gazelles (*Gazella* spp.), and pronghorn antelope (*Antilocapra americana*), and close handling in small enclosures (unless specially designed) can be traumatic. Other species such as bison (*Bison bison*), African buffalo (*Syncerus caffer*), and muskoxen may be unpredictable and sufficiently dangerous that they should be separated from people with fences or other barriers especially during the rut. The design of facilities must further account for their gregariousness, tractibility, size, strength, and agility.

Ecological constraints

Since seasonal resources of forage and water often are geographically separated, questions have been raised about restricting normal migratory movements of grazing animals (Sinclair & Fryxell, 1985). Under such circumstances, the choices are to use either extensive unfenced systems which accommodate far-ranging movements or, alternatively, more intensive systems which provide year-round water supplies and extend the grazing season with cultivated forages and supplemental feed. The aims of conservation are best served by the first but patterns of land tenure and other sociopolitical and economic factors often make the second more practical.

Ecological constraints also may influence the composition of the grazing community. The appropriate mix of herbivore species depends on the balance of forage resources and habitats as well as the threat of cross-transmission of parasites and diseases. Despite their complementary grazing habits, joint stocking of wildebeest and cattle, or domestic sheep and deer, may be limited by the threat of malignant catarrhal fever. *Parelaphostrongylus* may prevent coexistence of moose and white-tailed deer (*Odocoileus virginianus*) where suitable intermediate hosts are present (Anderson, 1972). *Eleophora* may limit opportunities for running deer and wapiti (*Cervus elaphus*) (Hibler & Adcock, 1971). Occasionally, animals interact more directly; wapiti appear behaviourally dominant to white-tailed deer and moose and may displace them from certain habitats (Telfer & Scotter, 1975).

Managerial constraints

Technological sophistication and its demand for equipment, infrastructure, services, and skilled (in the conventional sense) manpower has a powerful impact on appropriate designs. Generally, game production is considered to make fewer demands than conventional livestock husbandry although the methods may be less familiar. But there are exceptions; the logistics of game cropping to serve urban or export markets can be staggering, and intensive game farming differs little from conventional livestock husbandry in terms of its needs. Consequently, commercial game production is not always easily implemented, particularly in developing nations where opportunities for serving subsistence or informal rural markets should be fully explored (Chapter 25).   Game production as a form of land use must compete with alternative demands on time and resources. Particularly important is the heavy commitment necessary for pastoralism and intensive farming which may mean forgoing opportunities for wage employment (Ingold, 1980). Low input systems such as ranching are attractive where labour costs are high and opportunities for supplementary income are available. Such forces are noticeable in recent transformations of the reindeer industry.

Economic constraints

Some of the most harsh constraints are imposed by economic factors. Not everyone has the luxury of planning activities to maximise economic efficiency. Subsistence hunters and herders are necessarily motivated by risk aversion and are forced to plan mainly from day to day (Widstrand, 1975). A main thrust of development programmes presumably has been to help people enter cash economies so they can plan in the longer term. Even producers in developed nations are not purely profit-motivated, but to survive they are induced to maximise the disparity between cash costs and revenue, and to minimise investments per stock unit.

One of the heaviest development costs for containment systems is fencing (approximately US$ 4000/km). Fencing costs per hectare (hence per stock unit at an equivalent density) can be diluted by enclosing larger areas; unit costs decline roughly in proportion to the square root of the area. If cross-fencing is not considered, reasonable economy of scale is achieved for areas larger than 2000 ha. However, total development costs are high and such large contiguous parcels are not always easily found.

Relative investments also can be diluted by increasing stocking rates. One way is to stock an appropriate mix of browsers, grazers, and mixed feeders which would provide fuller and more balanced use of range resources (Owen-Smith & Cooper, 1985). Vegetation can be manipulated with fire, logging, or fertilising, or native vegetation may be replaced with higher yielding tame forages. The bottleneck of minimum seasonal forage supplies can be ameliorated by supplemental feeding, increasing carrying capacity many-fold.

Beyond increasing stocking rates, there are ways to improve productivity/stock unit. The simplest is to judiciously manage the sex and age structure of the animal population (Beddington, 1975a,b; Caughley, 1977). Nutritional manipulation can be used to enhance conception rates and other productive functions. Physiological control of growth and reproduction is possible by manipulating photoperiod or more directly by influencing endocrine status (Kay, Milne & Hamilton, 1984; Kay, 1985). This has already found some application in advancing calving dates of red deer to more closely match pasture growth. There are also opportunities for genetic selection and crossbreeding which already have made important contributions to the deer farming industry (Chapter 16).

As in other branches of the economy, the optimal scale and intensity of a production unit depends upon many factors. One of the most important is land value both in terms of cost and inherent productivity. Another factor is the availability and costs of other inputs including labour. The design also depends on the relative market value of breeding stock, antlers, venison, and hunting opportunities (Berry, 1986).

Social constraints

The design of production systems also may depend on the reactions of consumers, competitors, and even observers (the general public). Consumers express preferences for various products and dictate standards of hygiene. Competitors that either share the market (e.g., other sectors of the meat industry) or the land base (e.g., recreational public) may urge tight controls to protect their interests. There also is a growing public concerned with the conditions under which animals are raised and the purposes to which they are put. Legalisation of commercial game production often hinges on affirming that there is indeed a demand, product quality will be high, animals will be treated humanely, security of wild stocks can be maintained, and other land users will not be inconvenienced. Such requirements are more easily met by some production systems than others.

Consumer demand

The demand for recreational hunting opportunities has been one of the more stable demands, stimulating annual hunter expenditures of US$ 7.15 billion in the United States alone (Prescott-Allen & Prescott-Allen, 1986). Although absolute demand has kept pace until recently, the proportion of the public that chooses to hunt has declined and recreational hunting is viewed with increasing suspicion (Baker, 1985). Nevertheless, demands remain high for quality hunts in exotic places and, recognising this, some countries have increased foreign exchange earnings by offering trophy hunting to non-nationals. At an international level, demand is related more to political stability or costs of international travel than the general popularity of the sport.

Demand for special products such as velvet antler, horn, musk, urine, and tails depends both on traditions and disposable income. Interest in traditional folk medicines remains high and purchasing power continues to rise. Although the principal markets are considered to be Korea, Hong Kong, Taiwan, and the People's Republic of China, much of the processed material is exported to serve Oriental communities in North America and Europe (Kong & But, 1985). Traditional international markets seem to have become saturated at just under 100 metric tonnes (Hughes, 1986) in the absence of a major marketing initiative.

Traditional demands for venison in Europe, especially Germany, have been supplemented by growing demands in new markets where venison is considered a healthy alternative to beef and other fat meats. Whereas there are few cultural or religious food aversions to most game meats (Harris, 1987), there is some resistance to marketing meats from elephants, zebras, giraffe, and kangaroos. The tendency to market a variety of game meats under a common banner (venison, antelope, impala, etc.) is being reversed to placate consumers angered by this deception. Insects, rodents, primates, and carnivores which are important in the diets of local people are unlikely to stimulate much interest on international markets. Less differentiated market preferences exist for certain ungulate species, for small versus large cuts, and for wild versus farm origin.

Product standards

Although there are still few tariffs in the international venison trade, regulation is achieved indirectly by imposing certain standards of slaughter and meat handling. There was an early trend toward relaxing requirements to permit field slaughter. However, greater market opportunities are opening for game meats from slaughter facilities which provide both *ante* and *post-mortem* inspection. In addition to providing higher standards of hygiene and uniform quality, such systems provide greater security against illegal entry of poached venison.

Animal welfare

Development of husbandry systems for new species provides an opportunity to throw off entrenched but inhumane practices. It may be a necessity as well as an opportunity since the animal rights movement is gaining popular and political support (Regan, 1983). Attempts are being made to devise an international code of animal rights as well as standards of ethical or humane practice for specific circumstances. Generally, these codes relate to hunting ethics, products which can be harvested, slaughter methods, and husbandry systems.

Each society has its own perception of hunting ethics. Some jurisdictions ban sport hunting in fenced reserves or specify their minimum size. This restriction may have more to do with the quality of the sport than humaneness since it does not apply to commercial hunting which also uses such 'unsporting' practices as night-lighting. The most striking cultural contrast is the selection of sporting equipment (Chapter 6). North American hunters see primitive weapons (black powder shooting, archery) as more sporting than modern firearms whereas in most parts of Europe they are banned on humane grounds.

Many countries do not allow cutting of velvet antlers from living animals, which is considered unnecessarily cruel for what is perceived to be a frivolous purpose. Most Western countries that do permit velvetting require supervision by licensed veterinarians, and use of general and local anaesthetics although this may conflict with provisions for drug residues. Most other by-products are not taken from living animals so concerns for animal welfare do not arise although some products may be viewed with some disdain.

There is still some disagreement on standards of humane transport and commercial slaughter. In general, measures for domestic stock seem adequate for farmed deer (Farm Animal Welfare Council, 1984; Seamer, 1986). However, some people argue that it is more humane for deer to be shot on pasture although this may make it more difficult to meet standards of hygienic slaughter.

Although few game animals are raised in confinement, intensive management at high stocking rates has disturbed those who would frankly prefer a more park-like setting in which animals could consumate a normal range of behaviours. The public generally demands more than for domestic animals which are perceived to have been selected for contentment in simple surroundings. However, difficulties in devising objective and appropriate criteria have obviated implementation of standards. One indirect attempt has been to specify the minimum size of game farms, not to regulate stocking rates but rather to minimise spurious entry and short-term profiteering.

Protection of wild populations

Most jurisdictions have imposed various restrictions on commercial production to protect wild populations. In general, the price of tighter controls is more intensive husbandry so opportunities for landscape conservation may be forgone. However, there is a need to establish rules for determining where and how game will be raised and to monitor the production chain in the field, at the slaughter plant, and on retail markets. This can be achieved in several ways.

One of the easiest is to license the location and conditions under which animals will be raised. In some places, commercial production is zoned to separate wild and commercial herds as insurance against inadvertent introduction of diseases or genetic contamination. Licensing also may be contingent on land ownership to prevent conflicts in the public domain.

Adequate fencing often is a condition of licensing. For individual famers, fences clarify ownership, prevent crop damage, and increase productivity. For governments, fences prevent mixing of wild and tame stocks and impose a financial commitment which prevents spurious entry to the industry. Further controls on illegal movement of animals to or from farms can be imposed by stock inventories, permanent marking of animals, and issuance of transport permits.

Meat inspection regulations are designed to ensure product quality but they also provide a measure of protection against illegal entry of meat. To prevent trafficking at the retail level, retailers are often licensed and monitored. Distinctive sealed packaging provides a practical safeguard which increases costs somewhat but has the advantage of preventing the typically high moisture loss of lean meats.

Future of wildlife production

The long-term future of wildlife production (as well as the long-term future of wild ungulates) is uncertain. One view is that humans always have and always will have an economic dependence on wildlife. Another view questions whether future generations will necessarily see flesh as food or animal production as moral. Most likely, commercial game production will wax and wane in relation to changing attitudes and opportunities.

The heyday of commercial wildlife utilisation lies in the past when democratisation of access to wildlife throughout the Western world saw the depletion of many species largely for commercial purposes. But the most appalling slaughter occurred as colonial powers mined the wildlife resources of their new dominions in America, Asia, and Africa. This trade was curtailed only when stocks were profoundly depleted, for some species too late to avert extinction.

Current trade volumes are still too small to interpret short-term trends since they are so strongly tied to the fates of individual systems. For example, the world trade is presently dominated by wild hares exported by Argentina but with tremendous production capacity building in Oceania, venison from intensive deer farms is likely to dominate international trade within a decade. Momentary declines (Chapter 2) seem not related to declining interest but rather to a tightening of various controls on diseases and hygiene.

Clearly, conservation objectives are best served by extensive systems (Chapter 23) but they may continue to lose their market share to more intensive systems which can monitor disease more closely, standardise products, and provide superior standards of hygiene. Wild game may be favoured on traditional markets but uniformity and quality will have an edge in the penetration of new ones.

These considerations seem to have initiated a transformation of the industry. Whereas most game products once came from commercial hunting, there seems to be a continuing move to farm game. For example, New Zealand expects to supply 20,000 tonnes by 1994, almost two-thirds of the current world trade. Although traditional markets seem to be fully served, the long-term prospects for market development are good with ample opportunities to enter new areas, especially in the Pacific Rim, and to move down to lower-priced mass markets.

These changes in the size and specification of the market change the competitive edge for various wildlife production systems. The following sections speculate on the prospects for each of the main forms of wildlife utilisation.

Hunting

Opportunities for subsistence hunting are likely to decline as hunting cultures are swept into the cash economy and wildlife populations decline through habitat loss and non-sustainable utilisation. Traditional pursuits are also being transformed into more commercial endeavours based on more intensive production methods. Nevertheless, bushmeat on informal rural markets remains surprisingly high (de Vos, 1977; Sale, 1983). For example, the total harvest of wild meat (mostly bearded pigs (*Sus barbatus*)) in Sarawak (Malaysia) has been estimated at 10,000–30,000 tonnes/annum which almost equals the current international trade (Caldecott, 1986). But as human populations grow in developing countries, relative nutritional dependence will continue to decline.

Commercial hunting for formal markets has a less certain future. Reduction cropping as practised in eastern Africa in recent decades seems to have been a fleeting opportunity although certain parks and reserves may require harvest to control over-population despite the controversy it generates. Sustained yield cropping has a brighter future especially in centrally planned economies or where rights-of-access to wildlife are vested in landholders.

The main contribution to meat supplies continues to be made by sport hunters whether or not game meat enters commercial markets. Nevertheless, interest in sport hunting is declining in the face of dwindling opportunities, urbanisation of human populations, and growing public resistance to killing animals for sport. These trends are not new but their cumulative effects are significant.

Herding
Reindeer husbandry, because of harsh living conditions and high labour costs, has undergone massive transformation. In some places more extensive systems have been adopted (Ingold, 1980), but in others more intensive fenced systems have been tried (Chapter 10). In general, markets for antlers have breathed new, if not momentary, life into the industry and rising meat schedules have made reindeer husbandry much more profitable and attractive to young herders. Nevertheless, the carrying capacity of arctic ranges is limited and prospects for major increases in production are not great.

In southern latitudes, several recent attempts have been made to rationalise pastoral production. The best known is the Galana experiment, which attempts to insert new animals (eland, oryx) into traditional pastoral systems. However, even the future of traditional pastoralism is in question and the system is unlikely to attract much interest.

Containment systems
Containment systems resolve the problem of demonstrating ownership and preventing conflicts with adjacent land uses. Although movements of animals are limited by fences, extensive ranching systems offer an ecologically sound and practical opportunity although they may need the higher revenues

afforded by sport hunting (Berry, 1986). Ultimately, game ranchers may lose certain markets to game farmers who can provide better consistency and quality of meat, and higher standards of hygiene and disease control.

One of the remarkable success stories is the rise of intensive deer farming particularly in New Zealand. Traditional markets may discriminate against meat from farmed game since importance is attached to the concepts of wildness and the hunt. However, the greater control afforded by intensive production will allow penetration of new markets where high standards of quality are critical. Another reason for the rising popularity of deer farming is the fact that it is widely accessible since existing farm holdings can be diversified from conventional livestock to game. Game production need not be limited to large individual, corporate, or state landholders. Although game farming may not provide the same conservation benefits as extensive game ranching, it is often better than the agricultural land uses it replaces and adds welcome variety to the rural landscape.

References

Anderson, R. C. (1972). The ecological relationships of meningeal worm and native cervids in North America. *Journal of Wildlife Diseases*, 8, 304–10.

Arman, P. & Hopcraft, D. (1975). Nutritional studies on East African herbivores. 1. Digestibilities of dry matter, crude fibre and crude protein in antelope, cattle, and sheep. *British Journal of Nutrition*, 33, 255–64.

Baker, R. (1985). *The American Hunting Myth*. New York: Vantage Press.

Beddington, J. R. (1975a). Economic and ecological analysis of red deer harvesting in Scotland. *Journal of Environmental Management*, 3, 91–103.

Beddington, J. R. (1975b). Age structure, sex ratios and population density in the harvesting of natural animal populations. *Journal of Applied Ecology*, 11, 915–24.

Bell, R. H. V. (1971). A grazing ecosystem in the Serengeti. *Scientific American*, 225, 86–93.

Bell, R. H. V. (1982). The effect of soil nutrient availability on community structure in African ecosystems. In *Ecology of Tropical Savannas*, ed. B. J. Huntley & B. H. Walker, pp. 193–216. New York: Springer-Verlag.

Berg, R. T. & Butterfield, R. M. (1976). *New Concepts in Cattle Growth*. Sydney: Sydney University Press.

Berry, M. P. S. (1986). Comparison of different wildlife production enterprises in the northern Cape Province, South Africa. *South African Journal of Wildlife Research*, 16, 124–8.

Bigalke, R. C. (1986). Utilizing wild species. In *Bioindustrial Ecosystems*, Ecosystems of the World 21, ed. D. J. A. Cole & G. C. Brander, pp. 255–74. Amsterdam: Elsevier.

Brown, L. P., Chandler, W. U., Flavin, C., Pollock, C., Postel, S., Starke, L. & Wolf, C. (1986). *State of the World 1986*. New York: W. W. Norton & Co.

Caldecott, J. O. (1986). *Hunting and Wildlife Management in Sarawak*. Kuala Lumpur: World Wildlife Fund – Malaysia.

Caughley, G. (1977). *Analysis of Vertebrate Populations*. New York: Wiley & Sons.

Christopherson, R. J., Hudson, R. J. & Christophersen, M. K. (1979). Seasonal energy expenditures and thermoregulatory responses of bison and cattle. *Canadian Journal of Animal Science*, 59, 611–17.

Clutton-Brock, J. (1981). *Domesticated Animals from Early Times*. London: British Museum (Natural History).

Coe, M. J., Cumming, D. H. & Phillipson, J. (1976). Biomass and production of large African herbivores in relation to rainfall and primary production. *Oecologia*, 22, 341–54.

Crawford, M. A., Gale, M. M., Woodford, M. H. & Casperd, N. M. (1970). Comparative studies on fatty acid composition of wild and domestic meats. *International Journal of Biochemistry*, 1, 295–305.

Dahl, G. (1981). Production in pastoral systems. In *The Future of Pastoral Peoples*, ed. J. G. Galaty, D. Aronson, P. C. Salzman & A. Chouinard, pp. 200–9. Ottawa: International Development Research Centre.

Darling, F. F. (1960). Wildlife husbandry in Africa. *Scientific American*, 203, 123–34.

Dasmann, R. F. (1964). *African Game Ranching*. Oxford: Pergamon.

de Vos, A. (1977). Game as food. A report on its significance in Africa and Latin America. *Unasylva*, 29, 2–12.

Demment, M. W. & van Soest, P. (1985). A nutritional explanation for body-size patterns of ruminant and non-ruminant herbivores. *American Naturalist*, 125, 641–72.

Dezhkin, V. V. (1975). Ecologo-economic foundations of game management. In *Okhotovedeniye*. Moscow: Lesnaya Promyshlennost Publishers (in Russian).

Dezhkin, V. V. (1978). Production of game management. In *Okhotnichiye khozyaistvo RSFSR*. Moscow: Lesnaya Promyshlennost Publishers (in Russian).

Dezhkin, V. V. & Menkova, N. V. (1981). Comparative aspects of the use of resources of wild and domestic ungulates. In *Ekonomika, organizatsyia i ispolzovaniye resursov okhotnichyego khotyaistva RSFST (sbornik nauchnykh trudov TzNIL Glavokhoty RSFSR)*. Moscow (in Russian).

Drew, K. R. (1985). Meat production from farmed deer. In *Biology of Deer Production*, ed. P. F. Fennessy & K. R. Drew, Bulletin 22, pp. 285–90. Wellington: Royal Society of New Zealand.

East, R. (1984). Rainfall, soil nutrient status and biomass of large African savanna mammals. *African Journal of Ecology*, 22, 245–70.

Edwards, M. A. & McDonnell, U. (eds.) (1982). *Animal Diseases in Relation to Animal Conservation*, Symposium of the Zoological Society of London, 50. New York: Academic Press.

Eltringham, S. K. (1984). *Wildlife Resources and Economic Development*. New York: Wiley & Sons.

Falkow, L. P. & Mercer, J. B. (1986). Partition of heat loss in resting and exercising winter- and summer-insulated reindeer. *American Journal of Physiology*, 251, R32–R40.

Farm Animal Welfare Council. (1984). *Report on the Welfare of Farmed Deer*. Alnwick (Northumberland): MAFF Publications.

Fennessy, P. F. & Drew, K. R. (eds). (1985). *Biology of Deer Production*, Bulletin 22. Wellington: Royal Society of New Zealand.

Field, C. R. (1979). Game ranching in Africa. *Applied Biology*, 4, 63–101.

Foose, T. (1982). *Trophic Strategies of Ruminant Versus Nonruminant Ungulates*. Ph.D. Thesis. Chicago: University of Chicago.

Forss, D. A., Manley, T. R., Platt, M. P. & Moore, V. J. (1979). Palatability of venison from farmed and feral red deer. *Journal of Science Food and Agriculture*, 30, 932–5.

Gebczynska, Z., Kowalczyk, J., Krasinska, M. & Ziolecka, A. (1974). A comparison of the digestibility of nutrients by European bison and cattle. *Acta Theriologica*, 19, 283–9.

Georgiadis, N. J. (1985). Growth patterns, sexual dimorphism and reproduction in African ruminants. *African Journal of Ecology*, 23, 75–83.

Hammond, J. A. & Branagan, D. (1973). The disease factor in plans for domestication of wild ruminants in Africa. *Veterinary Record*, 92, 367–9.

Hansen, R. M., Mugambi, M. M. & Bauni, S. M. (1985). Diets and trophic ranking of ungulates of the northern Serengeti. *Journal of Wildlife Management*, 49, 823–9.

Harris, M. (1987). *The Sacred Cow and the Abominable Pig: Riddles of Food and Culture.* New York: Simon & Schuster.

Hawley, A. (1985). Commercial meat production from wild cervids. In *Biology of Deer Production*, ed. P. F. Fennessy & K. R. Drew, Bulletin 22, pp. 327–37. Wellington: Royal Society of New Zealand.

Hawley, A. W. L. (1986). Carcass characteristics of bison (*Bison bison*) steers. *Canadian Journal of Animal Science*, 66, 293–5.

Hawley, A. W. L., Sylven, S. & Wilhelmsen, M. (1983). Commercial moose meat production in Sweden. *Livestock Production Science*, 10, 507–16.

Hemmer, H. (1985). The aptitude and selection of large mammals for game farming and domestication. *Acta Zoological Fennica*, 172, 233–6.

Hibler, C. P. & Adcock, J. L. (1971). Elaeophorosis. In *Parasitic Diseases of Wild Mammals*, ed. J. W. Davis & R. C. Anderson, pp. 263–78. Ames: Iowa State University Press.

Hofmann, R. R. (1973). *The Ruminant Stomach: Structure and Feeding Habits of East African Game Ruminants*. Nairobi: East African Literature Bureau.

Hofmann, R. R. (1985). Digestive physiology of the deer – their morphophysiological specialization and adaptation. In *Biology of Deer Production*, ed. P. F. Fennessy & K. R. Drew, Bulletin 22, pp. 393–407. Wellington: Royal Society of New Zealand.

Hudson, R. J. (1985). Body size, energetics and adaptive radiation. In *Bioenergetics of Wild Herbivores*, ed. R. J. Hudson & R. G. White, pp. 1–24. Boca Raton: CRC Press.

Hudson, R. J. & Christopherson, R. J. (1985). Maintenance metabolism. In *Bioenergetics of Wild Herbivores*, ed. R. J. Hudson & R. G. White, pp. 121–42. Boca Raton: CRC Press.

Hughes, R. (1986). Velvet marketing: long-term Korean prospects poor. *The Deer Farmer*, 31, 33–5.

Ingold, T. (1980). *Hunters, Pastoralists and Ranchers: Reindeer Economies and Their Transformations*. Cambridge: Cambridge University Press.

Kay, R. N. B. (1970). Meat production from wild herbivores. *Proceedings of the Nutrition Society*, 29, 271–8.

Kay, R. N. B. (1985). Body size, patterns of growth, and efficiency of production in red deer. In *Biology of Deer Production*, ed. P. F. Fennessy & K. R. Drew, Bulletin 22, pp. 411–21. Wellington: Royal Society of New Zealand.

Kay, R. N. B., Milne, J. A. & Hamilton, W. J. (1984). Nutrition of red deer for meat production. *Proceedings of the Royal Society of Edinburgh*, 82B, 231–42.

Kay, R. N. B., Sharman, G. A. M., Hamilton, W. J., Goodall, E. D., Pennie, K. & Coutts, A. (1981). Carcase characteristics of young red deer farmed on hill pasture. *Journal of Agricultural Science (Cambridge)*, 91, 513–22.

King, J. M. (1979). Game domestication for animal production in Kenya: field studies of the body-water turnover of game and livestock. *Journal of Agricultural Science (Cambridge)*, 93, 71–9.

King, J. M. & Heath, B. R. (1975). Game domestication for animal production in Africa. *World Animal Review*, 16, 23–30.

King, J. M., Heath, B. R. & Hill, R. E. (1977). Game domestication for animal production in Kenya: theory and practice. *Journal of Agricultural Science (Cambridge)*, 89, 445–57.

King, J. M., Kingaby, G. P., Colvin, J. G. & Heath, B. R. (1975). Seasonal variation in water turnover by oryx and eland on the Galana Game Ranch Research Project. *East African Wildlife Journal*, 13, 287–96.

Kong, Y. C. & But, P. P. H. (1985). Deer – The ultimate medicinal animal (antler and deer parts in medicine). In *Biology of Deer Production*, ed. P. F. Fennessy & K. R. Drew, Bulletin 22, pp. 311–24. Wellington: Royal Society of New Zealand.

Kyle, R. (1987). *A Feast in the Wild*. Oxford: Kudu.

Lamprey, H. F. (1983). Pastoralism yesterday and today: the over-grazing problem. In *Tropical Savannas*, ed. F. Bouliere, Ecosystems of the World, Vol. 13, pp. 643–66. Amsterdam: Elsevier.

Ledger, H. P. (1968). Body composition as a basis for a comparative study of some East African mammals. In *Comparative Nutrition of Wild Animals*, ed. M. A. Crawford, Symposium of Zoological Society of London 21, pp. 289–310. New York: Academic Press.

Lewis, J. G. (1977). Game domestication for animal production in Kenya: activity patterns of eland, oryx, buffalo and zebu cattle. *Journal of Agricultural Science (Cambridge)*, 89, 551–63.

Manley, T. R. & Forss, D. A. (1979). Fatty acids of meat lipids from young red deer (*Cervus elaphus*). *Journal of Science of Food and Agriculture*, 30, 927–31.

Mason, I. L. (ed.) (1984). *Evolution of Domesticated Animals*. London & New York: Longman.

McNaughton, S. J. & Georgiadis, N. J. (1986). Ecology of African grazing and browsing mammals. *Annual Reviews of Ecology and Systematics*, 17, 39–66.

Mentis, M. T. (1977). Stocking rates and carrying capacities for ungulates on African rangelands. *South African Journal of Wildlife Research*, 7, 89–98.

Moss, R. (1975). Different roles of nutrition in domestic and wild game birds and other animals. *Proceedings of the Nutrition Society*, 34, 95–100.

Owen-Smith, N. & Cooper, S. M. (1985). Comparative consumption of vegetation components by kudus, impalas and goats in relation to their commercial potential as browsers in savanna regions. *South African Journal of Science*, 81, 72–6.

Parker, K. L. & Robbins, C. T. (1984). Thermoregulation in mule deer and elk. *Canadian Journal of Zoology*, 62, 1409–22.

Parker, K. L. & Robbins, C. T. (1985). Thermoregulation in ungulates. In *Bioenergetics of Wild Herbivores*, ed. R. J. Hudson & R. G. White, pp. 161–82. Boca Raton: CRC Press.

Peters, R. H. (1983). *The Ecological Implications of Body Size*. Cambridge: Cambridge University Press.

Prescott-Allen, C. & Prescott-Allen, R. (1986). *The First Resource: Wild Species in the North American Economy*. New Haven: Yale University Press.

Price, M. A. & White, R. G. (1985). Growth and development. In *Bioenergetics of Wild Herbivores*, ed. R. J. Hudson & R. G. White, pp. 183–213. Boca Raton: CRC Press.

Regan, T. (1983). *The Case for Animal Rights*. Berkeley: University of California Press.

Renecker, L. A. & Hudson, R. J. (1986). Seasonal energy expenditures and thermoregulatory responses of moose. *Canadian Journal of Zoology*, 64, 322–7.

Richmond, R. J., Hudson, R. J. & Christopherson, R. J. (1977). Comparison of forage intake and digestibility by American bison, yak and cattle. *Acta Theriologica*, 22, 225–30.

Ringberg, T. (1979). The Spitzbergen reindeer – a winter-dormant ungulate? *Acta Physiologica Scandinavica*, 105, 268–73.

Robbins, C. T. (1983). *Wildlife Feeding and Nutrition*. New York: Academic Press.

Rogerson, A. (1968). Energy utilization by the eland and the wildebeest. In *Comparative Nutrition of Wild Animals*, ed. M. A. Crawford, Symposium of the Zoological Society of London, Vol. 21, pp. 153–61. New York: Academic Press.

Sale, J. (1983). *The Importance and Values of Wild Plants and Animals in Africa*. Gland: IUCN.

Schaefer, A. L., Young, B. A. & Chimwano, A. M. (1978). Ration digestion and retention times of digesta in domestic cattle (*Bos taurus*), American bison (*Bison bison*), and Tibetan yak (*Bos grunniens*). *Canadian Journal of Zoology*, 56, 2355–8.

Schmidt-Nielsen, K. (1984). *Scaling: Why is Animal Size so Important?* Cambridge: Cambridge University Press.

Seamer, D. J. (1986). The welfare of deer at slaughter in New Zealand and Great Britain. *Veterinary Record*, 118, 257–8.

Simpson, A. M., Webster, A. J. F., Smith, J. S. & Simpson, C. A. (1978). The efficiency of utilization of dietary energy for growth in sheep (*Ovis ovis*) and red deer (*Cervus elaphus*). *Comparative Biochemistry and Physiology*, 59A, 95–9.

Sinclair, A. R. E. (1985). Does interspecific competition or predation shape the African ungulate community? *Journal of Animal Ecology*, 54, 899–918.

Sinclair, A. R. E. & Fryxell, J. M. (1985). The Sahel of Africa: ecology of a disaster. *Canadian Journal of Zoology*, 63, 987–94.

Talbot, L. M. (1966). *Wild Animals as a Source of Food*. US Bureau of Sport Fisheries and Wildlife, Special Scientific Report – Wildlife 98. Washington: US Government Printing Office.

Taylor, C. R. (1968). Hygroscopic food: a source of water for desert antelopes. *Nature*, 219, 181–2.

Taylor, C. R. (1969a). The eland and the oryx. *Scientific American*, 220, 89–95.

Taylor, C. R. (1969b). Metabolism, respiratory changes and water balance of an antelope, the eland. *American Journal of Physiology*, 217, 317–20.

Taylor, C. R. (1970a). Strategies of temperature regulation: effect on evaporation in East African ungulates. *American Journal of Physiology*, 219, 1131–5.

Taylor, C. R. (1970b). Dehydration and heat: effects on temperature regulation of East African ungulates. *American Journal of Physiology*, 219, 1136–9.

Taylor, R. D. & Walker, B. H. (1978). A comparison of vegetation use and condition in relation to herbivore biomass on a Rhodesian game and cattle ranch. *Journal of Applied Ecology*, 15, 565–81.

Teer, J. G., Burger, G. V. & Deknatel, C. Y. (1983). Commercial hunting in the United States. *Transactions of the North American Wildlife Natural Resources Conference*, 48, 445–56.

Telfer, E. S. & Scotter, G. W. (1975). Potential for game ranching in the boreal aspen forests of western Canada. *Journal of Range Management*, 28, 172–80.

Tennessen, T. & Hudson, R. J. (1981). Traits relevant to the domestication of herbivores. *Applied Animal Ethology*, 7, 87–102.

Tessaro, S. V. (1986). The existing and potential importance of brucellosis and tuberculosis in Canadian wildlife: A review. *Canadian Veterinary Journal*, 27, 119–24.

Topps, J. H. (1975). Behavioural and physiological adaptation of wild ruminants and their potential for meat production. *Proceedings of the Nutrition Society*, 34, 85–93.

von la Chevallerie, M. (1972). Meat quality of seven wild ungulate species. *South African Journal of Animal Science*, 2, 101–3.

Walker, B. H. (1979). Game ranching in Africa. In *Management of Semi-Arid Ecosystems*, ed. B. H. Walker, pp. 55–81. Amsterdam: Elsevier.

Walther, F. (1984). *Communication and Expression in Hoofed Mammals*. Bloomington: Indiana University Press.

Wenham, G. & Pennie, K. (1986). The growth of individual muscles and bones in the red deer. *Animal Production*, 42, 247–56.

White, R. J. (1986). *Big Game Ranching in the United States*. Mesilla (New Mexico): Wild Sheep & Goat International.

Widstrand, C. (1975). The rationale of the nomad economy. *Ambio*, 4, 146–53.

Yerex, D. (ed.) (1982). *The Farming of Deer: World Trends and Modern Techniques*. Wellington: Agricultural Promotion Associates.

25

Appropriate technology for rural development

ARCHIE S. MOSSMAN

Abstract

Subsistence game production refers to broadly based integrated utilisation of resources to serve informal local markets. It requires technology appropriate to local people and their culture as well as to the needs of wildlife management. Consequently, the best technology for developed nations may be nearly the worst for aid recipients. Development should be something people want, something they can be intimately involved with, and something they can and will do themselves after a short training period. If appropriate technologies are applied, subsistence game production offers an opportunity to improve the welfare of people while securing and improving that of wildlife and habitats.

Introduction

Human populations have increased and technologically impacted their surroundings to the extent that natural resources are inadequate to consistently feed us or provide us with safe drinking water. In many areas, this condition is being rapidly approached and the quality of life is already unacceptable. Frequent warfare exacerbates these situations. The poorest people and especially their children are most seriously affected.

Technological impacts can be reduced and appropriate technologies used to improve human conditions, but there is only one morally acceptable way to correct overpopulation, and that is to reduce birthrates allowing natural attrition to reduce populations to what Garrett Hardin (1986) terms *cultural carrying capacity*. Although stable populations may be a realistic goal in the long term, we must do what we can in the interim both to provide for people and to maintain the capacity to do so in the future. Subsistence game production is one possible way to help.

What I will call *subsistence game production* differs from managed subsistence hunting in that a broader approach is taken toward meeting local human needs. It attempts to carefully integrate use of all local resources and is not restricted to harvest of species usually considered game. It may include harvesting of small rodents and their food caches or edible insects such as termites, locusts, and beetles; protecting and using of medicinal plants and animals; protecting religious sites and organisms; or burning to promote production of plants used in basketry.

Understandably, most people from industrialised nations are culturally ill-prepared to anticipate what people living in subsistence economies are likely to perceive as their needs and priorities. This is especially so because the variation in the details of these economies is great; and that implies that assistance needs to be closely tailored for each situation.

This chapter attempts to bridge the gap between subsistence cultures and people interested in providing rural assistance. It was prompted by three concerns: (1) the tendency toward monoculture and domestication of presently wild species rather than toward more efficient multiculture in natural environments, (2) the failure of game ranching projects to involve and benefit local people, and (3) the inaccurate impression that improved use of wildlands is too complicated for rural people thus killing their interest and participation. As Schumacher (1973) observed ' ... *the apparent shortage of entrepreneurs in developing countries ... is precisely the result of the negative demonstration effect of a sophisticated technology infiltrated into an unsophisticated environment'*.

These concerns are linked to the reasons for our early efforts to implement game ranching in Africa; namely, to protect and improve the integrity of rangeland and their biota as much as possible by enhancing what they can contribute to people on a sustained basis (Dasmann & Mossman, 1961). This was not and is not an antihuman approach but rather rests on sound information that in the long-term, non-arable lands are more usefully productive through multiculture than monoculture. It also is based on evidence that many agricultural systems are probably not sustainable.

Appropriate technology

Schumacher (1973) provides the theoretical framework for the vision of appropriate technology that follows. Such ideas have to a considerable extent been formalised for aid purposes in the Cocoyoc Declaration of the United Nations Environment Program and the United Nations Conference on Trade and Development (UNEP/UNCTAD, 1974; United Nations, 1978). We can paraphrase Schumacher by saying that appropriate technology empowers the users, making their hands and brains more productive than ever before. It should not make them dependent upon a supplier of technology. It means that at least some among them must know how the technology works, can fix it if it breaks, if necessary make a usable substitute, and even improve upon it.

Why emphasise appropriate technology?

Most developing societies are so closely integrated that any disturbance of one segment reverberates throughout the societal structure; any transfer of technology into a society will certainly alter it. The less the mores of the society are altered, the greater the likelihood that the technology will be accepted. Some societies are robust, accepting many changes and successfully integrating them into their enriched societal structure (Reynolds & Doll, 1984). Others suffer and some disintegrate (Helm, 1961, 1976; Nelson, 1969; Turnbull, 1972).

Recipients must be intimately involved in early stages of the project so that the transfer of technology will not be seriously disruptive. Unfortunately, this is unlikely to be as easy as it seems (Jequier & Blanc, 1983). Because of the political and monetary power structures and their sensitivities, it is usually necessary to clear any project at high levels within the power structures before substantial contact with the intended recipients is advisable. Therefore, careful diplomacy can be necessary to leave room for essential adjustments when intensive direct contact with the aid recipients becomes possible.

How to determine what is technologically appropriate

The determination of what is technologically appropriate requires intimate knowledge of the people who will use it as well as the resources it will harness. The essence of what is appropriate can be appreciated by paraphrasing Schumacher (1973: 197): *give a person fishing tackle and the results are in doubt for even if the person catches fish, that person is still dependent; but teach a person how to make fishing tackle and how to fish and you have helped him to become not only self-supporting, but also self-reliant and independent.* This gives an idea of what appropriate or intermediate technology is all about. However, it is not sufficient. Part of the idea is to build on existing local technology with an emphasis on education, and to do so in such a way as to integrate in a socially positive fashion.

There are many examples of appropriate technology. Snaring is a common and effective way that fish, reptiles, birds, and mammals are obtained all over the world. Snares are often inhumane and nontarget individuals or species may be killed inadvertently. One possible way to minimise such problems is to use stopped snares (Mossman *et al.*, 1963, see Fig. 25.1). Where selection for males is desired, the stops can be located so that horned or antlered males are held whereas females are automatically released. When set and tended properly, animals unintentionally captured can be released.

In Zimbabwe lives a large, green, longhorned grasshopper (Tettigoniidae) that is attracted to lights at night and is sought for food. If there is some way to predict when they will swarm, it might be useful to devise a light trap. There are other edible insects that might be similarly obtained. It would be useful to develop efficient and appropriate methods of mass capture for agricultural pests that also serve as food. Quelea (*Quelea* spp.) and certain locusts are examples.

Where it could be done inexpensively and would not have unwanted ecological or sociological effects, the carrying capacity for water-dependent terrestrial species often can be increased by improving the distribution of water. People also can be taught how to make use of naturally occurring miniwatersheds, how to improve wildlife access to the water, how to increase

Fig. 25.1. Cable clamp on a wire cable snare that prevents complete closure. Captured animals can be released unharmed if the snare is properly set. (Photograph by the author.)

the water holding capacity of existing waterholes and, perhaps in the process, some may learn to more frequently think in innovative ways.

Among subsistence agriculturalists, loss of crops to wildlife is a serious concern. Where socially feasible, methods of re-siting gardens, fencing, and repelling wildlife or attracting them elsewhere could be identified and then the principles and methods taught. Where the wildlife can be utilised, the problem is greatly reduced and, in some cases, the crops may be maintained partly to feed and to attract them.

Ingenuity needs to be applied to solving the problems of subsistence economies in appropriate ways. There probably are no purely technological answers in the sense of mechanical contrivances or chemical substances. Every potential answer has sociological and educational components that cannot be ignored if success is to be achieved.

Subsistence game production

Why consider subsistence game production for aid purposes? The short answer is that people living in largely subsistence economies are in most need of assistance and many occupy lands that are marginal for intensive agriculture. Furthermore, game production that emphasises mulitculture and the maintenance and enhancement of the ecosystem can produce most for people on a sustained basis.

Subsistence economies

There is no clear line separating subsistence and other economies, only a gradation from one to the other. One dictionary defines subsistence farming as *'farming whose products are intended to provide for the basic needs of the farmer, with little surplus for marketing'* (Stein & Urdang, 1967). The question is always how much marketing is possible if the economy is still to be considered a subsistence one. People define subsistence in a way that suits their own interests and perceptions. This is not surprising because if a person has ever grown or obtained from wild sources any food that they or their family have consumed, or any product that they have put to some other beneficial use, they have practised subsistence. Likewise, anyone who has merely exchanged gifts, let alone sold something for cash, has participated in a market. Subsistence, like many other terms, becomes absurd when pushed to the possible but unreasonable limits. Clearly, some trade is a normal and necessary part of subsistence economies.

Goals

Our primary goal must be to protect and improve the soil. Our concern is that the soil be protected by vegetation and litter and that a favourable mix of animal and plant life not only protects and improves the soil, but also provides for human welfare. Human welfare is related to the amounts and kinds of use an area can provide. We therefore need to consider the intensity with which we will crop the biota.

Amounts of use

For years, maximum sustained yield was the accepted goal for fish and wildlife management. It also has been part of the legal framework for resource use, but maximum sustained yield is clearly inappropriate as a goal for subsistence game production. In fact, it is not appropriate in most other circumstances (Mossman, 1974; Holt & Talbot, 1978; Bailey, 1982). Many of the reasons are the same, but one is especially important to subsistence game production; namely, that to approach maximum sustained yield, the technology is both formidable in itself and also formidably expensive. The overall management goal for subsistence game production or any game production systems lies somewhere short of maximum sustained yield.

Kinds of use

Game ranges offer both renewable and non-renewable resources which can be used in consumptive or non-consumptive ways. Non-renewable resources for consumptive uses include everything from clay for locally made pottery to oil and titanium for industry. The decision to use these needs to be locally determined and frequently local people may need some help in evaluating, directing, and possibly resisting attempts to extract non-renewable resources from their lands. More commonly, consumptive uses are made of renewable resources as when wildlife is cropped for food and hides.

Much of the income made by some game ranches is earned through trophy hunting. These animals are worth many times more than their meat value, and the meat can still be retained or sold profitably to the hunter or to others. If rural people decide to sell hunting opportunities, they can do so directly by providing the infrastructure for handling the clients, or they can contract with a safari or guide service to manage the hunting part of the enterprise. In addition to possible profits, other financial, social, and psychological effects need to be considered. This caveat also applies to extraction of non-renewable resources and to non-consumptive uses of resources on a wildlife production area.

The infrastructure needed for trophy hunting is not greatly different from that needed for photo safaris, a potential nonconsumptive use. In fact, provision of a unique but safe experience of living in an indigenous culture would be especially easy, and such experience could be mutually beneficial. One indication that such an experience is saleable is the fact that the opportunity to stay in traditional Swazi housing is sought by some tourists in Mlilwane Game Reserve, Swaziland (S. Ellis & T. Reilly, personal communications). Similar experiences could be provided by other cultural groups (Mossman & Mossman, 1976).

When we are considering the amounts of each kind of use, we need a rationale for deciding the intensity of management. Generally, we should aim for a moderate but recognisable improvement of present conditions achieved in a manner that will not preclude additional improvements in the future. Such goals have four important attributes: they are usually quickly attainable, they can be improved in the future, the improvement can be recognised, and the cultural changes can be relatively small, gradual, and hence acceptable.

Achieving the goals

Participation by aid recipients largely determines both what is feasible and how it needs to be achieved. There are three related considerations. One is that interest in education and willingness to educate and be educated are essential for the providers of aid. The second is the need to realise that much more sophisticated things can be accomplished by people lacking formal education than most aid providers probably realise. The third is that the ways aid providers normally go about solving similar problems in their own culture are probably the worst for assistance.

We have already discussed education as an integral part of development. The potential abilities of people lacking formal education need to be kept firmly in mind. If they are, better communication and hence better results will be achieved. Often there are traditional ways of subsistence that are rooted firmly in ecological functioning. Some of these may be quite sophisticated ecologically and it is the responsibility of the aid provider to ensure that he or she is properly educated by the recipients.

One area that can often be improved is estimation of the size of the crop to be taken. To save space, I will simply say that people lacking formal education and modern technology, and having very little cash can, with assistance, gather data and use it to allow fairly sophisticated management.

Why not just set an arbitrary conservative harvesting rate of 5, 10, or 15% depending on the species? Such an approach has been used and it can result in both under and overharvesting, both of which may be more deleterious than one would think. To illustrate, the government of Zimbabwe established a uniform 10% harvest rate for species on game ranches in 1972–73. Impala (*Aepyceros melampus*) and others were severely underharvested with the expected ecological consequences but also with the consequence that game ranching was made to appear uneconomic. At the same time, elephant (*Loxodonta africana*) were overharvested making game ranching appear conservationally unsound as well (Mossman & Mossman, 1976). Admittedly, this was an extreme example of 'just setting an arbitrary conservative harvesting rate', but the principles apply whenever the rates are 'wide of the mark'. We need to do better and we can.

Management errors and appropriate technology

We will always run the risk of making mistakes. There is no avoiding this because natural ecosystems, even simple ones, are so complex that we can never know all the factors and interactions. However, three things let us keep the inevitable errors small and manageable. One is to know as much as possible about factors such as weather variations, fire, human activities, and epidemics that can drastically alter biotic responses and thus be prepared to counter or adapt management to them if necessary. The second is to be in nearly constant contact with what is going on in the field so that corrections can be made in a timely manner. The third is to manage conservatively.

Some see the need for management conservatism as justifying data collection almost without end before undertaking management. They thus may act wrongly through failure to manipulate when manipulation is needed. Such errors of omission are especially serious when they result in land conversion to less desirable use.

Part of the reason for the frequent reluctance of wildlife professionals to actually manage probably comes from our focus on managing wildlands in the most precise way possible given present knowledge. In fact, meaningful improvements in the use of wildlife and wildlands can be made with very unsophisticated techniques and in the absence of a large part of the data collection that we biologists usually consider necessary. The insistence on the application of our most powerful wildlife management methods is, for many places, not only unnecessary but actually disadvantageous.

That statement is probably going to elicit a strong negative reaction from many wildlife biologists. I even found myself bothered 'at the gut level' upon realising that useful improvements in the management of wildlife, especially in developing countries, can be made with very little recourse to the scientific background we have spent our lives acquiring. I still find it a little hard to accept that 'doing it right' in a biological sense is actually 'doing it wrong' given the social realities in a great many places in the world.

Rural employment

There appears to be a misapprehension that subsistence modes of production offer insufficient employment and thereby account for the influx of people into urban areas. However, there are many reasons for urban immigration such as starvation resulting from 'land shortage', land erosion, drought, need for money to pay taxes and school fees, hope for a way to get food, for social advancement and so on, but not for lack of employment in subsistence economies.

Decreased reliance on subsistence may be necessary as populations burgeon, but there is at least a faint possibility that this may not always be so. Historically, the shift to agriculture from subsistence on wild resources probably did not occur primarily because of shortage of wild subsistence resources, but rather because of the ease of plant domestication and the opportunities that offered for greater sedentariness coupled with the security of knowing where one's food would be obtained. As reliance on cultivation increased so did the need to remain with the crops to tend and protect them. That in turn provided the circumstances for enlarged governance systems and other specialisations of human effort that were not possible in a society entirely dependent upon wild resources. This scenario, of course, flies in the face of the usual contention that it was the increased food and fibre supplies provided by agriculture that allowed urbanisation and the other characteristics of modern cultures. Why doubt such a generally accepted explanation?

The human population of the area now encompassed by the contiguous 48 states of the United States has been estimated to have been 846,000 at the time of settlement by the first European immigrants (Stirling, 1955). In 1981, the population was 225,740,000 (National Geographic Society, 1981). Three per cent of these people now farm about 20 per cent of the land and not only feed the nation but in 1980 exported $41 billion worth of agricultural products (National Geographic Society, 1981). However, it is generally accepted that America's soils have deteriorated, land available for food production has decreased, and both salt and freshwater fisheries have deteriorated. And some authorities contend that agricultural land actually produces less than wild land (see Atjay *et al.*, 1979; Olson *et al.*, 1983; Vitousek *et al.*, 1986). We should at least consider the possibility that shifts from reliance upon wild resources to cultivation were not caused by overall shortage of wild subsistence resources. We also need to consider the corollary that for some lands subsistence game production may prove superior to other agricultural land uses.

With our present technology, the disadvantages of having to travel to exploit resources are much reduced. Other technologies such as modern firearms, and for fish, large nylon nets allow the exploitation of sparse or highly aggregated populations. A few radio collars attached to caribou (*Rangifer tarandus*) can tell us where the herds are at almost any moment, and dogsleds, snowmobiles, and aircraft can get us to them and get the meat and by-products out. No longer need people starve because migratory herds alter their movement patterns.

With the arrival of the horse, the agricultural Indians of the eastern Great Plains turned from farming to hunting (Stirling, 1955). That 'technological breakthrough' allowed them to exploit the great migratory herds in a manner impossible when all travel had to be accomplished on foot. We ought to explore the feasibility of following their lead when we are trying to decide how to manage the remaining marginal lands of the world. I suspect that wild game and plant production will usually prove to be the best use of those lands.

Conclusions

The use of wildland products possibly offers a more reliable means of subsistence than does agriculture for people living at the subsistence level (Lee, 1979). When drought ruined the crops of their neighbours, the !Kung San of southern Africa allowed them to exploit the same subsistence resources upon which they, themselves depended. Certainly, in the vicinity of Lee's study area wild resources were more reliable. That wild resources can support more people is also suggested by the finding that acacia savannas of East Africa could support domestic cattle at the rate of 19.6 to 28.0 kg/ha whereas comparable areas supported 65.5–157.6 kg/ha of wild ungulates (Talbot, 1963).

If we are to realise the potentials of wildland resources we need to put much more effort into understanding wild biotas and how to improve wild land utilisation. Domestication, and so on, certainly should and will continue, but that is not where the greatest potential lies for helping the people most in need of help and for improving wild land productivity. We also need to be very careful to truly involve the aid recipients and to keep our promises so that not only is the immediate effort successful but also so it is possible to start other similar ones, preferably with the help of people assisted in earlier projects.

If we are to assist rural people, we need to do much more than give lip service to applying appropriate technology and involving the recipients, and it is not easy. Subsistence game production as an effective means of rural assistance need be no more sophisticated than successful pastoralism and subsistence agriculture. The animals and plants may be different but the principles are the same. By remembering that, we can guard against 'the negative demonstration effect of a sophisticated technology infiltrated into an unsophisticated environment' that Schumacher (1973) warns about. If a relatively low key, appropriate technology approach is used, there is a considerable opportunity to improve the welfare of people while securing and improving that of wildlife and wild areas. However, if we do not keep the philosophy behind appropriate technology firmly in mind and act accordingly, I see little long-term hope for effective rural assistance.

References

Ajtay, G. L., Ketner, P. & Duvigneaud, P. (1979). Terrestrial primary production and phytomass. In *The Global Carbon Cycle*, ed. B. Bolin, E. T. Degens, S. Kempe & P. Ketner, pp. 129–82. New York: John Wiley & Sons.

Bailey, J. A. (1982). Implications of 'muddling through' for wildlife management. *Wildlife Society Bulletin*, 10, 363–9.

Dasmann, R. F. & Mossman, A. S. (1961). Commercial use of game animals on a Rhodesian ranch. *Wild Life*, 3, 6–14.

Hardin, G. (1986). Cultural carrying capacity: a biological approach to human problems. *BioScience*, 36, 599–606.

Helm, J. (1961). *The Lynx Point People: The Dynamics of a Northern Athapaskan Band*. National Museum of Canada Bulletin Number 176. Anthropological Series Number 53. Canada Department of Northern Affairs and National Resources.

Helm, J. (1976). *The Indians of the Subarctic, a Critical Bibliography*. Bloomington: Indiana University Press.

Holt, S. J. & Talbot, L. M. (1978). New principles for the conservation of wild living resources. *Wildlife Monographs*, 59. Washington: The Wildlife Society.

Jequier, N. & Blanc, G. (1983). *The World of Appropriate Technology: a Quantitative Analysis*. Paris: Development Centre of the Organization for Economic Co-operation and Development.

Lee, R. B. (1979). *The !Kung San: Men, Women and Work in a Foraging Society*. Cambridge: Cambridge University Press.

Mossman, A. S. (1974). *Conservation*. New York: Intext Educational Publishers.

Mossman, A. S., Johnstone, P. A., Savory, C. A. R. & Dasmann, R. F. (1963). Neck snare for live capture of African ungulates. *Journal of Wildlife Management*, 27, 132–5.

Mossman, S. L. & Mossman, A. S. (1976). *Wildlife Utilization and Game Ranching*. International Union for the Conservation of Nature and Natural Resources Occasional Paper Number 17. Morges: IUCN.

National Geographic Society. (1981). *Atlas of the World*. 5th edn. Washington: National Geographic Society.

Nelson, R. K. (1969). *Hunters of the Northern Ice*. Chicago: University of Chicago Press.

Olson, J. S., Watts, J. A. & Allison, L. J. (1983). *Carbon in Live Vegetation of Major World Ecosystems*. ORNL–5862. Oak Ridge: Oak Ridge National Laboratory. Environmental Science Division.

Reynolds, B. & Doll, D. (1984). Eskimo hunters of the Bering Sea. *National Geographic*, 165, 814–34.

Schumacher, E. F. (1973). *Small is Beautiful*. Perennial Library edition. New York: Harper & Row Publishers Inc.

Stein, J. & Urdang, L. (eds.) (1967). *The Random House Dictionary of the English Language*, unabridged edition. New York: Random House.

Stirling, M. W. (1955). Indians of our western plains. In *Indians of the Americas*, ed. M. W. Stirling, pp. 73–98. Washington: National Geographic Society.

Talbot, L. M. (1963). Comparison of the efficiency of wild animals and domestic livestock in utilization of East African rangelands. In. *Conservation of Nature and Natural Resources in Modern African States*. New Series 1. Morgues: International Union for the Conservation of Nature and Natural Resources.

Turnbull, C. M. (1972). *The Mountain People*. New York: Simon & Schuster.

UNEP/UNCTAD. (1974). *The Cocoyoc Declaration*. Adopted by the participants in the Symposium on Patterns of Resource Use, Environment and Development Strategies. Cocoyoc, Mexico, 8–12 October 1974. New York: United Nations.

United Nations. (1978). *Transfer of Technology. Its Implications for Development and Environment*. Secretariat of the United Nations Conference on Trade and Development. New York: United Nations.

Vitousek, P. M., Ehrlich, P. R., Ehrlich, A. H. & Matson, P. A. (1986). Human appropriation of the products of photosynthesis. *BioScience*, 36, 368–73.

INDEX